NANOSENSORS

Editors

Dimitrios Nikolelis
Laboratory of Environmental Chemistry
Department of Chemistry
University of Athens
Panepistimiopolis-Kouponia
15771 Athens, Greece

Georgia Paraskevi Nikoleli
Department of Biomedical Sciences
Faculty of Health and Care Sciences
University of West Attica
Egaleo Park Campus
12243 Athens, Greece

CRC Press
Taylor & Francis Group
Boca Raton London New York

CRC Press is an imprint of the
Taylor & Francis Group, an **informa** business
A SCIENCE PUBLISHERS BOOK

Cover images provided by the first editor, Dimitrios Nikolelis.

First edition published 2023
by CRC Press
6000 Broken Sound Parkway NW, Suite 300, Boca Raton, FL 33487-2742

and by CRC Press
4 Park Square, Milton Park, Abingdon, Oxon, OX14 4RN

Library of Congress Cataloging-in-Publication Data (applied for)

ISBN: 978-0-367-36985-9 (hbk)
ISBN: 978-1-032-48030-5 (pbk)
ISBN: 978-0-367-82228-6 (ebk)

DOI: 10.1201/9780367822286

Typeset in Times New Roman
by Radiant Productions

Preface

Nanotechnology and biosensors show how nanotechnology is used to create affordable, mass-produced, portable, small sized biosensors to directly monitor a wide range of chemical compounds, i.e., toxicants such as environmental pollutants. In addition, it provides information on their integration into components and systems for mass market applications in food analysis, environmental monitoring and health diagnostics. Nanotechnological advancements have led to an improvement in the characteristics of biosensors such as lifetime, costs, performance, detection limits, sensitivity and selectivity of biosensors. This text covers a lot of ground and will be of interest to those wishing for a general overview of the range of nanomaterials used for biosensing in environmental, food and medical sensing applications.

Nanotechnology plays an important role in the development of biosensors. Application of nanomaterials in biosensors allows the use of many new signal transduction technologies. Because of their small size, nanosensors, nanoprobes, and other nanosystems are revolutionizing the fields of chemical analysis, to enable rapid detection of multiple chemicals in food and environmental samples. Recent progress in nanotechnology has finalized the production of low cost, mass-produced nanosensors and in integrating these nanosensors into components and systems (ca. portable ones) for market applications. Sensing includes chemical food toxicants, such as toxins, insecticides, pesticides, herbicides, microorganisms, bacteria, viruses, and other microorganisms, phenolic compounds, allergens, genetically modified foods, hormones, and dioxins. The limit of detection, costs and performance of these devices is being improved by the use of nanomaterials for their construction. The use of these nanomaterials has permitted the introduction of many new methods of signal transduction in biosensors. Because of their small size, nanosensors, nanoprobes, and other nanosystems have permitted simple and rapid analyses *in vivo*. Portable devices which are able to analyse multiple components are becoming available. This book reviews the status of the various nanostructure-based nanosensors and investigates prototype nanosensing devices specially in the design and microfabrication which now is based on novel nanotechnological tools providing nanosensors that are suitable for the rapid and in the field detection of food toxicants and environmental pollutants.

Contents

1

Applications of Sensitive Electrode Surfaces; Determination of Different Classes of Antibiotics

*Stella Girousi** and *Christina Sarakatsanou*

1. Introduction

Antibiotics are among the drugs most commonly administered to patients with marked inter-individual pharmacokinetic variability. Non-invasive monitoring of antibiotic levels in biological fluids contributes to the evaluation of the body's adequate exposure to them. In this way, we can understand the relationship between the dose of antibiotics and their clinical effects [1].

There is a wide variety of antibiotics that differ in use and mechanism of action. The following figure (Figure 1) summarizes the main classes of antibiotics and provides a brief overview of how they fight bacterial infections. We will also look at each of the groups, in this chapter, along with some antibiotics that belong to each category [2].

Bacteria can be divided in two categories—gram-positive and gram-negative. These categories are named by the Gram test, which involves adding a violet dye to the bacteria. Gram-positive bacteria retain the violet color of the dye, while Gram-negative bacteria do not; the latter are instead colored red or pink. Gram-negative bacteria are more resistant to antibodies and antibiotics than Gram-positive bacteria [2].

2. β-lactams

The first antibiotic to be discovered in the β-lactams class of antibiotics was penicillin, which Alexander Fleming identified in 1928. The antibiotics in this class contain a β-lactam ring and include penicillins, such as amoxicillin and cephalosporins. They

Analytical Chemistry laboratory, School of Chemistry, Faculty of Sciences, 54124 Thessaloniki, Greece.
* Corresponding author: girousi@chem.auth.gr

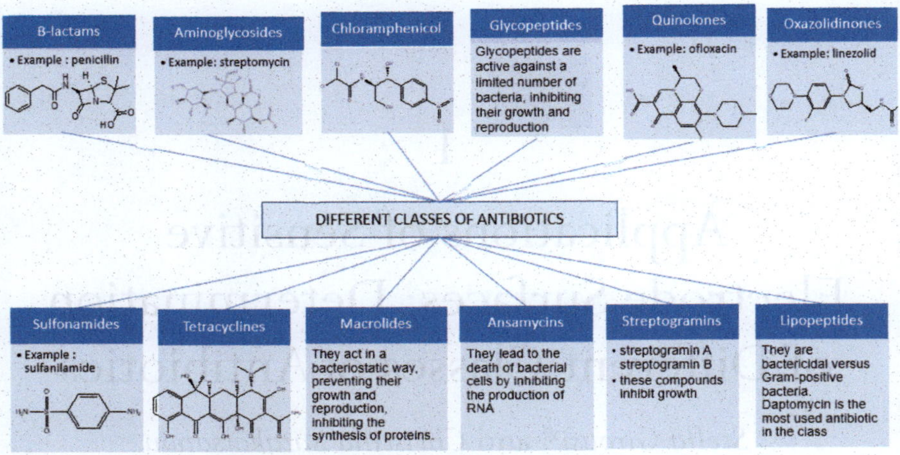

Figure 1. Different classes of antibiotics.

work by interfering with the synthesis of peptidoglycan, a component of the bacterial cell wall, and are mainly used against gram-positive bacteria [2].

2.1 Penicillin

In order to detect penicillin V, Švorc et al. developed a simple, sensitive and selective differential pulse voltammetric method on a bare boron diamond electrode. The oxidation peak of penicillin V was reproducible and well-defined irreversible peak, at a very positive potential of +1.6 V (vs. Ag/AgCl). Optimal experimental conditions for penicillin V oxidation were achieved in an acetate buffer solution (pH 4.0). These optimal conditions were: modulation amplitude of 0.1 V, modulation time of the pulse as 0.05 s, and a scan rate 0.05 V s^{-1}, selected for differential pulse voltammetry. Linear response of peak current on the concentration ranged from 0.5 to 40 μM, with a coefficient of determination of 0.999, good repeatability (RSD of 1.5%) and a detection limit of 0.25 μM; these were observed without any chemical modification or electrochemical surface pretreatment. The effect of possible interferents such as stearic acid, glucose, urea, uric acid and ascorbic acid appeared to be negligible, proving the good selectivity of the method. The above method was applied for the detection of penicillin V in pharmaceutical formulations (tablets) and human urine samples, with satisfactory recoveries (98 to 101% for tablets and 97 to 103% for human urine) [3].

2.2 Ceftazidime

Ceftazidime (CFZ) is one of the third-generation cephalosporin antibiotics and is widely used against both gram-positive and gram-negative bacteria. CFZ is not metabolized and is excreted unchanged (mainly through urine). It is an electroactive compound and therefore electrochemical techniques offer a good opportunity for its detection. Thus, a simple, sensitive and selective sensor based on carbon paste electrode modified by functionalized mesoporous silica material was developed by

Dehdashtian and Abdipur for the electrochemical detection of ceftazidime (CFZ). The oxidation of CFZ was studied on modified carbon paste electrode using cyclic voltammetry, differential pulse voltammetry and electrochemical impedance spectroscopy (EIS). The working electrode was an unmodified or modified carbon paste electrode, and the auxiliary electrode and the reference electrode were platinum wire and SCE respectively. The results showed that the oxidation peak current of CFZ on the modified carbon paste electrode was improved compared to that obtained on the bare carbon paste electrode. Under optimum conditions, the sensor exhibited a linear response over the CFZ concentration range of 1–2500 nM, with a detection limit of 0.3 nM. The proposed sensor was successfully applied for monitoring of CFZ in the pharmaceutical and biological samples, and satisfactory results were obtained [4].

2.3 Amoxicillin

The first study for the detection of amoxicillin (AMX) developed an electrochemical sensor based on a glassy carbon electrode (GCE) modified by three-dimensional graphene (3D-GE) and polyglutamic acid (PGA). AMX response at PGA/3D-GE/GCE involving the transfer of one electron and an equal number of protons was determined using electrochemical approaches. By optimizing the experimental conditions, the performance of AMX was linear in the range of 2–60 μM. In addition, the detection limit was found to be 0.118 μM (S/N = 3). The modified electrode could determine the concentration of AMX in human urine samples. Overall, the developed PGA/3D-GE/GCE for the detection of AMX showed great potential in practice [5].

According to a second research, Cu(II)-exchanged clinoptilolite nanoparticles (Cu(II)-NCL) were prepared and characterized by FT-IR, BET, XRD and TEM techniques. The obtained Cu(II)-NCL was then used to form a carbon paste electrode (CPE). The resulting Cu(II)-NCL/CPE was finally used for the voltammetric determination of amoxicillin (AMX). The best voltammetric response was obtained by the electrode containing 20% of the modifier (copper–doped nano-clinoptilolite) in 0.05 mol L^{-1} NaCl at 2.2 and 7.2 pH levels. The electrode showed a linear response in the concentration range of $4.0 \times 10^{-8} - 1.0 \times 10^{-4}$ M AMX, with a detection limit of 2.0×10^{-8} M, in square wave voltammetry. The electrode showed good repeatability, reproducibility and long lifetime as confirmed by statistical tests. The electrode also had good selectivity and applicability in the detection of AMX in urine and pharmaceutical tablets [6].

Furthermore, a new electrochemical device based on a combination of nanomaterials such as Printex 6L Carbon and cadmium telluride quantum dots within a poly(3,4-ethylenedioxythiophene) polystyrene sulfonate film was developed by Wong et al. for the sensitive detection of amoxicillin [7]. The electrochemical determination of AMX was carried out using square-wave voltammetry. Under optimized experimental conditions, the proposed sensor showed high sensitivity, repeatability and stability to AMX detection, with an analytical curve in the AMX concentration range from 0.90 to 69 μM, and a low detection limit of 50 nM. No significant interference in the electrochemical signal of the antibiotic was observed due to possible biological factors or drugs such as uric acid, paracetamol, urea, ascorbic acid and caffeine. It turned out that without any sample pre-treatment and with the use of a simple measurement device, the sensor could prove to be an effective alternative method for not only the analysis of pharmaceutical products (commercial tablets) and clinical samples (urine), but also the examination of food quality (milk samples) [7].

Finally, a new electrochemical method for indirect chronoamperometric detection of amoxicillin (AMX) in the presence of copper ions Cu (II) at a carbon paste electrode (CPE) was suggested by Hrioua et al. [8]. The method was based on the interaction between the β-lactam antibiotic and copper ions. This interaction was initially studied by square wave voltammetry (SWV), wherein the peak corresponding to the oxidation of AMX at 0.7 V decreased significantly in the presence of copper ions. The product of the reaction was characterized by UV-visible spectrophotometry and infrared spectroscopy (IR). Moreover, when AMX was added to the electrolyte solution containing copper ions, the chronoamperometric current from the Cu(II)/Cu(I) redox system was increased. This increase in current was used for the indirect detection of amoxicillin. By selecting an appropriate concentration of copper (II), the calibration curve for AMX was established in the concentration range of 1.95×10^{-7} to 1.46×10^{-5} M. The detection limit was found to be 8.84×10^{-8} M. The effect of different organic compounds on the detection of AMX and the effect of coexistence of other metal ions in the electrolytic solution were studied. The applicability of the proposed method was tested in human blood and pharmaceutical tablets [8].

2.4 Ampicillin

The design and construction of a "signal-on" electrochemical sensor based on aptamer E-AB for the detection of ampicillin was reported by Yu and Lai, and the sensor's response was reportedly fast [9]. Since all the sensor components were surface-immobilized, it was regenerable and could be reused for a minimum of three times. The sensor demonstrated good specificity, and distinction could be made between ampicillin and structurally similar antibiotics such as amoxicillin. This sensor can be used directly in complex samples, including serum, saliva and milk. Although both alternating current voltammetry (ACV) and square wave voltammetry (SWV) are suitable sensor characterization techniques, the results showed ACV to be more suitable for target analysis in this case. Even under optimal experimental conditions, the detection limit of the sensor obtained in ACV (1 μM) was significantly lower than that obtained in SWV (30 μM) [9].

2.5 Cefadroxil

The first study in this section presents a novel nano-silver amalgam paste microelectrode (nano-Ag-APME) for the detection of cefadroxil in pharmaceutical formulations and biological specimens [10]. The proposed electrode, prepared in the plastic tip of a micropipette tied with the nano-silver amalgam paste, mainly consists of silver nano-powder and pure mercury (1:9 w/w). The silver nanopowder and the nanosilver amalgam paste were characterized using scanning electron microscope, energy dispersive X-ray spectrometer and X-Ray Diffraction (XRD). The proposed method, based on square-wave stripping voltammetry for the measurement of cefadroxil at pH 4 in Britton–Robinson buffer, showed a well-defined reduction peak at −0.160 V. The comparative study of SW-AdSV and spectrophotometric methods showed the linear response for CFL in a range from 0.033–0.304 μM and 10–70 μM, respectively. The analytical performance for the quantification of CFL by the SW-AdSV method was significantly better than the spectrophotometric method ($p < 0.01$). Moreover, it was indicated that SW-AdSV is more sensitive and selective for routine analysis of CFL, compared to the spectrophotometric method [10].

The electrochemical behavior of cefadroxil has also been studied at glassy carbon electrodes in aqueous media over a wide pH range [11]. The first oxidation process is of the phenol moiety and follows an ECE mechanism, generating catechol and resorcinol derivatives as sub-products, which are subsequently reduced and oxidized in subsequent cycles. The sulfur heteroatom present in the cyclic structure near the β-lactam moiety is oxidized in two steps to form sulfoxide and sulfone. This process was identified from direct comparison with amoxicillin, which has a similar molecular structure, although they belong to different classes of β-lactam antibiotics. For amoxicillin, the oxidation of the sulfur heteroatom occurs at more positive potentials, most likely due to structural difficulties in stabilizing the charged oxidized species. The formation of a complex between copper (II) and the above antibiotics was studied by cyclic voltammetry in commercial samples [11].

2.6 Cefixime

Afkhami et al. introduced a new and efficient electrochemical sensor for the detection of cefixime (CEF) in 0.04 M buffer solution (pH 3.0), using gold nanoparticles electrodeposited on a modified multi-walled carbon paste electrode (GNPs/MWCPE). The voltammetric behavior of cefixime on this modified electrode was studied using cyclic and square wave voltammetric techniques. The oxidation of cefixime was irreversible and showed a pH-controlled diffusion process. The effect of various experimental parameters, including pH, scan rate, accumulation potential and time on the voltammetric response of CEF, was investigated. The results indicated that the process of oxidation of this compound was irreversible and adsorption-controlled. Under optimal conditions, the CEF concentration was determined using square wave voltammetry (SWV) in a linear range of 1.0×10^{-2} to 2.0×10^{2} μM, with a correlation coefficient of 0.9996, and the detection limit was found to be 3.0×10^{-9} M. Square wave voltammetric method was successfully applied to the detection of cefixime in a pharmaceutical dose and a urine sample. GNPs/MWCPE shows excellent analytical performance for the detection of cefixime, with a very low detection limit, high sensitivity, fantastic repeatability and reproducibility, compared to other methods reported in the literature [12].

The voltammetric behavior of cefixime has also been studied using cyclic, linear sweep, differential pulse and square wave voltammetric techniques [13]. The oxidation of cefixime was found to be irreversible and showed a pH-controlled diffusion process. In the study, various parameters were tested to optimize the conditions for the detection of cefixime. Furthermore, the dependence on pH, concentration, scan rate and nature of the buffer was also investigated. According to the linear relationship between the peak current and concentration, differential pulse (DPV) and square wave (SWV) voltammetric methods were used to detect cefixime in pharmaceutical forms and biological fluids. An acetate buffer at pH 4.5 was also proposed, allowing for quantitation over $6 \times 10^{-6} - 2 \times 10^{-4}$ M range in supporting electrolyte and spiked serum sample, $8 \times 10^{-6} - 2 \times 10^{-4}$ M range in urine sample and $6 \times 10^{-6} - 1 \times 10^{-4}$ M range in breast milk samples for both the techniques. The repeatability, reproducibility and accuracy of the methods in all media were investigated. No electroactive interferences from the excipients and endogenous substances were found in the pharmaceutical dosage forms and biological samples respectively [13].

In another work, a novel and cost-effective electrochemical sensor was developed for the electrochemical investigation of cefixime [14]. SEM, CV and EIS were used for the characterization of electrodes. At CTAB/AuNPs/PGE, the oxidation of cefixime was found to be irreversible and diffusion-controlled. The linearity was established in the range of 10×10^{-9} to 30×10^{-8} M, with a detection limit of 1.21×10^{-10} M. The studies involved pharmaceutical formulations and biological fluids [14].

2.7 Cephalosporins

In a study by Feier et al. the electrochemical oxidation of seven cephalosporins (ceftriaxone, cefotaxime, ceftazidime, cefadroxil, cefuroxime, cefaclor, cefalexin) was evaluated at high potentials, using a bare boron-doped diamond electrode, and the influence on the analytical response of the side chains was investigated. According to the anodic oxidation of the cephalosporin nucleus, a simple and sensitive method was developed for the electrochemical detection of cefalexin by differential pulse voltammetry. After optimization of the experimental conditions, a linear correlation was achieved between the peak height and the molar concentration of cefalexin in the range of 0.5 μM–700 μM, with a detection limit of 34.74 ng mL^{-1} [15]. The anodic peak for cefalexin was evaluated in the presence of other cephalosporin molecules and other common interferents. The developed method was applied to detect cefalexin in environmental, biomedical and pharmaceutical samples, and the results were found to be satisfactory. The electrochemical oxidation of cephalosporins was successfully adapted for flow injection analysis, with sensitive and reproducible successive analysis of cefalexin in different concentrations. Flow analysis also made it possible to determine the total amount of cephalosporins present in the sample [15].

3. Sulfonamides

Sulfonamides belong to a large class of antibiotics and are capable of acting on both gram-positive and gram-negative bacteria. Unlike β-lactams, they do not act by directly killing bacteria, but by inhibiting the bacterial synthesis of Vitamin B or folic acid, thus preventing the growth and reproduction of bacteria. Today, sulfonamides are rarely used, partially because of the development of bacterial resistance, and also because of concerns regarding their side effects (e.g., hepatotoxicity) [2].

3.1 Sulfamethoxazole

For the detection of sulfamethoxazole (SMX) and trimethoprim (TMP), a new electrochemical sensor modified with multi-walled carbon nanotubes decorated with Prussian blue nanocubes (MWCNT/PBnc) was developed by Sgobbi et al. [16]. MWCNT/PBnc composite combines the high presence of catalytically active sites on the PBnc edges with the MWCNT platform to enhance PBnc stability at neutral pH levels. The detection of these antibiotics was performed by differential pulse voltammetry. The MWCNT/PBnc film exhibited high selectivity and sensitivity to both the antibiotics. A linear detection range of 0.1–10.0 μM, with the limit of detection (LOD) as 60 nM was achieved for TMP; for SMX, a linear detection range of 1.0–10.0 μM, with an LOD of 38 nM was observed. The sensor allowed for the simultaneous detection of SMX and TMP in artificial urine, with recovery between 91.3 to 101% [16].

3.2 Sulfanilamide

Ferraz et al. detected Sulfanilamide (SFD) in otologic solution, human urine and serum by electroanalytical techniques on glassy carbon electrode (GCE) [17]. The cyclic voltammetry (CV) experiments showed an irreversible oxidation peak at +1.06 V in 0.1 M BRBS solution (pH 2.0) at 50 mV/s. Different voltammetric scan rates (from 10 to 250 mV/s) were applied, and it was evident that the oxidation of SFD on the GCE was controlled by diffusion. Square-wave voltammetry (SWV) under optimized conditions showed a linear response to SFD from 5.0 to 74.7 μM (R = 0.999), with detection and quantification limits of 0.92 and 3.10 μM respectively. The SWV method proved to yield better results for the detection limit and linear range, compared to the chronoamperometry method. This technique has been successfully deployed to determine SFD concentration in pharmaceutical formulation, human urine sample and serum samples, with recovery close to 100% [17].

3.3 Sulfasalazine

Sadhegi et al. developed an electrochemical sensor to detect the antibiotic sulfasalazine, based on molecularly imprinted polymer (MIP) as the recognition element [18]. A sulfasalazine selective MIP and a non-imprinted polymer (NIP) were synthesized and then incorporated into carbon paste to prepare the modified carbon paste electrodes. The MIP-based sensor showed high recognition ability towards sulfasalazine, compared to that of the NIP-based sensor. The influence of experimental parameters such as carbon paste composition, pH, incubation time, potential scan rate and potential amplitude were also optimized. Under optimal conditions, the MIP-based sensor performed well for sulfasalazine over the concentration range of 1.0×10^{-8} to 1.0×10^{-6} M, with a detection limit of 4.6×10^{-9} M and a sensitivity of 1.11×10^{7} μA L mol^{-1}. The MIP-based sensor was successfully used in a commercial pharmaceutical formulation and a human serum [18].

4. Aminoglycosides

Aminoglycosides inhibit protein synthesis in bacteria, eventually leading to cell death. They are effective against some gram-negative bacteria and some gram-positive bacteria, but are not absorbed during digestion and should hence be injected. Streptomycin was the first drug found to be effective in the treatment of tuberculosis. However, due to problems with aminoglycoside toxicity, their use is currently limited [2].

4.1 Streptomycin

Streptomycin (STR) is used in humans and animals to treat gram-negative bacterial infections. In the present study, gold nanoparticles (AuNPs) and thiol graphene quantum dots (GQD-SH) were used as the nanomaterial for ultrasensitive detection of streptomycin (STR). GQD-SH was immobilized onto the surface of a glassy carbon electrode (GCE). AuNPs were immobilized on SH groups of GQDs through bond formation of Au\S, and Apt was loaded on the electrode surface through interactions between thiol groups of the aptamer. By incubating STR as a target on the surface of the prepared Apt/AuNPs/GQD-SH/GCE as a proposed nanoaptasensor, the Apt/STR complex was formed and changes in the electrochemical signal were evaluated using EIS. The proposed nanoaptasensor showed wide linear detection range from 0.1 to 700 pg ml^{-1}. Consequently, the proposed nanoaptasensor was successfully used for the detection of STR in real samples, yielding satisfactory results [19].

5. Tetracyclines

Tetracyclines are antibiotics that work against both gram-positive and gram-negative bacteria. They inhibit protein synthesis, thus preventing the growth and reproduction of bacteria. The increased resistance of bacteria towards them has reduced their usage; however, they still find use in the treatment of acne, urinary tract infections (UTI), respiratory tract infections and chlamydial infections [2].

5.1 Tetracycline

The first research we discuss under this section dealt with an electrochemical sensor based on a carbon paste electrode modified with a combination of multi-walled carbon nanotubes functionalized with carboxyl groups (MWCNT-COOH), together with graphene oxide (GO), for the detection of tetracycline. Such sensors were constructed using carbon paste modified with 2.6% (w/w) of MWCNT-COOH and 3.1% (w/w) of GO. Under optimal conditions, using adsorptive stripping differential pulse voltammetry (AdSDPV), the sensor showed a linear response for tetracycline concentrations between 2.0×10^{-5} and 3.1×10^{-4} M, a sensitivity of 1.2×10^{4} µA L mol^{-1}, and a detection limit of 3.6×10^{-7} M. The sensor showed great sensitivity, selectivity and stability. The MWCNT-COOH-GO/CPE sensor was successfully used for the detection of tetracycline in river water, artificial urine and pharmaceutical samples, without sample pretreatment. The relative standard deviation (RSD) of the electrochemical measurements was less than 6.0% (n = 3) [20].

Another study in this domain reported the construction of an electrochemical sensing platform based on platinum nanoparticles supported on carbon (PtNPs/C) for the detection of the antibiotic tetracycline (TTC). Initially, the PtNPs/C was synthesized and characterized by powder X-ray diffraction spectroscopy and transmission electron microscopy. Different experimental parameters for the electrochemical behavior of the PtNPs/C-coated glassy carbon electrode (PtNPs/C/GCE) were evaluated. Under the optimal experimental conditions, the linear calibration range, detection limit, quantification and sensitivity of the PtNPs/C/GCE were 9.99–44.01 μM, 4.28 μM, 14.3 μM and 3.32 μAL μmol^{-1} cm^{-2} respectively. Eventually, the proposed electrochemical sensing platform was successfully deployed to detect low TTC concentrations in urine samples [21].

6. Chloramphenicol

Chloramphenicol is an antibiotic which acts by inhibiting protein synthesis and, in turn, the growth and reproduction of a few bacteria. It is used to treat typhoid fever and infections including salmonellosis. Due to potential severe toxic effects, it is generally used in developed countries only in cases where infections are considered life-threatening. Nevertheless, it is a very common antibiotic in developing countries, due to its low cost. Thus, the use of CP in pharmaceuticals, food-producing animals and agriculture can pose a serious threat to both human and animal health. Therefore, the development of a highly selective CP detection method has become essential for the protection of human health [2, 22, 23].

6.1 Chloramphenicol

In a particular study for the detection of CPL, well dispersive palladium nanoparticles decorated with graphene oxide (Pd NPs/GO) nanocomposite were developed using facile ice bath method. The electrochemical behavior of the electrochemical sensor was investigated by cyclic voltammetry and amperometric method. The Pd NPs/GO nanocomposite showed an excellent electrocatalytic behavior in terms of higher cathodic peak current and lower peak potential. Furthermore, the Pd NPs/GO-modified GCE displayed an excellent linear response range (0.007 to 102.68 μM), good sensitivity (3.0479 μA μM^{-1} cm^{2}) and a low detection limit (0.001 μM). The Pd NPs/GO/GCE sensor exhibited an excellent selectivity for CPL sensing in the presence of other interfering compounds. Additionally, the nanocomposite was efficaciously applied to biological and food samples for the detection of CPL and demonstrated good recoveries values [22].

7. Macrolides

Macrolides, as antibiotics, are mainly effective against gram-positive bacteria. They act in a bacteriostatic way, preventing their growth and reproduction, inhibiting the synthesis of proteins. These antibiotics have shown to be effective against numerous types of bacteria. While some bacterial species have developed macrolide resistance, they are still the second most commonly prescribed antibiotic in the NHS, with erythromycin being the most commonly prescribed antibiotic in this class [2].

7.1 Azithromycin

Azithromycin (AZM) is a widely used macrolide antibiotic; it is used to treat or prevent multiple bacterial infections [2]. As a semi-synthetic derivative of macrolide antibiotics, it has significant tissue pharmacokinetic properties and antibiotic activity with rapid absorption, long half-life (68 hours), very large volume of distribution and low plasma protein binding [2, 24]. However, AZM also has some side effects on the body's gastrointestinal tract. High levels of AZM accumulate in ocular tissues, with approximately 10% of the administered AZM excreted unchanged in the urine [2]. Therefore, a need arises for the administration of the appropriate dose of AZM [24].

In one of the studies, a selective and sensitive molecularly imprinted polymer (MIP)-based electrochemical sensor was fabricated for the detection of azithromycin from various biological samples (urine, tears, plasma). The reversible boronate ester bond-mediated, thin (~ 75 nm) MIP-based biomimetic recognition layer was electrodeposited in non-aqueous media onto the surface of a glassy carbon electrode. The surface morphology and the analytical performance of the developed sensor were assessed using scanning electron microscope (SEM), atomic force microscope (AFM), electrochemical impedance spectroscopy (EIS) and cyclic voltammetry (CV). The sensor showed a wide dynamic range (13.33 nM–66.67 μM) for AZM detection, and an estimated detection limit in the subnanomolar range (0.85 nM). The sensor was simple to build, reusable and had a good shelf-life, exhibiting remarkable selectivity over a wide number of antibiotics, commonly associated drugs and endogenous compounds [1].

7.2 Erythromycin

Erythromycin has a similar antimicrobial spectrum as penicillin and is widely used especially for the treatment of patients allergic to penicillin. In a research towards detection of Erythromycin, Vajdle et al. used a renewable silver-amalgam film electrode (Hg(Ag)FE) for the characterization and determination of erythromycin ethylsuccinate (EES), a widely used esterified form of the antibiotic, by means of cyclic voltammetry (CV) and square wave voltammetry (SWV). In the aqueous Britton-Robinson buffer (pH 5.0–9.0), one reduction peak of EES was observed in the potential range of –0.75V to –1.80 V vs. SCE, with peak potential maxima ranging from –1.59 V to –1.70 V, depending on the applied pH as well as the peak shape. The optimal conditions resulted in stable SWV responses with a decent linear correlation in the EES concentration range from 4.53 to 29.8 μg mL^{-1} (LOD = 1.36 μg mL^{-1}), and from 0.69 μg mL^{-1} to 2.44 μg mL^{-1} (LOD 0.21 μg mL^{-1}) in the case of optimized SW-AdSV. The relative standard deviation was found to be below 1.5%. The reliability of the elaborated procedures and thus the accuracy of the results were validated by comparing them with those obtained by means of

HPLC-DAD measurements. The direct cathodic SWV method was successfully applied for the detection of EES in the pharmaceutical preparation Eritromicin1, while SW-AdSV was applied in spiked urine sample. In both cases, the standard addition method was used [25].

8. Glycopeptides

Glycopeptide antibiotics are active against a limited number of bacteria, inhibiting their growth and reproduction, rather than killing them immediately [2]. Glycopeptides include the drug vancomycin—commonly used when other antibiotics fail.

8.1 Vancomycin

Gill et al. modified a copper(II) benzene-1,3,5-tricarboxylate (BTC) metal-organic framework (MOF) with poly(acrylic acid) (PAA) and used it in a vancomycin electrochemical sensor. The MOF, which was synthesized through a single-pot method, improved MOF solubility and dispersibility in water, without compromising its crystallinity and porosity. It showed improved electrocatalytic properties when placed on a glassy carbon electrode (GCE). This led to better dispersion of the MOF and improved interaction between MOF and vancomycin. Furthermore, the structural, spectral and electrochemical properties of the MOFs and their vancomycin complexes was characterized. The modified GCE proved to be a viable tool for electrochemical detection (best at a working potential of 784 mV vs. Ag/AgCl) of the antibiotic vancomycin in spiked urine and serum samples. Response was linear in the 1–500 nM vancomycin concentration range, and the detection limit was found to be 1 nM, with a relative standard deviation of ±4.3% [26].

9. Oxazolidinones

Oxazolidinones are active against gram-positive bacteria and act by inhibiting protein synthesis, and their growth and reproduction in turn. They are a completely synthetic class of antimicrobial agents. Two less toxic oxazolidinone derivatives were developed in the 1990s, one of which is linezolid morpholine. It binds to the subunit of ribosomes in bacteria and inhibits protein synthesis. The pharmacokinetic properties of Linezolid (LNZ), along with its effective ability to enter and concentrate in the tissue in many parts of the body, make it an effective drug for use in surgical infections [2, 27].

9.1 Linezolid

Aydin et al. investigated the electrochemical properties of linezolid, using unmodified renewable pencil graphite and carbon paste electrodes, in a Britton Robinson buffer solution (pH 2–10) [27]. The experimental studies were examined by square-wave (SW) and cyclic voltammetry (CV). Initially, the parameters affecting the performance of both the electrodes, such as electrode activation, support electrolyte selection, pH effect, interval of potentials, scan rates and voltammetric parameters were investigated with square wave voltammetry. Linezolid was not reversibly oxidized in supporting electrolyte solutions, yielding well-defined peaks in its oxidation potential range. Using the pencil graphite electrode (PGE), a linear relationship between peak current and concentration was obtained between 0.01 and 0.2 μg mL^{-1} ($2.96 \times 10^{-8} - 5.93 \times 10^{-7}$ M); at the carbon paste electrode (CPE), the range was 0.1 and 7.5 μg mL^{-1} ($2.96 \times 10^{-7} - 2.22 \times 10^{-5}$ M), as obtained by SWV. The selectivity, sensitivity, precision and other validation parameters of the developed voltammetric method was also evaluated. Eventually, the concentration of linezolid was examined in pharmaceutical and biological samples using the standard addition technique [27].

10. Ansamycins

Ansamycins are effective as antibiotics against a few gram-positive and gram-negative bacteria. They lead to the death of bacterial cells by inhibiting the production of RNA, which is critical for bacterial cells [2].

11. Quinolones

Quinolones are bactericidal compounds that interfere with the replication and transcription of DNA in bacterial cells. They are broad-spectrum antibiotics and are used for treating urinary tract infections as well as for veterinary purposes [2].

11.1 Levofloxacin

Rkik et al. developed a method to detect the presence of Levofloxacin (LEV) antibiotic in human serum and urine samples, using cyclic voltammetry (CV) and square wave voltammetry (SWV). Boron-doped diamond (BDD) was used as a sensor for these voltammetric methods. This electrode was used as the anode, the electrochemical response of which showed three irreversible well-defined peaks. The oxidation mechanism of the molecule included the transfer of 2 electrons and 2 protons, leading to LEV N-oxide. Under optimized conditions, CV was less sensitive than SWV. At length, the SWV method gave the lowest limit of detection and quantification. LOD and LOQ were found to be 2.88 and 9.60 μM for SWV respectively [28].

11.2 Ofloxacin

In a study by Elfiky et al. a new sensor was prepared using a carbon paste electrode mixed with flake graphite and oxidized multi-walled carbon nanotubes. The electric conductivity and electrocatalytic activity of this sensor towards the electrochemical oxidation of ofloxacin were better than those of modified graphite flake-shaped carbon paste electrodes or oxidized multi-walled carbon nanotubes. Under optimal conditions, the modified carbon paste electrode achieved detection limits of 0.18 nM and 0.24 nM in pharmaceutical samples and spiked human urine samples respectively. The sensor proved to be extremely stable, reusable and free of interference from other common excipients in drug formulations [29].

11.3 Ciprofloxacin

Zhang et al. prepared a glassy carbon electrode modified with poly(alizarin red)/electrodeposited graphene (PAR/EGR) composite film for the detection of ciprofloxacin (CIP), in the presence of ascorbic, uric acid and dopamine [30]. The electrocatalytic oxidation of CIP on AR/EGR was investigated by cyclic voltammetry (CV) and differential pulse voltammetry (DPV). A linear detection range of 4×10^{-8} to 1.2×10^{-4} M was observed, with a detection limit (S/N = 3) of 0.01 μM. The modified electrode could be applied to the detection of CIP individually as well as in the presence of ascorbic acid, uric acid and dopamine. In conclusion, the method proved to be very simple and selective for the detection of CIP in pharmaceutical preparation and biological media [30].

In another study, ciprofloxacin (CIP) was detected using an indirect electrochemical method based on the complexation of CIP with Cd^{2+} [31]. On a graphene-modified electrode, Cd^{2+} showed a strong anodic stripping peak current response, which was efficiently prohibited in the presence of CIP. Thus, the anodic stripping peak current of Cd^{2+} was considered as the indicative signal for CIP detection. Some parameters such as the modified volume of graphene, solution pH, accumulation time and complexation reaction time were also investigated. Under optimum conditions, the concentration range of CIP was found to be 1.0×10^{-7} to 1.0×10^{-5} M, with a detection limit of 5.9×10^{-8} M. The method was very selective and reproducible, and could be successfully used to detect CIP in pharmaceutical formulations and human urine [31].

Voltammetric methods for the detection of ciprofloxacin have been further developed. In one such study, a cathodically pre-treated boron-doped diamond (BDD) electrode coupled with square-wave voltammetry (SWV) and differential pulse voltammetry (DPV) was used [32]. Analytical curves were obtained for CIP concentrations from 2.50 to 50.0 μM for SWV, and from 0.500 to 60.0 μM for DPV, with detection limits of 2.46 and 0.440 μM respectively. Moreover, sufficient recovery values were obtained for the detection of CIP in synthetic urine samples by DPV. On the other hand, SWV was employed to evaluate the interaction between CIP and double-stranded dsDNA. The results showed that CIP binds to dsDNA by intercalation, with a binding constant calculated as 5.91×10^{5} L mol^{-1}. Thus, the cathodically pre-treated BDD electrode was successfully deployed for the detection of CIP in biological samples and for studies on the interaction of this fluoroquinolone with dsDNA [32].

In yet another study, a new and efficient electrochemical sensor was developed to detect ciprofloxacin, using a composite of magnetic multi-walled carbon nanotubes (MMWCNTs) and molecularly imprinted polymer (MIP) [33]. A magnetic MIP (MMIP) was synthesized by simple procedure; MMIP can respond selectively to ciprofloxacin. The performance of the carbon paste electrode modified with MMIP was investigated using cyclic voltammetry and differential pulse voltammetry. The detection limit of this method was found to be 0.0017 μM, with a linear detection range (3Sb/m) of 0.005–0.85 μM. The detection of ciprofloxacin was possible in pharmaceutical samples and biological fluids. This method proved to have potential applications in routine analysis, given its high specificity, excellent reproducibility and good stability [33].

The aim of the final paper discussed under this section was to use boron-doped diamond electrodes for a sensitive, simple and reliable voltammetric detection of ciprofloxacin in human urine samples. Initially, an optimal level of boron doping was determined to achieve the highest sensitivity. Ciprofloxacin provided a well-defined irreversible oxidation peak at a potential of +1.15 V, using cyclic voltammetry, in an ammonium acetate buffer (pH 5). Optimal experimental conditions were achieved and the concentration range obtained by square wave voltammetry was linear from 0.15 to 2.11 μM ($R^2 = 0.9974$), while the limit of detection (0.05 μM) was very low. The developed square wave method was successfully used to detect ciprofloxacin in human urine samples, with a decent recovery (97 to 102%) [34].

12. Streptogramins

Streptogramins are usually given as a combination of two antibiotics from different classes of antibiotics: streptogramin A and streptogramin B. Individually, these compounds inhibit growth, but when used in combination, they have a synergistic effect and can directly kill bacterial cells by inhibiting protein synthesis. They are often used to treat resistant infections; however bacterial resistance has developed for these antibiotics as well [2].

13. Lipopeptides

Lipopeptides are bactericidal towards gram-positive bacteria. Daptomycin is the most widely used antibiotic in this class and has a unique mechanism of action. It disrupts the function of the cell membrane in bacteria, and is administered through injection, treating infections of the skin and tissues. Hence, this mechanism of action of daptomycin is significantly beneficial and the cases of drug resistance appear to be rare [2].

14. Future prospects

Electroanalysis may hold excellent prospects for detecting antibiotics in biological fluids, given their high sensitivity and accuracy. However, electrochemical techniques have a limited field of application when compared with optical and separation methods due to their relatively lower selectivity for some techniques and significant interference. Thus, it is proposed to develop new selective electrodes, as well as conduct additional research on 26 molecularly imprinted polymers (MIPs) and aptamers—which are distinguished for their high specific recognition capacity for the respective target molecules, while at the same time being unaffected by any interference that may be present in the respective substrates. Furthermore, modified electrodes can be an important object of research due to the difficulty in renewing their surface, which in turn reduces their repeatability. Ultimately, it would be interesting to further explore, study and develop electrochemical DNA biosensors that would make a significant contribution towards improving health.

Summary Table

Antibiotics	Electroanalytical Technique	Electrode	Detection Limit (μM)	Linear Range (μM)	Substrate	References
Penicillin	Differential Pulse Voltammetry (DPV)	Boron-Doped Diamond Electrode (BDDE)	0.25 μM	0.5–40 μM	Human urine and pharmaceutical formulations	[3]
Amoxicillin	Square Wave Voltammetry (SWV)	PGA/3D-GE/GCE [glassy carbon electrode (GCE) modified by three-dimensional graphene (3D-GE) and polyglutamic acid (PGA)]	0.118 μM	2–60 μM	Urine	[5]
	Square Wave Voltammetry (SWV)	Cu(II)-NCL/CPE (carbon paste electrode modified with Cu(II)-exchanged clinoptilolite nanoparticles)	0.02 μM	0.04–100 μM	Urine and pharmaceutical tablets	[6]
	Chronoamperometry	Carbon paste electrode with copper ions Cu(II)	8.84×10^{-2} μM	0.195–14.6 μM	Human blood and pharmaceutical tablets	[7]
	Square Wave Voltammetry (SWV)	Nanomaterials such as Printex 6L Carbon and cadmium telluride quantum dots within a poly(3,4-ethylenedioxythiophene) polystyrene sulfonate film	0.05 μM	0.90–69 μM	Urine, pharmaceutical and milk samples	[8]

Table contd. ...

...Table contd.

Antibiotics	Electroanalytical Technique	Electrode	Detection Limit (μM)	Linear Range (μM)	Substrate	References
Cefadroxil	Square-wave Stripping Voltammetry (SW-AdSV)	Nano-silver amalgam pastes microelectrode	0.011 μM	0.033–0.304 μM	Pharmaceutical formulations and biological specimen	[10]
	Cyclic Voltammetry	Glassy carbon electrode (GCE)	0.3 μM	1–5 μM	Commercial samples	[11]
Ampicillin	Alternating Current Voltammetry (ACV) & Square Wave Voltammetry (SWV)	"Signal-on" electrochemical aptamer-based sensor	**ACV:** 1 μM **SWV:** 30 μM	**ACV:** 5–5000 **SWV:** 100–5000 μM	Serum, saliva, and milk	[9]
Cefixime	Cyclic Voltammetry & Square Wave Voltammetry	Multi walled carbon paste electrode modified with gold nanoparticles GNPs/MWCPE	3×10^{-3} μM	$1 \times 10^{-2} - 2 \times 10^{2}$ μM	Urine and pharmaceutical samples	[12]
	Cyclic, linear sweep, differential pulse and square wave voltammetric techniques	Glassy Carbon Electrode	**DPV:** in **Serum** sample: 0.428 μM **Urine:** 5.31×10^{-2} μM **Breast milk:** 0.177 μM **SWV:** in **Serum** sample: 0.541 μM **Urine:** 0.245 μM **Breast milk:** 0.186 μM	For **Serum** sample: 6–200 μM **Urine:** 8–200 μM **Breast milk:** 6–100 μM for DPV and SWV techniques	Biological fluids (urine, serum, breast milk)	[13]
	Differential Pulse Voltammetry (DPV)	Modified pencil graphite electrode (PGE) using gold nanoparticles (AuNPs) and cetyltrimethyl ammonium bromide (CTAB)	1.21×10^{-4} μM	0.01–0.3 μM	Pharmaceutical formulations and biological fluids	[14]

Ceftazidime	Cyclic & differential pulse voltammetry, electrochemical impedance spectroscopy (EIS)	Carbon paste electrode modified by functionalized mesoporous silica material	0.30×10^{-3} μM	0.001–2.5 μM	Biological and pharmaceutical samples	[4]
Cephalosporins	Differential Pulse Voltammetry (DPV)	Bare boron-doped diamond electrode (BDDE)	34.74 nM (0,099 μM)	173.80–243313.00 μg/L (0.5–700 μM)	Enviromental, biomedical and pharmaceutical samples	[15]
Sulfamethoxazole & Trimethoprime	Differential Pulse Voltammetry (DPV)	Electrochemical sensor modified with multiwalled carbon nanotubes decorated with Prussian blue nanocubes (MWCNT/PBnC)	**TMP:** 60×10^{-3} μM **SMX:** 38×10^{-3} μM	**TMP:** 0.1–10 μM **SMX:** 1–10 μM	Artificial urine	[16]
Sulfanilamide	Square-wave voltammetry	Glassy carbon electrode (GCE)	0.92 μM	5–74.7 μM	Otologic solution, human urine and serum	[17]
Sulfasalazine	Differential Pulse Voltammetry (DPV)	Carbon paste electrode modified with molecularly imprinted polymer (MIP)	4.6×10^{-3} μM	0.01–1 μM	Human serum and pharmaceutical sample	[18]
Streptomycin	Electrochemical impedance spectroscopy (EIS)	Apt/AuNPs/GQD-SH/GCE Nanoaptasensor	5.67×10^{-8} μM (0.033 pg/ml)	$1.72 \times 10^{-7} - 1.2 \times 10^{-3}$ μM	Human serum	[19]

Table contd. ...

...Table contd.

Antibiotics	**Electroanalytical Technique**	**Electrode**	**Detection Limit (µM)**	**Linear Range (µM)**	**Substrate**	**References**
Tetracycline	Adsorptive stripping differential pulse voltammetry (AdSDPV)	Carbon paste electrode modified with multiwalled carbon nanotubes functionalized with carboxyl groups (MWCNT-COOH), together with graphene oxide (GO)	0.36 µM	20–310 µM	River water, artificial urine, and pharmaceutical samples	[20]
	Cyclic Voltammetry & Differential Pulse Voltammetry (DPV)	Platinum nanoparticles supported on carbon (PtNPs/C/GCE)	4.28 µM	9.99–44.01 µM	Urine	[21]
Linezolid	Square-wave (SW) and cyclic voltammetry (CV)	Pencil graphite and carbon paste electrodes (PGE/CPE)	BR buffer at **pH = 4:** 1.39×10^{-3} µM BR buffer at **pH = 7:** 9.25×10^{-3} µM	At **pH = 4 (PGE):** 0.0296–0.593 µM At **pH = 7 (CPE):** 0.296–22.2 µM	Pharmaceutical and biological samples	[27]
Chloramphenicol	Amperometry	PdNPs/GO/GCE	0.001 µM	0.007–102.68 µM	Biological and food samples	[22]
Ciprofloxacin	Cyclic voltammetry (CV) and differential pulse voltammetry (DPV)	Modified glassy carbon electrode	0.01 µM	0.04–120 µM	Biological and pharmaceutical samples	[30]
	Anodic stripping voltammetry	Graphene-modified electrode	59×10^{-3} µM	0.1–10 µM	Pharmaceutical formulation and human urine	[31]

	Square-wave voltammetry (SWV) and differential pulse voltammetry (DPV)	Boron doped diamond electrode (BDD)	**SWV:** 2.46 µM **DPV:** 0.440 µM	**SWV:** 2.50–50.0 µM **DPV:** 0.500–60.0 µM	Urine	[32]
	Cyclic voltammetry (CV) and differential pulse voltammetry (DPV)	Electrochemical sensor MMWCNTs/MIP	0.0017 µM	0.005–0.85 µM	Biological fluids & pharmaceutical samples	[33]
	Cyclic voltammetry (CV) & square wave voltammetry (SWV)	Boron doped diamond electrode	0.05 µM	0.15–2.11 µM	Human urine	[34]
Azithromycin	Cyclic voltammetry (CV) and differential pulse voltammetry (DPV)	Electrochemical sensor MMWCNTs/MIP	0.0017 µM	0.005–0.85 µM	Biological fluids & pharmaceutical samples	[1]
Erythromycin	Cyclic voltammetry (CV) & square wave voltammetry (SWV)	Boron doped diamond electrode	0.05 µM	0.15–2.11 µM	Human urine	[25]
Ofloxacin	Square wave adsorptive anodic stripping voltammetry (SWAdASV)	Carbon paste electrode blended with flake graphite and oxidized multiwall carbon nanotubes ([FG_2/MW_2]/CPE)	In **urine:** 0.24×10^{-3} µM In **medicines:** 0.18×10^{-3} µM	**urine:** 0.0008–0.02 µM **medicines:** 0.0006–0.015 µM	Pharmaceutical and human urine samples	[29]

Table contd. ...

...Table contd.

Antibiotics	Electroanalytical Technique	Electrode	Detection Limit (μM)	Linear Range (μM)	Substrate	References
Levofloxacin	Cyclic voltammetry (CV) and square wave voltammetry (SWV)	Boron doped diamond electrode (BDD)	**CV:** 10.01 μM **SWV:** 2.88 μM	**CV:** 48–100 μM **SWV:** 10–80.9 μM	Urine and human serum	[28]
Vancomycin	Cyclic Voltammetry	Glassy carbon electrode modified with MOF [copper(II) benzene-1,3,5-tricarboxylate (BTC) metal-organic framework]	10^{-3} μM	10^{-3}–0.5 μM	Urine & serum	[26]

Literature

[1] Stoian, I.A., Bogdan-Cezar Iacob, Cosmina-Larisa Dudaş, Lucian Barbu-Tudoran, Diana Bogdan, Iuliu Ovidiu Marian, Ede Bodoki and Radu Oprean. 2020. Biomimetic electrochemical sensor for the highly selective detection of azithromycin in biological samples. Biosens. Bioelectron. 155(February): 112098.

[2] Compound-Interest. A Brief Overview of Classes of Antibiotics. [Online]. Available: http://www.compoundchem.com/2014/09/08/antibiotics/.

[3] Švorc, Ľ., J. Sochr, M. Rievaj, P. Tomčík and D. Bustin. 2012. Voltammetric determination of penicillin V in pharmaceutical formulations and human urine using a boron-doped diamond electrode. Bioelectrochemistry 88: 36–41.

[4] Dehdashtian, S. and Z. Abdipur. 2017. Fabrication of an ultrasensitive electrochemical sensor based on a mesoporous silica material functionalized by copper ion (SBA-15-Cu(II)) modified carbon paste electrode for determination of antibiotic ceftazidime and its application in pharmaceutical an. J. Iran. Chem. Soc. 14(8): 1699–1709.

[5] Chen, C. et al. 2019. Amoxicillin on polyglutamic acid composite three-dimensional graphene modified electrode: Reaction mechanism of amoxicillin insights by computational simulations. Anal. Chim. Acta 1073: 22–29.

[6] Nosuhi, M. and A. Nezamzadeh-Ejhieh. 2017. Comprehensive study on the electrocatalytic effect of copper–doped nano-clinoptilolite towards amoxicillin at the modified carbon paste electrode–solution interface. J. Colloid Interface Sci. 497: 66–72. Doi: 10.1016/j.jcis2017.02.055. Epub 2017 Feb 24.

[7] Wong, A., A.M. Santos, F.H. Cincotto, F.C. Moraes, O. Fatibello-Filho and M.D.P.T. Sotomayor. 2020. A new electrochemical platform based on low cost nanomaterials for sensitive detection of the amoxicillin antibiotic in different matrices. Talanta 120252.

[8] Hrioua, A., A. Farahi, S. Lahrich, M. Bakasse, S. Saqrane and M.A. El Mhammedi. 2019. Chronoamperometric detection of amoxicillin at graphite electrode using chelate effect of copper(II) Ions : Application in human blood and pharmaceutical tablets. Chemistry Select 4(28): 8350–8357.

[9] Yu, Z. gang and R.Y. Lai. 2018. A reagentless and reusable electrochemical aptamer-based sensor for rapid detection of ampicillin in complex samples. Talanta 176: 619–624. Doi: 10.1016/j.talanta.2017.08.057. Epub 2017 Aug 19.

[10] Atif, S., J.A. Baig, H.I. Afridi, T.G. Kazi and M. Waris. 2020. Novel nontoxic electrochemical method for the detection of cefadroxil in pharmaceutical formulations and biological samples. Microchem. J. 154(December 2019): 104574.

[11] Sanz, C.G., S.H.P. Serrano and C.M.A. Brett. 2019. Electrochemical characterization of cefadroxil β-lactam antibiotic and Cu(II) complex formation. J. Electroanal. Chem. 844(January): 124–131.

[12] Afkhami, A., F. Soltani-Felehgari and T. Madrakian. 2013. Gold nanoparticles modified carbon paste electrode as an efficient electrochemical sensor for rapid and sensitive determination of cefixime in urine and pharmaceutical samples. Electrochim. Acta 103: 125–133.

[13] Golcu, A., B. Dogan and S.A. Ozkan. 2005. Anodic voltammetric behavior and determination of cefixime in pharmaceutical dosage forms and biological fluids. Talanta 67(4): 703–712.

[14] Manjunatha, P. and Y.A. Nayaka. 2019. Cetyltrimethylammonium bromide-gold nanoparticles composite modified pencil graphite electrode for the electrochemical investigation of cefixime present in pharmaceutical formulations and biology. Chem. Data Collect. 21: 100217.

[15] Feier, B., A. Gui, C. Cristea and R. Săndulescu. 2017. Electrochemical determination of cephalosporins using a bare boron-doped diamond electrode. Anal. Chim. Acta 976: 25–34.

[16] Sgobbi, L.F., C.A. Razzino and S.A.S. Machado. 2016. A disposable electrochemical sensor for simultaneous detection of sulfamethoxazole and trimethoprim antibiotics in urine based on multiwalled nanotubes decorated with Prussian blue nanocubes modified screen-printed electrode. Electrochim. Acta 191: 1010–1017.

[17] Ferraz, B.R.L., T. Guimarães, D. Profeti and L.P.R. Profeti. 2018. Electrooxidation of sulfanilamide and its voltammetric determination in pharmaceutical formulation, human urine and serum on glassy carbon electrode. J. Pharm. Anal. 8(1): 55–59.

[18] Sadeghi, S., A. Motaharian and A.Z. Moghaddam. 2012. Electroanalytical determination of sulfasalazine in pharmaceutical and biological samples using molecularly imprinted polymer modified carbon paste electrode. Sensors Actuators, B Chem. 168: 336–344.

[19] Ghanbari, K. and M. Roushani. 2018. A novel electrochemical aptasensor for highly sensitive and quantitative detection of the streptomycin antibiotic. Bioelectrochemistry 120: 43–48.

[20] Wong, A., M. Scontri, E.M. Materon, M.R.V. Lanza and M.D.P.T. Sotomayor. 2015. Development and application of an electrochemical sensor modified with multi-walled carbon nanotubes and graphene oxide for the sensitive and selective detection of tetracycline. J. Electroanal. Chem. 757: 250–257.

[21] Kushikawa, R.T., M.R. Silva, A.C.D. Angelo and M.F.S. Teixeira. 2016. Construction of an electrochemical sensing platform based on platinum nanoparticles supported on carbon for tetracycline determination. Sensors Actuators, B Chem. 228: 207–21.

[22] Kokulnathan, T., T.S.K. Sharma, S.M. Chen, T.W. Chen and B. Dinesh. 2018. *Ex-situ* decoration of graphene oxide with palladium nanoparticles for the highly sensitive and selective electrochemical determination of chloramphenicol in food and biological samples. J. Taiwan Inst. Chem. Eng. 89: 26–38.

[23] Yadav, M., V. Ganesan, R. Gupta, D.K. Yadav and P.K. Sonkar. 2019. Cobalt oxide nanocrystals anchored on graphene sheets for electrochemical determination of chloramphenicol. Microchem. J. 146: 881–887.

[24] Hu, L., Tingting Zhou, Jingwen Feng, Hua Jinc, Yun Tao, Dan Luo, Surong Mei and Yong-Ill Lee. 2018. A rapid and sensitive molecularly imprinted electrochemiluminescence sensor for Azithromycin determination in biological samples. J. Electroanal. Chem. 813(November 2017): 1–8.

[25] Vajdle, O., Valéria Guzsvány, Dušan Škorić, Jasmina Anojčić, Pavle Jovanov, Milka Avramov-Ivić, János Csanádia Zoltán Kónya, Slobodan Petrović and Andrzej Bobrowski. 2016. Voltammetric behavior of erythromycin ethylsuccinate at a renewable silver-amalgam film electrode and its determination in urine and in a pharmaceutical preparation. Electrochim. Acta 191: 44–54.

[26] Gill, A.A.S., Sima Singh, Nikhil Agrawal and Rajshekhar Karpoormath. 2020. A poly(acrylic acid)-modified copper-organic framework for electrochemical determination of vancomycin. Microchim. Acta 187(1).

[27] Aydin, I., H. Akgun and P.T. Pınar. 2019. Analytical determination of the oxazolidinone antibiotic linezolid at a pencil graphite and carbon paste electrodes. Chemistry Select 4(34): 9966–9971.

[28] Rkik, M., M. Ben Brahim and Y. Samet. 2017. Electrochemical determination of levofloxacin antibiotic in biological samples using boron doped diamond electrode. J. Electroanal. Chem.

[29] Elfiky, M., N. Salahuddin, A. Hassanein, A. Matsuda and T. Hattori. 2019. Detection of antibiotic Ofloxacin drug in urine using electrochemical sensor based on synergistic effect of different morphological carbon materials. Microchem. J. 146: 170–177.

[30] Zhang, X., Y. Wei and Y. Ding. 2014. Electrocatalytic oxidation and voltammetric determination of ciprofloxacin employing poly(alizarin red)/graphene composite film in the presence of ascorbic acid, uric acid and dopamine. Anal. Chim. Acta. 835: 29–36. Doi: 10.1016/j.aca.2014.05.020.

[31] Shan, J., Y. Liu, R. Li, C. Wu, L. Zhu and J. Zhang. 2015. Indirect electrochemical determination of ciprofloxacin by anodic stripping voltammetry of Cd(II) on graphene-modified electrode. J. Electroanal. Chem. 738: 123–129.

[32] Garbellini, G.S., R.C. Rocha-Filho and O. Fatibello-Filho. 2015. Voltammetric determination of ciprofloxacin in urine samples and its interaction with dsDNA on a cathodically pretreated boron-doped diamond electrode. Anal. Methods 7(8): 3411–3418.

[33] Bagheri, H., H. Khoshsafar, S. Amidi and Y. Hosseinzadeh Ardakani. 2016. Fabrication of an electrochemical sensor based on magnetic multi-walled carbon nanotubes for the determination of ciprofloxacin. Anal. Methods 8(16): 3383–3390.

[34] Radičová, M., Miroslav Behúl, Marián Marton, Marian Vojs, Róbert Bodor, Robert Redhammer and Andrea Vojs Staňová. 2017. Heavily boron doped diamond electrodes for ultra sensitive determination of ciprofloxacin in human urine. Electroanalysis 29(6): 1612–1617.

2

Electrochemical DNA Sensors Based on Nanomaterials for Pharmaceutical Determination

Anna Porfireva,[1] *Veronika Subjakova*,[2] *Gennady Evtugyn*[1] and *Tibor Hianik*[2,]*

1. Introduction

An electrochemical biosensor is a self-contained integrated device which can provide specific quantitative or semi-quantitative analytical information using a biological recognition element (biochemical receptor) retained in direct spatial contact with an electrochemical transduction element [1]. This IUPAC definition elaborated in numerous discussions sheds light on the main peculiarities of biosensors' assembly and behavior that set them apart from other chemical sensors and provide undisputable advantages and a few drawbacks. The source of the recognition element plays a crucial role in the performance of a biosensor—in terms of both extraordinary selectivity and sensitivity of recognition event and modest stability of the signal, but with a rather short lifetime of the biosensor. Although some new receptors mimicking nature have been recently introduced in biosensor assembly (aptamers [2], molecularly imprinted polymers [3] and DNA chimeras [4]), they are yet to surpass the numerous properties and opportunities that natural components offer—in terms of analysis, especially in medicine, pharmacy, toxicology and environmental monitoring. From the very beginning of biosensor history, they were considered as a tool of diagnostics, which were simpler in operation, and faster and more reliable

[1] A.M. Butlerov' Chemistry Institute of Kazan Federal University, 18 Kremlevskaya Street, Kazan, 420008, Russian Federation.

[2] Department of Nuclear Physics and Biophysics, Comenius University, Mlynska dolina F1, 842 48 Bratislava, Slovakia.

* Corresponding author: Tibor.Hianik@fmph.uniba.sk

than traditional assays based on "wet chemistry". Biochemical reactions, mainly those catalyzed by enzymes, were considered as simple models for the same process in human beings. Complication of biochemical parts of biosensors and introduction of biochemical reactants other than enzymes made it possible to extend the potential application of biosensors.

This is especially true for deoxyribonucleic acids (DNAs) introduced in biosensor investigations later than enzymes. On one hand, DNAs have a unique structure of a huge size and considerable regularity, offering many opportunities for molecular recognition. On the other hand, this molecule is too big and natural functions of DNAs are very complicated; they demand much longer periods than those covered with biosensor signal measurement. For this reason, instead of whole DNA molecules, their parts related to some specific genes have been mainly used in biosensor assembly (DNA probes). The detection of hybridization event, i.e., the reaction of the DNA probe with the complementary sequence, has been used as an evidence of appropriate genetic material in samples tested though quantitative analysis of appropriated sequences. However, this was not sufficiently accurate or reproducible. Today, DNA sensors have considerably extended areas of application and have learned to mimic other functions like elongation and shortening of primary nucleotide sequences. This is compensated for by a much more complicated measurement protocol, which is sometimes far from the initial idea of a reagent-free sensor [5].

Nevertheless, in the area of DNA sensors, one topic remains mostly in trend for simple and fast analysis of small molecules. This is the investigation of DNA—drug interactions that cover both the theoretical aspects of kinetics exploration or affinity constant calculations, and the applied area of real sample assays (including screening of new drug formulations and individual dose assessment). Most of such investigations cover anti-cancer drugs that are either targeted at DNA or affect DNA-related biochemical functions. In this review, various aspects of such biosensors are discussed, along with the role and importance of nano-sized components that now offer new opportunities in the sensitivity of assays and microfabrication of DNA sensors as parts of microfluidics and lab-on-chip devices. Rather useful for biosensor assembly are sensing layers performed by electropolymerization that are characterized by a variety in the shape and size. All such materials are considered as a part of biorecognition system and their influence on the performance of DNA sensors has been illustrated through examples of drug detection or determination.

Recently, several comprehensive reviews associated with various aspects of biosensors used for drug analysis have been published [6, 7]. Most of them consider the classification of appropriate biosensors based on biochemical components, and a significant part of the reviews is devoted to enzyme-based and immune sensors. In this review, DNA-based sensors have been focused on and significant attention has been paid to the mechanism of signal generation and the role of microenvironment on the DNA immobilized on the transducer for the analytical characteristics of the response.

2. Nanomaterials used in the assembly of electrochemical DNA sensors

Current progress in the development of biosensors is, to a great extent, based on the synthesis and application of new nanomaterials. Depending on their nature, size, shape and functionality, nanomaterials can serve as universal platforms for immobilization of bioreceptors, or as signal generating elements or amplifiers intended for enhancing the signals of existing labels. Nanomaterials are sufficiently compatible with miniaturization and microfabrication trends and are considered as an indispensable part of future measurement devices to be used for biomedical applications within a paradigm of point-of-care platforms. Recently, several reviews have summarized the trends in the application of nanomaterials in electrochemical biosensors, with an emphasis on DNA-based devices [8].

2.1 Carbon nanomaterials

Carbon nanomaterials such as carbon nanotubes, graphene and its derivatives, and carbon black have garnered significant attention for offering unique chemical, physical and electrical properties, and due to their ability as nanomaterials to allow for tuning their size and shape for a high surface area. These properties have thus opened up the possibility of designing novel systems and fabrication of electrochemical sensors [9]. Electrochemical sensors provide several advantages in terms of cost, sensitivity, selectivity, reliability and size, and hold prospects for many promising applications in biomedical fields and analytical chemistry [10].

Carbon nanotubes

Since their discovery in 1991 [11], carbon nanotubes (CNTs) have attracted the attention of researchers and have come into use in a wide range of biosensing applications. Carbon nanotubes are cylindrical molecules made of sp^2 hybridized carbon atoms in a hexagonal lattice; their size is typically in the order of several nanometers in diameter and micrometers in length. CNTs are known in two structural compositions: single-wall carbon nanotubes (SWNTs) with one rolled-up graphene sheet (Figure 1A), and multi-wall carbon nanotubes (MWNTs) consisting of two or more rolled concentric graphene sheets (Figure 1B) [12–14]. The application of CNTs in electrochemical sensors is based on their remarkable electronic properties and high chemical stability [12]. Furthermore, for using CNTs in biosensors, functionalization of CNT surface is necessary to overcome their low dispersion in most solvents and to increase biocompatibility through covalent or non-covalent bonding [15].

Graphene and its derivatives

Graphene is a one atom thick, two-dimensional planar carbon material with notable chemical and physical properties. Reduced graphene can be produced by different methods—chemical, thermal and electrochemical; the resulting defects and oxygen groups help enhance electron transfer and make it more effective for further functionalization by biomolecules [13]. Graphene oxide (GO) contains hydroxyl and carboxyl groups, and can easily adsorb various functional molecules such as DNA through mainly π-π stacking and hydrogen bonding (Figure 1C). GO can be used in electrochemical sensors and as efficiently in optical sensors [16]. For using graphene in electrochemistry, it is necessary to study its electrochemical properties and behaviors [17]. Chemically reduced graphene on glassy carbon electrode has been used for studying the oxidation characteristics of free DNA bases. The electrochemical activity of DNA bases is greatly enhanced at GO-modified surface [18]. The enhancement of the activity of GO has also been confirmed for other species like dopamine, NADH, etc. Functionalized graphene-modified graphite electrode has been used for the detection dopamine to eliminate interference from ascorbic acid and uric acid. Three separate oxidation peaks were thence observed in the voltammogram. It has been found that functionalized graphene increases the electrocatalytic activity of graphite electrode [19]. Electrochemical graphene-based biosensors have also been reviewed in literature [20].

Carbon nanomaterials have been also used in many electrochemical sensors for monitoring DNA-drug interactions. The main principle is based on the electroactive nature of DNA bases and drugs [21]. For example, DNA-GO-modified glassy carbon electrode was investigated for studying the electrochemical signal of DNA depending on the concentration of diclofenac [22]. GO-modified electrode demonstrated enhanced electrochemical signal for the detection of the anti-cancer drug mitomycin C. 21.4% and 34.4% signal enhancement was observed for mitomycin C and guanine respectively, compared to an electrode without GO. Specific binding of mitomycin to DNA resulted in a suppressed electrochemical signal [23]. The guanine oxidation peak was examined by DPV as a reference for the detection of the anti-cancer drug methotrexate. DNA was immobilized on GO-modified glassy carbon electrode by electrodeposition. The applied potential and time duration affected the guanine oxidation peak. The presence of GO has shown higher sensitivity for the detection of methotrexate (limit of detection (LOD): 7.6 nM) [24].

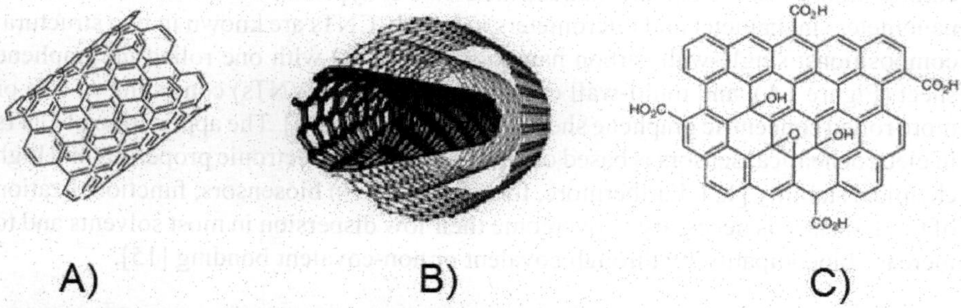

Figure 1. Examples of structures of (A) single-wall carbon nanotubes (SWNTs), (B) multi-wall carbon nanotubes (MWNTs), and (C) graphene oxide (GO).

SWCNTs and MWCNTs have been successfully employed for electrode modification and for the study of the interaction of drugs with DNA, indicating excellent electrocatalytic behavior [25], as well as for the development of aptasensors [26].

Carbon nanomaterials with other polymers and metal nanoparticles are very attractive in terms of increasing the active surface and electron transfer at the electrode, combining the benefits of nanomaterials. Composites of graphene and chitosan [27] and MWCNTS and chitosan [28] increase the active surface area significantly on a glassy carbon electrode, resulting in a higher oxidative peak in the presence of $K_3[Fe(CN)_6]$. The above-mentioned modified electrodes have been used for studying the interaction of dsDNA and the anti-cancer drug 6-mercaptopurine. Furthermore, dsDNA was immobilized on a pencil graphite electrode modified by a composite of polypyrrole/functionalized MWCNTs through external application of voltage. The decrease in the oxidation response of guanine and adenine after the interaction with 6-mercaptopurine was also examined. The response of guanine was proportional to the concentration of the analyte and an LOD of 0.08 μM was achieved [29].

Carbon black

Among carbon materials, carbon black is an extremely low-cost conductive material widely used in electrochemical sensors due to its morphological and electronic properties with a 3–100 nm size range of agglomerates. Fabrication of a carbon black electrode is based on a thin layer of composites [30]. A DNA biosensor based on a composite of acetylene black has been used for the detection of DNA hybridization, with an LOD as low as 0.12 fM [31]. In another study, a voltammetric DNA sensor was fabricated based on a composite of carbon black, pillar[5]arene molecules and poly (Neutral red), layer-by-layer on a glassy carbon electrode, for the detection of oxidative DNA damage [32]. The sensor allowed for the detection of oxidative damage of DNA in the presence of reactive oxygen species generated by Cu^{2+}/H_2O_2 mixture.

2.2 Gold nanoparticles

Gold nanoparticles (AuNPs) are small particles (less than 100 nm in diameter) with tunable optical and electrical properties with respect to size and shape. For the synthesis of AuNPs, various methods have been introduced; the most widely known process therein was developed by Turkevich et al. [33]. Their method was based on the reduction of hydrogen tetrachloroaurate ($HAuCl_4$) (heated till boiling) by citric acid. Other strategies have also been developed for the synthesis of AuNPs in controlled sizes and shapes [34]. There is a wide range of applications of AuNPs in sensing, targeted drug delivery and imaging, by virtue of their easy surface modification and biocompatibility [35].

Nanoparticles play an important role in electrochemical sensors for immobilization or labeling of biomolecules, catalysis of chemical reaction and enhancement of electron transfer [36]. In this light, AuNP-modified electrodes increase the active surface and conductivity, and can be prepared by electrodeposition, for controlled properties [37, 38].

Gold electrodes are very common as working electrodes in electrochemical measurements due to their high stability in air and high affinity to thiol groups. Molecules containing thiol groups can be chemisorbed on the gold surface and form self-assembled monolayers (SAM) (Figure 2) [39, 40]. The effect of surface pre-treatment is very crucial for the formation of well-ordered SAM as a support for further immobilization of biomolecules and for blocking non-specific interactions. Several cleaning techniques have been reported and compared in literature [41]. AuNPs can be used for increasing the effective area of the biosensor for the immobilization of DNA probes (Figure 2A) as well as for enhancing the detection of DNA hybridization (Figure 2B).

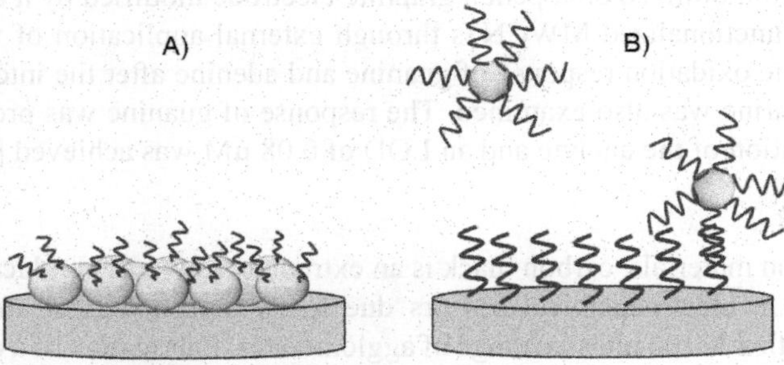

Figure 2. Schematic illustration of AuNPs in an electrochemical sensor: (A) AuNPs increase the active surface area of the electrode, (B) AuNPs as a label, for DNA hybridization detection, for signal amplification.

Antibiotic drugs play an important role in disease treatment; however, an overdose of antibiotics leads to bacterial resistance. Similarly, even anti-cancer drugs cause many side effects in the human body. Herein, several electrochemical DNA sensors based on AuNPs have been reported for the detection of antibiotic and anti-cancer drugs [42–47]. For example, electrodeposition of AuNPs on a flat gold electrode improves the sensitivity of the DNA sensor for antibiotic drug detection (nogalamycin, mithramycin and netropsin). Electron transfer was measured by impedance spectroscopy and LOD was enhanced by a factor of 15–40 with AuNPs [42]. Another strategy based on AuNPs, designed for the detection of ampicillin, another antibiotic drug, uses a glassy carbon electrode, wherein DNA functionalized AuNPs are immobilized on the electrode surface through hybridization with DNA aptamers. Therein, single-stranded DNA binding protein (ssDNA-BP) had a specific interaction with DNA-functionalized AuNPs, resulting in a drop in the signal from the redox probe. In the presence of the drug, the electrochemical signal increased due to the blocking interaction between DNA-functionalized AuNPs and ssDNA-BP. An LOD of 0.38 pM was achieved and the recovery in spiked milk samples was observed to be 95.5–105.5% [46]. Furthermore, addition of functionalized AuNPs amplifies the changes in the peak current after the specific interaction of DNA bases with the anti-cancer drug dacarbazine [43].

The presence of AuNPs increases the active surface area of a pencil graphite electrode by a factor of 5.1, for modification by dsDNA, and facilitates the detection of anti-cancer drugs, with an LOD of 1 nM in the linear response range of 5 nM–45 µM. Such a biosensor has shown good performance in terms of detection of drugs in real samples, such as capsules, urine and blood serum [47].

Fabrication of nanocomposites combines the unique characteristics of nanomaterials as well as polymers. AuNPs are often used in nanocomposites to increase the electrochemical active area of electrodes and their conductivity [48, 49].

In one study, a nanocomposite platform of AuNPs and polymer polyvinylferrocenium was used for analyzing the interaction of DNA and chemotherapeutic agent mitomycin C [50]. Therein, changes in the peak current of electroactive DNA bases (guanine and adenine) were investigated using DPV. The presence of AuNPs in polymer-coated Pt electrode enhanced the oxidation peak signal for both the DNA bases.

A team of researchers constructed an electrochemical sensor for the detection of dopamine on carbon ionic liquid electrode by step-by step electrodeposition of graphene, AuNPs and dsDNA [51]. The application of nanomaterials resulted in good sensor performance and well-defined redox peaks. The sensor had a linear response in the range 70–600 μM, with an LOD of 19 nM [51]. Other electrochemical sensors based on AuNP nanocomposites have been demonstrated for the detection of chemotherapeutic drug valrubicin [45] and for monitoring the interaction of isoproterenol-(β-adrenergic receptor agonist) induced DNA damage of cardiomyocyte [52]. Thus, nanoparticles are a rather useful tool for biosensor fabrication, increasing the sensitivity of response.

2.3 Metal organic frameworks structures

Metal organic frameworks (MOFs) are novel prospective molecular crystalline materials with a wide range of applications in biosensing, catalysis, separation and nanomedicine. MOFs are composed of metal ions or clusters linked by organic ligands. The organic linkers involved in MOFs are the source of a conjugated π-electron system and provide a basis for hydrogen bonds that can facilitate interactions between MOFs and single-stranded DNA (ssDNA). MOFs can recognize DNA and other biomolecules by means of electrochemical changes [53]. These materials can be compared with sponges with unique abilities—being able to absorb, hold and release molecules from their pores. With a highly-ordered framework of pores, MOFs exhibit the largest known surface areas per gram—one gram of an MOF can have a surface area of up to 7000 m^2. Therefore, MOFs are the fastest growing class of materials in chemistry today. More than 20000 MOFs have been found in the last 20 years. The scheme of formation of hemin@MOF composite is presented in Figure 3.

A unique feature of MOFs, owing to their high electron transfer ability and porous structure, is that they can be doped with various species such as nanoparticles, nucleic acids, protein receptors, etc. However, for electrochemical sensing the MOF must be doped with conducting materials like metallic nanoparticles or carbon nanomaterials. Among the several advantages of MOFs, the following are important: (1) the highly porous three-dimensional structure of MOFs, which enlarges the electroactive surface area (hence, more molecules can be involved in electrochemical response); (2) the hydroxyl and carboxyl groups in MOFs can interact with biomolecules via H-bonding, which also provides better sensitivity; and (3) the well-arranged pore structures of

MOFs can improve sensor selectivity [54]. Recent progress in research on MOFs has already been reviewed [55]. In this section, we will focus on the applications of MOFs in the development of electrochemical DNA sensors. Electrochemical methods are rather attractive due to their relative simplicity, fast response, high selectivity, low cost and easy operation. MOFs facilitate the amplification of the signal generated. The development of electrochemical DNA sensors based on MOFs is in premature stages; therefore, only a few papers have been published on this topic so far. Moreover, all of them invariably focused on gene detection. As of now, there have been no papers using MOFs aimed at the development of sensors for the detection of DNA-drug interactions.

Shao et al. reported an electrochemiluminiscence (ECL) biosensor for the detection of miRNA-141. Micro RNAs (miRNAs) are short non-protein coding strands (19–23 nucleotides) [56]. It has been seen that an increased concentration of miRNA can be associated with cancer diseases (for example, prostate cancer) [57, 58]. Therefore, miRNA is considered as an important cancer marker. Ruthenium-MOFs (Ru-MOFs) have been used as a platform for ECL signal detection. ECL is a rather sensitive method based on luminiscence induced by redox reaction at the electrode surface, initiated by an externally applied voltage [59]. The capture DNA (cDNA) is immobilized on the magnetic Fe_3O_4 nanoparticles to form a capture unit. Addition of miRNA-141 to the solution containing capture units results in hybridization with cDNA. Using magnetic glassy carbon electrode, it is possible to attach the nanoparticles containing miRNA-141 to the electrode surface. Addition of Ru-MOFs modified by signaling DNA allows for extending the double electric layer that improves the sensitivity of measurements. The ECL signal is provided by the oxidation of tripropylamine (TPrA) that serves as a co-reactant in the $Ru(bpy)_3^{2+}$/TPrA couple. The luminophore in the excited state (Ru^{2+}) relaxes to the ground state and emits a photon. The measurements performed in 0.1 M PBS (pH 7.4) containing 20 mM TPrA have shown the rather high sensitivity of this approach, with a detection limit of 0.3 fM and a wide dynamic range 1 fM–10 pM.

MOFs are quite useful in the development of non-enzymatic biosensors. In one case, hemin@MOF composite was used for the fabrication of a sensor for detecting hydrogen peroxide and DNA [54]. Hemin has been used as a catalyst in many reactions. It has been observed that MOFs improve the stability of hemin. The high surface area of MOFs helps in isolating hemin, thus eliminating its aggregation and preserving its catalytic activity [60]. The hemin@MOF composite mentioned earlier was prepared by doping hemin onto Cu-MOF-199. The composite was adsorbed at the surface of a glassy carbon electrode (GCE) and served as a catalyst for the detection of H_2O_2. This hemin-containing composite was used for detecting hydrogen peroxide in the concentration range of 0.1 μM to 2.2 mM, with an LOD of 0.07 μM. Furthermore, hemin@MOF was used as a platform for the development of DNA sensors. (The principles of DNA detection have been presented in Figure 3). When carboxyl functionalized graphene (CFGR-COOH), which possesses excellent conducting properties, was coated on the surface of GCE, 5'-amino and 3'-biotin DNA probes (pDNA) were immobilized at the said surface via amidation reaction. This fixed the stem-loop structure of pDNA. In the next step, streptavidin

(SA) functionalized hemin@MOF (hemin@MOF-SA) was prepared for recognizing biotin at the 3'end of pDNA. After the addition of the matched tDNA, the loop structure of pDNA unfolded due to hybridization with tDNA. Subsequently, hemin@MOF-SA could access the electrode through the binding of SA to the biotin of pDNA. The electrocatalytic properties of Hemin@MOF-SA were employed for the detection tDNA. This DNA sensor allowed the detection of DNA with an LOD of 6.9×10^{-16} M, with an excellent selectivity of one base mismatch.

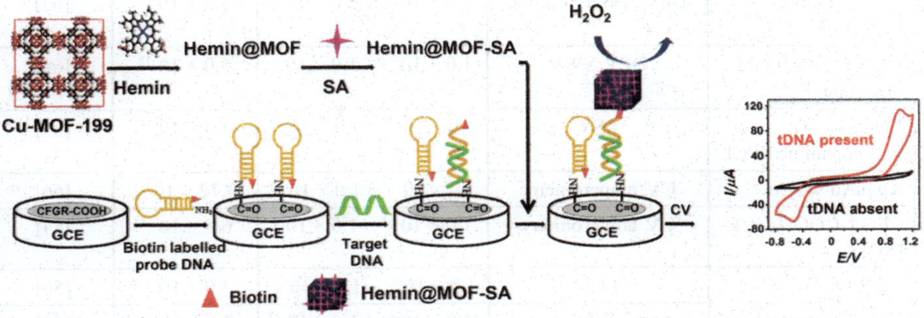

Figure 3. The scheme of detection of DNA through a biosensor based on the electrocatalytic activity of hemin@MOF composite towards H_2O_2. Adopted from [54] with permission of the Electrochemical Society.

NiCo-based metal-organic framework (NiCo-MOF) impedimetric DNA sensor for the detection HIV-1 virus was recently reported [53]. In the said case, EIS measurements were performed in 0.01 M PBS containing 0.1 M KCl, 5 mM $K_3[Fe(CN)_6]$ and 5 mM $K_4[Fe(CN)_6]$. It was shown that the hybridization of the target HIV-1-derived DNA with the probe DNA immobilized at the surface of the NiCo-MOF composite ($NiCo_2O_4$/CoO/CNTs) resulted in an increase in charge transfer resistance. This was because electron transfer was impeded due to the increased negative surface charge during the hybridization of the probe DNA with HIV-1 target DNA. The LOD for this detection was found to be 16.7 fM. The sensor was validated in HIV-1 DNA spiked healthy human blood serum samples diluted by 200 times, with recovery and relative standard deviation values in the range of 101.85–111.94% and 2.9–4.4% respectively. Table 1 compares the sensitivity of the DNA sensors based on various surfaces (including MOFs).

In the preparation of electrochemical DNA sensors, a crucial requirement is the stability of the DNA probe at the surface. Typically used thiol-based methods are relatively easy and facile for DNA immobilization; however, storage of such sensors in a buffer, results in the loss of DNA within several days after sensor preparation [75]. Recent studies have demonstrated that MOFs can be used for the immobilization of various biomolecules to protect them from degradation [76]. Ma et al. [77] proposed a simple method for the protection of DNA self-assembled monolayers on gold surface by covering them with MOFs. It was shown that zeolitic imidazolate framework-8 (ZIF-8) can be grown on a gold electrode modified with DNA probes and can be completely removed by rinsing, thus restoring the detection properties. ZIF-8 exoskeleton also protected DNA probes on the gold electrode surface from different

Table 1. Comparison of the sensitivity of various electrochemical DNA sensors.

Surface modification	Method of detection	Linear range (mol/L)	Detection limit (mol/L)	References
graphene-Nafion/GCE	EIS	1.0×10^{-13} - 1.0×10^{-10}	2.3×10^{-14}	[61]
Exo III-assisted biosensor	Amperometry	$1.0 \times 10^{-11} - 1.0 \times 10^{-6}$	1.18×10^{-11}	[62]
RCA-based biosensor	DPV, MB as a probe	$2.0 \times 10^{-14} - 5.0 \times 10^{-8}$	8.3×10^{-15}	[63]
ssDNA/MCH/gold electrode	DPV, SWV	$1.0 \times 10^{-12} - 1.0 \times 10^{-8}$	5.0×10^{-13}	[64]
Electro-polydopamine/SPCE	LSV	$1.0 \times 10^{-12} - 7.0 \times 10^{-11}$	3.0×10^{-13}	[65]
Ag@AuNP/Au	CV, amperometry	$1.0 \times 10^{-14} - 1.0 \times 10^{-12}$	3.72×10^{-13}	[66]
CFGR-COOH/GCE-MOFs	CV, amperometry	$1.0 \times 10^{-15} - 1.0 \times 10^{-6}$	6.9×10^{-16}	[54]
Ru-MOFs	ECL	$1.0 \times 10^{-15} - 1.0 \times 10^{-11}$	3.0×10^{-16}	[56]
AuNTs/nanoporous polycarbonate	EIS	$1.0 \times 10^{-14} - 1.0 \times 10^{-6}$	1.0×10^{-15}	[67]
PANI-NC/ITO	EIS	$1.0 \times 10^{-17} - 1.0 \times 10^{-6}$	7.0×10^{-18}	[68]
AuNPs-PAT/rGO/GCE	EIS, DPV	$1.0 \times 10^{-12} - 4.0 \times 10^{-10}$	7.0×10^{-14}	[69]
pDNA/PANI-MoS 2/ITO	EIS	$1.0 \times 10^{-17} - 1.0 \times 10^{-6}$	3.0×10^{-18}	[70]
NiCo/CNTs-MOF	EIS	$1.0 \times 10^{-13} - 2.0 \times 10^{-8}$	16.7×10^{-15}	[53]
Ferrocene-PAMAM dendrimers	CV, DPV	$1.3 \times 10^{-9} - 2.0 \times 10^{-8}$	3.8×10^{-10}	[71]
Te-ZnO nanowires	EIS	$1.0 \times 10^{-12} - 1.0 \times 10^{-6}$	1.0×10^{-13}	[72]
AuNPs/TiO/CdS/dendrimer	Photoelectrochemistry	$1.0 \times 10^{-15} - 1.0 \times 10^{-10}$	1.0×10^{-10}	[73]
AuNPs	DPV	$1.0 \times 10^{-17} - 1.0 \times 10^{-13}$	0.68×10^{-18}	[74]

Abbreviations: EIS – electrochemical impedance spectroscopy, CV – cyclic voltammetry, DPV – differential pulse voltammetry, LSV – linear sweep voltammetry, SWV – square wave voltammetry, ECL – electrochemiluminiscence, MB – methylene blue.

extreme environmental conditions such as heat and ionic strength, as well as from nuclease degradation. The DNA on the surface was protected for at least 30 days. The growth of MOFs on the DNA monolayer can be monitored by mere electrochemical measurement. The efficiency of the protection of the DNA layer by ZIF-8 has been demonstrated on DNA electrochemical sensors prepared using four types of DNA probes labeled by methylene blue (MB) redox marker—single-stranded DNA (ssDNA), hairpin DNA (hpDNA), double-stranded DNA (dsDNA) and tetrahedron skeleton (TDN). The current intensity of MB was monitored in various conditions. It was seen that without protection, all the types of sensors lost their stability rapidly, as indicated by a decrease in the MB signal. For example, after 15 days of storage, the signaling current fell to ~ 41% for ssDNA sensor, ~ 29% for hpDNA sensor, ~ 15% for dsDNA sensor and ~ 8% for TDN sensor. This was mainly due to the rupture of Au-S bonds in the buffer. It was also shown that ssDNA-, hpDNA- and dsDNA-based sensors underwent substantial current reduction when exposed to a solution containing DNAase I (0.5 U/μL) for merely 1 hour. However, TDN sensor exhibited better resistance against DNAase I; yet, even for this sensor, only 31% of the current signal was retained after 20 hours of incubation. The application of ZIF-8 substantially improves the stability of DNA layers. For example, the weakest TDN-based sensor retains over 68% of the MB signal for up to 30 days, versus a total loss without ZIF-8. By washing the substrate with an acidic buffer, the MOFs are completely removed and the electrode surface with immobilized DNA probes is recovered. The mechanism of protection herein is based on the *in situ* growth of ZIF-8 around DNA probes. ZIF-8, being small in size, tightly encapsulates DNA probes and protects them from the influence of environmental factors (including temperature) and DNAses. Recent advancements in MOF-based electrochemical aptasensors have also been reviewed in literature [78].

2.4 Electropolymerized supports

Electropolymerization involves the anodic oxidation of an appropriate monomer, resulting in the formation of radicals capable of undergoing oligomerization, and precipitation of insoluble products on an appropriate support, mainly on the electrode surface [79]. As per traditional organic chemistry, the reaction of electropolymerization belongs primarily to the category of polycondensation and is accompanied by a release of by-products with low molecular weights (usually water molecules). The phenomenon of formation of such oligomers though known for many years, was generally considered as a drawback of the electrolysis protocol as the deposition of oligomeric products reduced the active surface area and decreased the current and the yield of the main product. However, the situation has changed with the dawn of the biosensor era. Most electropolymerized products have found applications in biosensor assembly and their contribution to the progress of biosensor has not yet exhausted [80, 81].

From the point of view of electrochemical properties, all the electropolymerized products applied in DNA sensors can be divided into three groups [82]:

(1) Electroconductive polymers: They exhibit conductivity akin to many semiconductors, in a wide range of polarization potentials. Polyaniline, polypyrrole, polythiophene and their derivatives are the most frequently used polymers in this group [83].

(2) Redox active polymers: These polymers are formed mostly via polymerization of polyheteroaromatic structures like phenazine and phenothiazine dyes. They contain redox centers capable of reversible electron/hydrogen ion transfer [84]. Their behavior is like that of conventional mediators of electron transfer deposited on the electrodes. Redox active polymers can be easily combined

with electroconductive nanoparticles like carbon nanotubes, Au nanoparticles, etc. [85–87]. In such composites, a synergy of electrochemical properties takes place, resulting in current amplification and decrease in the potential of analyte oxidation.

(3) Electrochemically inactive polymers: These compounds are obtained by the oxidation of some phenols and aromatic amines, or from the overoxidation of conductive polymers [88]. Such polymers can serve as supports for the immobilization of DNA oligonucleotides or for mechanical protection of the biosensitive layer of a biosensor. Insulating layers can be also used for the formation of molecularly imprinted polymers selective towards the transfer of low-molecular species [89].

It should be noted that the mechanism of electropolymerization does not dramatically differ from that of chemical oxidation of appropriate monomers. It involves the formation of the cation radical, followed by its dimerization, hydrogen ion subtraction, and restoration of the neutral dimeric analog of the initial compound (Scheme 1). After that, the cycle repeats until the oligomer is deposited on the electrode. This means that coupling of each monomer unit requires the transfer of two electrons so that the electropolymerization does not correspond to the electrocatalysis and its efficiency can be characterized by the current yield [90].

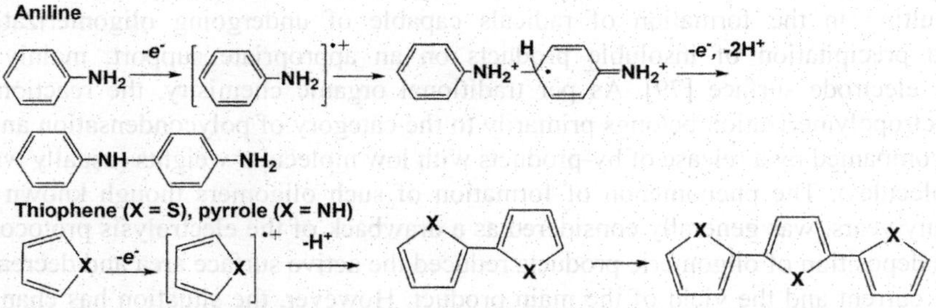

Scheme 1. Schematic outline of aniline, pyrrole and thiophene electropolymerization.

Although the described mechanism is rather common for most electrochemically active polymers, some peculiarities affecting the performance of polymeric products are worth mentioning. The oxidation of aniline results in the formation of a linear polymer with repeating fragments connected with imine bonds capable of both hydrogen ion and electron transfer. This results in a high dependence of the redox properties on the acidity of the environment. Consecutive electron/H^+ transfers result in the formation of three main forms of polyaniline—pernigraniline, emeraldine and leuco-emeraldine—in equilibrium, depending on the redox potential and pH. Out of these three forms, only the second one (emeraldine salt) shows electroconductivity. For this reason, polyaniline is better for synthesis in strongly acidic media and for application in biosensors that can be operated at pH < 3.0 [91]. For the same reason, polyaniline properties significantly depend on the nature of the acid used in its synthesis, i.e., on the counter-anion in the polymer matrix that controls equalization of the charge changes related to the redox conversion (Scheme 2) [92].

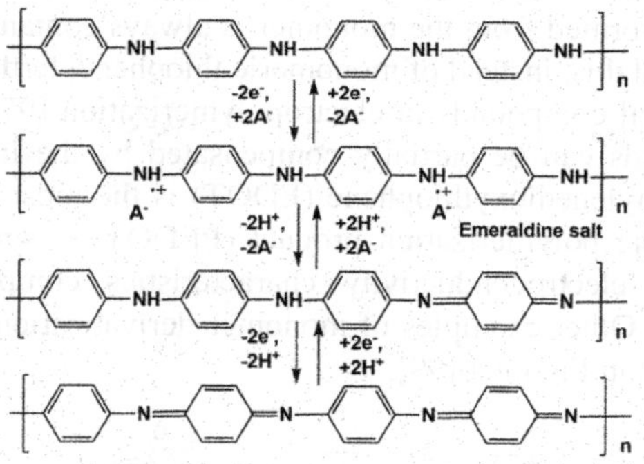

Scheme 2. Redox and acid-base equilibrium of polyaniline.

The involvement and release of acids, also known as doping and dedoping respectively, are followed by significant changes in polymer volume. DNA exerts a high negative charge of phosphate residues and hence forms complexes with polyaniline as a polyanionic substance. This spontaneous reaction is used for the amplified synthesis of polyaniline (DNA templating effect [93]) or for the detection of DNA-related reactions resulting in changes in the charge distribution of the biopolymer (hybridization effects, chemical DNA damage, DNA denaturing, etc.).

To suppress the pH sensitivity of polyaniline, acidic groups can be introduced in the monomer or the polymer structure. Such polymers formed from sulfanilic acid or anthranilic acid or from their mixtures with unsubstituted aniline are called 'self-doped' polyaniline [94]. Depending on the biosensor assembly, polyaniline can be used as a support for the immobilization of DNA probes or specific labels and for the generation of biosensor signal. In case of the latter, it is possible to monitor any biospecific reaction that alters the redox equilibrium in the DNA-polymer composite. This can be done through voltammetry or electrochemical impedance spectroscopy.

Unlike polyaniline, thiophene, polypyrrole and their derivatives contain only C-C bonds and are hence less sensitive to the acidity of the microenvironment and the acid-base equilibrium [95]. On the other hand, substituents at heteroaromatic rings can serve as anchors for DNA probes with minimal loss in the electroconductivity of the product. However, similar to polyaniline, low solubility of these polymers limits their use in DNA sensors. From this point of view, electropolymerization allows for the direct assembly of the sensing layer, with control over the film's thickness and its redox properties by choosing the regime of electrolysis. Polypyrrole and polythiophene are synthesized in neutral aqueous media but can be also obtained from polar organic solvents (like acetonitrile and dichloromethane).

Thiophene is oxidized to initiate electropolymerization at a potential higher than that required for the oxidation of its oligomers ("thiophene paradox", Scheme 3) [96].

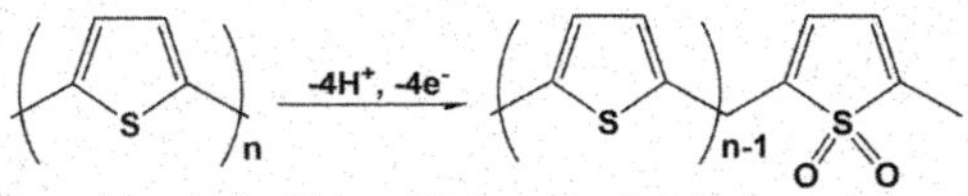

Scheme 3. Sulfone formation during the oxidation of polythiophene due to "thiophene paradox".

Polythiophene formed from the monomer is always contaminated with sulfone derivatives. To avoid this, instead of monomeric thiophene, terthiophene derivatives are used as the initial compounds in electropolymerization [97]. Low solubility of the above compounds can be partially compensated by the introduction of polar substituents. 3,4-Ethylenedioxythiophene (EDOT) is the most popular example of such derivatives. The polymerization product (PEDOT) is synthesized in neutral media and shows electroconductivity characteristics comparable with 'pure' polythiophene [98]. Other examples of monomer derivatization for attaching DNA probes are presented in Figure 4 [99].

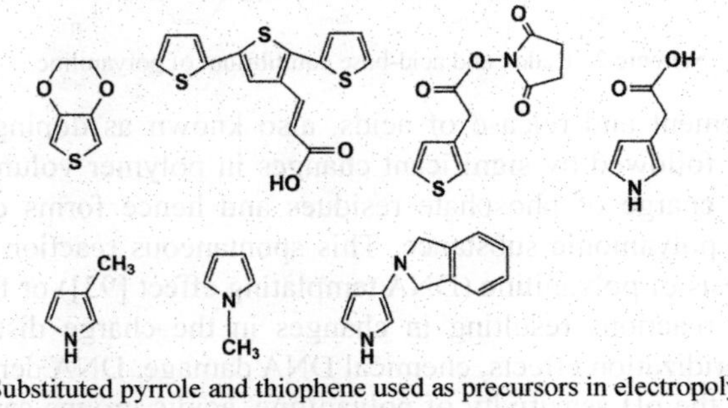

Figure 4. Substituted pyrrole and thiophene used as precursors in electropolymerization.

It should be mentioned that polypyrrole and polythiophene change their electrochemical properties as a response to target DNA interactions but to a lower extent in comparison with polyaniline. For this reason, they are mostly used for the electrical wiring of other mediators [100] and nanoparticles introduced in the sensing layer (for instance, Au nanoparticles, reduced graphene oxide nanosheets, carbon nanotubes, metal complexes, etc.).

Regarding redox active polymers, the mechanism of their polymerization has been explored to a much less extent, even for the most popular poly(Methylene blue) and poly(Neutral red) [101]. Polymerization of Methylene blue probably involves *N*-demethylation and formation of a covalent bond between the amino group of a monomer preliminarily oxidized to cation-radical and the aromatic ring belonging to another monomeric molecule of the dye (Scheme 4).

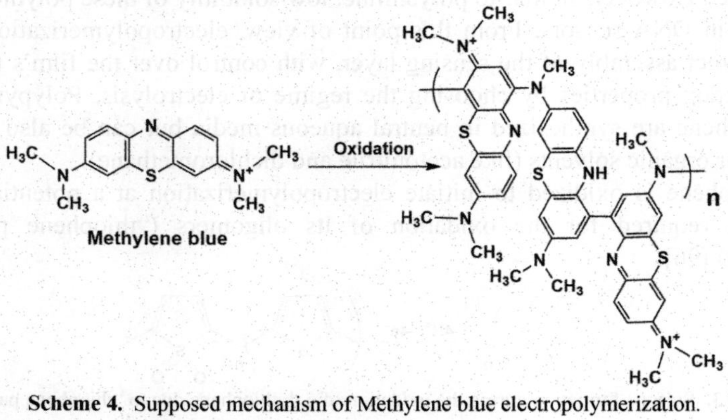

Scheme 4. Supposed mechanism of Methylene blue electropolymerization.

Appropriate products have been found among dimerization and trimerization products investigated by preparative electrolysis coupled with electrospray mass spectrometry [102]. A similar product has also been proposed by researchers for the oxidative coupling of Neutral red dye (Scheme 5) [101].

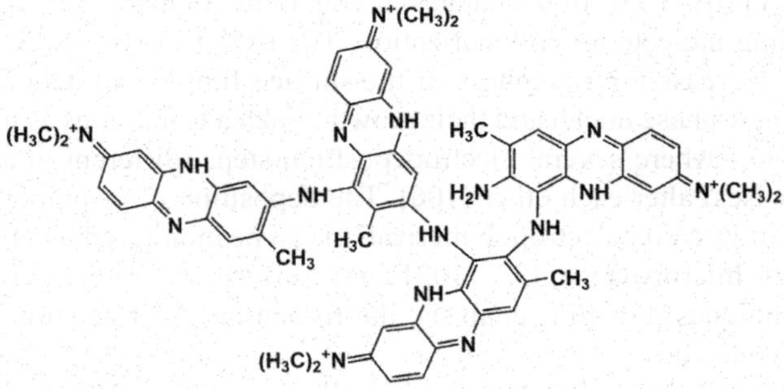

Scheme 5. Possible structure of Neutral red tetramer formed by electropolymerization.

Some polyheteroaromatic species with a primary amino group (thionine) are electropolymerized following the mechanism of polyaniline formation [103].

Most of the electropolymerized materials used in a DNA sensor assembly are formed by the anodic oxidation of monomers. Although this reaction can be performed in various regimes, multiple cycling of potential is the method mostly used in biosensor assembling. This protocol facilitates the formation of thin dense films with high adhesion to the electrode. Variation of scan rate, potential area and number of cycles makes it possible to influence the morphology of the surface and its roughness. In the case of polyaniline (and to a lesser extent for polypyrrole), the characteristics of the resulting product depend also on the nature of the counteranions present in the solution. In the case of electroconductive polymers, the peaks attributed to the products of polymerization tend to increase with the number of cycles. Phenazine (phenothiazine) dyes commonly show two pairs of reversible peaks related to the redox equilibrium of the monomeric and polymeric forms (Figure 5).

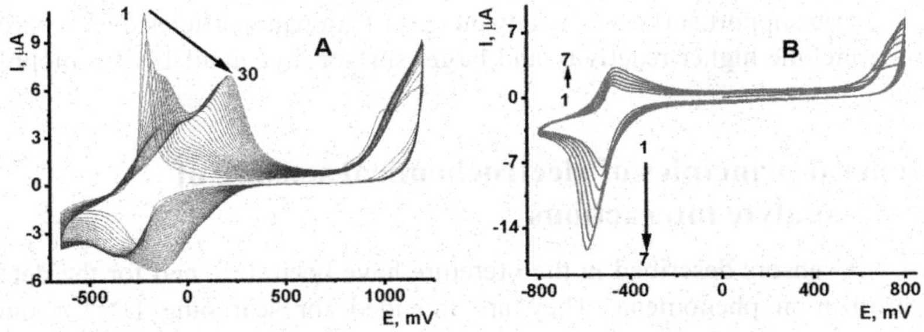

Figure 5. Electropolymerization of (A) Methylene blue and (B) Neutral red on a glassy carbon electrode by multiple cycling of potential. Reaction conditions: (A) Methylene blue: 0.1 M phosphate buffer + 0.1 M Na_2SO_4, pH = 8.2, 0.5 mM solution; (B) Neutral red: 0.025 M phosphate buffer + 0.1 M $NaNO_3$, pH = 5.5, 0.5 mM solution; scan rate: 50 mV/s. Arrows indicate changes in the voltammogram shape with increasing number of cycles.

The latter peaks are less reversible and are shifted to higher potentials, compared to the monomeric forms, due to slower electron exchange via polymeric film deposition on the electrode [103, 104]. Neutral red is the only exception herein. In this case, the peaks of the monomeric and polymeric forms are recorded at the same potential [101, 105]. Low changes in voltammograms or their stability during multiple cycling indicate no polymerization. The use of the pulse technique makes it possible to increase the roughness of the surface film by influencing the rate of formation of new phase nuclei and their growth. Such a technology is mostly applied for polypyrrole—where several electrodeposition steps (different in amplitude and duration of pulses) alter each other [106]. The deposition of insoluble products and the influence of DNA on electropolymerization can be monitored by electrochemical quartz crystal microbalance [32, 107] and surface plasmon resonance (SPR) [93, 108] techniques. For SPR analysis, the formation and reduction of Au oxides should be considered.

All electropolymerized materials are well compatible with nanomaterials that can be used during electropolymerization from their dispersions or deposited on the polymer film. Au nanoparticles and reduced graphene oxide can be obtained in one step by chemical or electrochemical oxidation of aniline [109]. This can be performed onto an electropolymerized layer or Au film to increase the thickness and stability of the layer. Covalent binding assumes preliminary functionalization of nanomaterials. Carbon nanodots can be doped with nitrogen to form amino groups on the surface [110]. Single-walled carbon nanotubes are oxidized to obtain carboxylic groups at the ends of the tube. Multiwalled carbon nanotubes are formed in addition to the carboxylic groups at the defects of the side walls. In both the cases, oxidation with a mixture of nitric acid and sulfuric acid is used. Oxidative treatment decreases the average molar mass, length and diameter of the carbon nanotubes. After that, carboxylated particles can be covalently attached to the amino groups of polymers (for example, thionine and Neutral red polymers) or can electrostatically hold onto positively charged polymers preferably existing in the oxidized cationic form [111]. In a similar manner, aminated DNA oligonucleotides are covalently bonded to the carboxylic groups of the carriers and thiolated derivatives to the Au nanoparticles on the polymeric support surface. The advantages of Au nanoparticles over a bare gold electrode include higher reactivity and larger surface area available for biopolymer binding.

3. General principles of electrochemical sensing of DNA-analyte interactions

Most DNA sensors described in the literature have been designed for the detection of hybridization phenomena. They are intended for searching DNA sequences complementary to those used in the biosensor assembly (DNA probes). The hybridization between complementary ssDNA strands is due to multiple weak molecular interactions by hydrogen bonds between the nucleobases. Being rather weak, they nevertheless provide high efficiencies in recognition of hybridization due to multiple new bonds established in the recognition event. The determination of drugs or the quantification of DNA-drug interactions significantly differs from the working and purpose of such biosensors.

The size of a drug is incomparably lower than that of the DNA probe attached to the electrode surface. This implies that a single binding of the analyte molecule does not remarkably alter the structure of the surface layer. This makes its difficult to recognize of target interactions. Furthermore, the specificity of DNA-analyte interactions influences the nature of the factors limiting the performance of the biosensor. For detection of the hybridization event, steric factors dominate to offer the access of the DNA sequence to the probe immobilized on the electrode. In such biosensors, covalent immobilization of DNA probe via terminal functional groups is preferred. In the case of low-molecular analytes, electrostatic factors mostly affect target interactions with the DNA sequence. Thus, with physical immobilization by entrapment in the growing polymer film, self-assembly or adsorption of polyelectrolye complexes on the polar sorbents can take precedence over covalent single point binding.

The methods for electrochemical sensing of DNA-analyte interactions are commonly divided into three groups (Figure 6): (1) application of redox signals to the groups covalently attached to the DNA molecule mostly via terminal functional groups; (2) label-free techniques, where diffusion-free indicators change their redox properties due to target interactions on the electrode interface; and (3) self signal of the drug accumulated on the DNA as specific adsorbent or entrapped molecule in the DNA-containing matrix. In the case of drugs, the first strategy of labeling should be extended. In these methods, all the types of biosensors are included (with redox response which does not involve diffusion as a mass transfer mechanism).

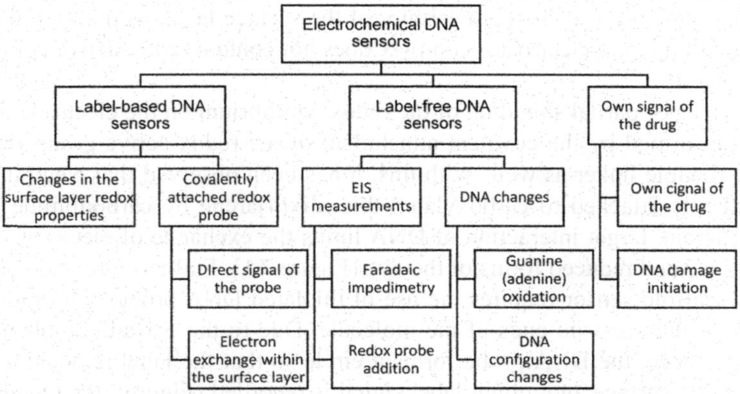

Figure 6. Classification of electrochemical DNA sensors based on signal measurement protocol.

Thus, changes in the intrinsic redox activity of the electropolymerized film are responsible for the signal of this group of methods.

3.1 Label-based techniques

Changes in the redox equilibrium involving DNA and redox active polymers on the electrode surface are mostly mentioned in this group of methods. Such changes can be observed for almost all polymers with redox activity—both electroconductive and redox active—but the signal value is higher for polyaniline and related derivatives, while the sensitivity of the assay is comparable with that of other polymers with less expressed charge sensitivity of their redox status.

There are two mechanisms that affect the redox equilibrium on the biosensor interface. The first one is associated with the partial elimination of the DNA's influence on the relation between the oxidized and reduced forms of the polymer. As has been established for polyaniline, the presence of DNA in close contact with polymer molecules increases the conductivity and extends the range of pH values for maintaining electrochemical activity of polyaniline, compared to the same polymer synthesized with no DNA. This is due to the ability of DNA to stabilize oppositely charged forms of polyaniline, i.e., emeraldine salt, by virtue of electrostatic interactions with phosphate residues of the DNA helix. In this case, any reactant capable of shielding (neutralizing) negative charge would diminish the electroconductivity of the surface layer. This results in synchronous increase of charge transfer resistance and capacitance, as measured by electrochemical impedance spectroscopy (EIS). Redox-active polymers also exert reversible changes in the peak currents recorded by direct current voltammetry or EIS parameters resulting from the electrostatic force shielding; this has been observed for poly(Neutral red) [112], poly(Azure B) [113] and polythionine [114].

The second mechanism deals with DNA intercalation and increases its volume and flexibility, affecting charge separation and redox equilibrium of the polymeric matrix. This mechanism has been indirectly confirmed by experiments performed with polyelectrolyte complexes with DNA as a polyanionic substance. Thermal DNA denaturation, its treatment with reactive oxygen species and interaction with doxorubicin affect the permeability of the surface layer, as monitored by EIS, even though the polyelectrolyte complex does not contain redox-active components [115].

The idea of partial shielding of the redox equilibrium on the electrode interface can be performed by the covalent attachment of the redox-active group (label) on a rather flexible linker as well. With this, it has been observed that Neutral red can be covalently attached to carboxylated thiacalix[4]arene by carbodiimide binding [112]. Therein, target interaction of DNA limits the exchange of electrons between the oxidized and reduced forms of the dye (Figure 7A). Furthermore, the concept of E-DNA scaffold sensor requires the use of thiolated DNA probes with redox label attached to the opposite ends of the molecule. Due to the partial complementarity of the sequence, the DNA probe forms a circle so that the label is positioned near the electrode surface and a high label signal is recorded (Figure 7B). Owing to the reaction with the target (e.g., a short complementary sequence), the probe is converted to the linear configuration and the distance between the label and the electrode dramatically increases. The E-Sensor protocol has been elaborated primarily for hybridization detection [116]. However, there are some examples of its application in drug detection [117].

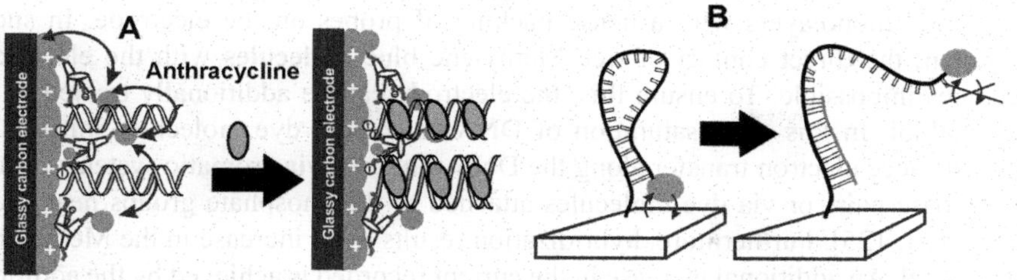

Figure 7. Mechanism of signal generation for (A) a DNA sensor based on controlled intrinsic electron exchange, and (B) an E-DNA scaffold biosensor).

3.2 Methylene blue case

Methylene blue is a phenothiazine drug that exhibits photosensitizing properties and can promote oxidative DNA damage due to the generation of singlet oxygen [118]. Methylene blue has found enormous application in DNA sensors due to the numerous advantages it offers – like photobleaching stability, high cell stain abilities and reversible redox behavior, to name a few (Scheme 6).

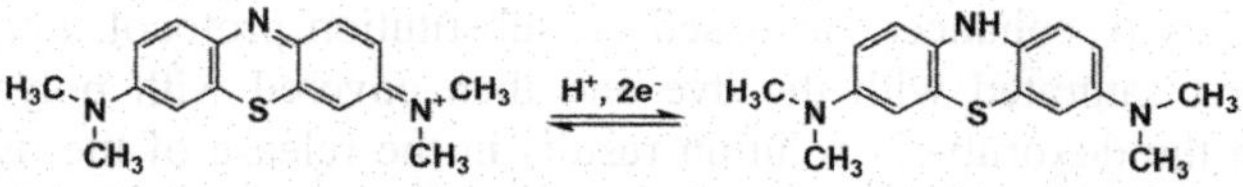

Scheme 6. Reversible redox conversion of Methylene blue used for amperometric detection of DNA-specific interactions.

The reaction of Methylene blue with dsDNA in a solution results in a partial decrease of the signal measured by DPV [119]. The estimation of the stoichiometry of interaction performed by UV spectroscopy and DPV techniques gives the ratio 1:2, indicating maximal binding of one dye molecule per two nucleobases [120]. This reaction has been considered as a multipoint interaction with partial intercalation of the dye in the guanine-cytosine-rich parts of the DNA molecule [121]. Moreover, it has been established that Methylene blue can alternatively bind to the minor grooves in DNA fragments with alternating adenine-thymine base sequences [122].

The transfer of DNA molecules on the electrode interface complicates the observation of their interaction with Methylene blue. In the case of physical adsorption on carbonaceous materials, including that electrostatically enforced by electrode-positive polarization, DNA molecules are randomly positioned along the surface of the electrode. Methylene blue shows a higher affinity towards ssDNA than towards dsDNA; hence, its reduction signal increases against a bare electrode by up to 40% of the initial value. Hybridization performed on the same electrode converts an ssDNA probe into a dsDNA helix. Intercalation of the dye decreases its signal on cyclic voltammogram or DPV to the extent depending on the degree of hybridization. This makes it possible to use Methylene blue as an effective indicator of hybridization event, capable of distinguishing between single and double mismatches [123].

The use of covalent immobilization of DNA probes on Au electrodes or Au nanoparticles via terminal –SH groups results in the formation of regular self-assembled monolayers with a dense packing of probes on the electrode. In such a system, the direct contact of free Methylene blue molecules with the electrode becomes impossible. To ensure this, the electrode can be additionally treated with hexanethiol. In this case, saturation of DNA helix with dye molecules initiates a long-distance electron transfer along the DNA sequence via aromatic systems of the nucleobase pairs or via dye molecules attached to the phosphate groups near each other [124, 125]. Furthermore, hybridization results in an increase in the Methylene blue signal. An additional increase in the current recorded is achieved by the addition of ferricyanide ions involved in the mediated oxidation of the dye molecules [126, 127].

Finally, Methylene blue can serve as a label in the DNA-A scaffold biosensors described above (Figure 7B). Additional modification of the dye with carboxylic groups neutralizes its positive charge and excludes non-specific electrostatic interaction with negatively charged DNA sequences. This increases the specificity of interaction. Thus, depending on the biosensor format, Methylene blue provides signal-on or signal-off sensing of a specific interaction.

In the domain of drugs, diffusion-free Methylene blue has been successfully used for sensitivity enhancement based on substitution protocol. For this reason, dsDNA is first saturated with the dye and then covered with polyaniline layer. Incubation in the doxorubicin solution results in the release of the dye. However, it cannot leave the surface layer due to the upper polyaniline film and is involved in an electron exchange chain providing sensitive response towards anthracycline preparation inactive at working potential [128].

3.3 Label-free DNA sensors

In most cases, specific binding of DNA with drug molecules starts with the formation of molecular complexes with phosphate residues at minor and major grooves of the DNA helix or by incorporation of drug molecules between the pairs of nucleobases. The latter is promoted by hydrophobic interactions of aromatic electron systems so that appropriate drugs have a planar polyaromatic structure in a rather large area [129]. Intercalation process mostly results in the shielding of drug molecules which lose their redox activity. Some extreme cases that contradict this common mechanism have been discussed below. Intercalation can be used in DNA sensors to control the conformation of double-stranded (ds) DNA molecules in extreme environments or after chemical/thermal damage, or to quantify the hybridization of complementary sequences. Detection of hybridization, which is more popular, is out of the scope of this review. Nevertheless, the interaction of intercalators with DNA retains interest for drug design and monitoring of pharmacokinetics.

Direct measurement of permeability is performed in a manner common for monitoring the surface coating of modified electrodes. In such protocols belonging to Faradaic EIS, an equimolar mixture of $K_3[Fe(CN)_6]$ and $K_4[Fe(CN)_6]$ is added to the solution. Using direct current voltammetry, equilibrium potential is determined as the half-sum of the peak potentials. At this potential, small alterations in the applied potential in EIS do not shift the equilibrium to anodic or cathodic reaction;

hence, the Faradaic current does not disturb the measurement results. Subsequently, charge transfer resistance and constant phase element (close to the pure capacity for an ideal electrode reaction) are calculated from the Nyquist diagram, reflecting the ratio between the imaginary and real parts of impedance. This approach has become widely prevalent, although the interpretation of the EIS data is sometimes far from theory. In actuality, redox activity and reversible electron exchange typical for DNA sensors based on appropriate polymers pose difficulties towards describing the processes within the layer as the existing equilibrium compensates for the changes caused by external stimuli (like polarization of the electrode interface). Regarding the use of nanomaterials, changes in charge transfer resistance can result from increased electrode area but not from DNA-specific reactions.

For inert carriers, Faradaic EIS offers results that are closer to theory and DNA-drug target interactions. The use of negatively charged $[Fe(CN)_6]^{3-/4-}$ probe is preferable because of electrostatic repulsion from phosphate residues that prevent non-specific accumulation of the probe. On the other hand, electron exchange with the participation of this redox probe is independent of pH. Other candidates like hydroquinone or Ru hexamine complex are less suitable for such DNA sensors. Redox conversion of hydroquinone is complicated due to the formation of 1:1 molecular complex (quinhydrone) and pH-dependence of the reaction. Consequently, the redox equilibrium is sensitive to the dissolved molecular oxygen, as opposed to the ferricyanide/ferrocyanide reaction.

Irrespective of the redox probe used, drug accumulation results in an increased charge transfer resistance due to the lower rate of diffusional transfer of the ferricyanide ion in the layer after drug binding. Hydrophobic interactions mentioned above can also decrease the signal due to higher hydrophobicity of the DNA-drug complex against the initial DNA sensor and slower transfer of a small but trivalent/tetravalent-charged anion of the redox probe. If aptamers are used as recognition elements, even small quantities of drugs can provoke conformational reorganization and formation of G4 quadruplexes (see E-DNA scaffold sensor described above) with a denser packing of the layer. Although there are a few examples of DNA aptamers used for drug determination, they are beyond the scope of this review.

Another instance of label-free sensing is based on DNA changes that follow the formation of a DNA-drug complex. This is typical for cytostatic drugs like anthracyclines (Section 4.1). Their incorporation in between planar pairs of complementary nucleobases is anchored by amino groups capable of protonation and interaction with phosphate residues [130]. Such a process disturbs the configuration of the DNA helix and opens guanine (and, to a lower extent, adenine) residues for attack from reactive oxygen species. As a result, oxo-guanine and oxo-adenine signals appear on DPVs as a sign of those reactions. The aforementioned nucleobases also yield irreversible oxidation peaks at high anodic potentials so that the synchronous changes attributed to nucleobases and their oxo-products are observed. Except for a high working potential, such signals are quite convenient for the detection of intercalation event. To avoid DNA oxidative damage following drug intercalation, voltammetric signals should be measured in de-aerated media as the product of a single electron transfer on the dioxygen molecule initiates the same reactions as direct oxidation of native DNA. Similar signals can be obtained for dehybridization reactions and other cases of alteration of native DNA configuration. To distinguish between the various possible causes of guanine (or adenine) oxidation, EIS measurements should be performed. Intercalation increases the average distance between the phosphate residues of the DNA strands, resulting in increased charge separation and decreased effective charge of the DNA complex. Meanwhile, intercalation results in consolidation of the DNA layer of the biosensor. In other words, charge transfer resistance and interface capacitance change in the opposite directions for DNA intercalation and both tend to increase in case of severe damage to the DNA structure.

4. Electrochemical DNA sensors for drug determination

Although there are many targets for DNA sensors, anti-tumor drugs have, for a long period time, remained a priority due to the severe consequences of cancer on the human population, the rapid growth in the number of affected patients, and the lethal outcomes. Some recent reviews have been devoted to the determination of drugs by electrochemical sensors and biosensors using nanomaterials and bioreceptors [6, 131, 132]. They show opportunities, achievements and prospects of these biomedical devices for preliminary diagnostics, dose control and new pharmaceuticals research.

4.1 Anthracyclines

Adriamycin (doxorubicin chloride) is one of the most widely researched anthracycline drugs and has been known for about 50 years (Scheme 7). It strongly intercalates DNA in guanine-cytosine rich areas and interferes with DNA regulation processes.

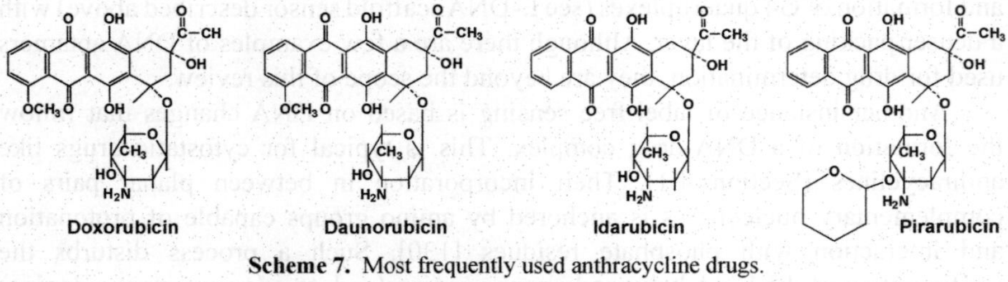

Scheme 7. Most frequently used anthracycline drugs.

Furthermore, adriamycin, like other anthracyclines, promotes the generation of reactive oxygen species and chemical DNA damage. In aqueous solutions, adriamycin shows a pair of reversible peaks associated with the reduction of 5,12-diquinone groups, and two irreversible peaks in the anodic area at 0.5 and 0.6 V—attributed to the oxidation of the 6,11-dihydroquinone unit [133]. All the peaks are pH-dependent. The contact of adriamycin with immobilized DNA produces a short-lived radical at –0.6 V, causing damage to the DNA and appearance of a new peak, associated with

8-oxoguanine (a product of oxidative DNA damage), at 0.4 V. The relation between the cathodic reduction of adriamycin and the formation of 8-oxoguanine has been confirmed by experiments in Ar atmosphere and by transfer of electrodes with a physically adsorbed thick DNA layer into a buffer solution containing oxygen in the absence of the drug [134]. Also, anthracyclines can involve Fe(II) ions in oxidative mutagenesis due to the formation of the 1:1 complex exerting a toxic effect on the native DNA [135]. Conversely, addition of antioxidants to anthracycline drugs allays their damaging effect on DNA; this happens due to the scavenging radical species formed in by reaction between the drug and oxygen [136]. Reduced working potential of doxorubicin oxidation can be achieved by using other nanosized modifiers like multiwalled carbon nanotubes with Pt [137] or Ag nanoparticles [138], polystyrene-Fe_3O_4-sulfonated graphene oxide nanomagnetic particles [139], or thiacalix[4]arene bearing Neutral red terminal fragments [112]. In the case of thiacalix[4]arene bearing Neutral red terminal fragments, interaction with DNA hinders electron exchange between the reduced and the oxidized forms of the dye, hence increasing charge transfer resistance and decreasing the reversibility of the cyclic voltammograms recorded with the fericyanide pair added to the solution.

In a similar manner, DNA-polyaniline composites obtained with DNA added to the solution or drop-casted on polyaniline surface appears to be sensitive to the DNA configuration, including changes in the native shape by intercalation. Direct addition of DNA in the reaction medium may result in DNA damage because of the strong mineral acids used to obtain an electroconductive polymer. However, substitution of sulfuric acid with oxalic acid makes it possible to increase the pH to 3.0 and retain the DNA structure and its accessibility to external stimuli [140]. Appropriate biosensors respond to incubation in anthracycline preparations by increasing the charge transfer resistance and decreasing the current of $[Fe(CN)_6]^{3-}$ as the diffusion-free redox indicator. The following LODs have been achieved in optimal conditions: 0.01 nM for doxorubicin, 0.1 nM for daunorubicin and 0.2 nM for idarubicin. The associated signals have been found to be quite selective and only slightly altered in the presence of serum proteins, sulfanilamide antibiotics and plasma electrolytes.

In later studies, it was proposed to introduce DNA between two polyaniline layers to minimize the negative effects of the high acidity of the polymerization media. Intrinsic redox activity of the polyaniline layer was determined through constant current voltammetry. The oxidation peak current increased with the number of potential cycles applied during the stage of electropolymerization and with the quantity of dsDNA on the polymer surface. For increased sensitivity towards doxorubicin, the DNA was saturated with Methylene blue (see previous Section). Its substitution with doxorubicin resulted in larger changes in the permeability and redox activity of the composite layer [128]. 0.6 pM was the minimum LOD achieved.

Some researchers have used adriamycin as the redox indicator of hybridization event though its oxidation potential is rather high. Thus, subsequently, a glassy carbon electrode consecutively modified with carboxylated multi-walled carbon nanotubes, electropolymerized 3-(3-pyridyl)acrylic acid and silver nanoparticles, followed by the attachment of thiolated DNA probe was used [141]. The signal of hybridization resulted in an increase in the signal associated with adriamycin. A possible cause of the opposite direction of the signal change is long distance electron transfer via anthracycline aromatic systems after intercalation of the hybridization product. The linear range of concentrations is determined from 0.009 to 9.0 nM (LOD 3.2 pM).

Doxorubicin forms with DNA in homogeneous solutions the complexes of 1:1 to 1:6 stoichiometry. These complexes can be identified by the hypochromic shift of the peak at 240–260 nm in the UV spectra. Doxorubicin forms a pair of reversible peaks at –0.55 and –0.80 V and one irreversible anodic peak at 0 to –0.1 V on appropriate voltammograms recorded in the absence of dissolved oxygen. The addition of dsDNA results in lower reversible peaks associated with anthracycline moiety [142].

Epirubicin is structurally related to doxorubicin cytostatic drug; however, it shows fewer side effects. Its reduced toxicity is attributed to the spatial orientation of hydroxyl groups at 4' position in the sugar moiety, making it responsible for faster elimination of the drug from the organism. Epirubicin is used against breast and ovarian cancer, gastric cancer, lung cancer and lymphomas. It is oxidized on a glassy carbon electrode, forming a single peak near 0.4 V; this peak is lowered and shifts to higher potentials in the presence of dsDNA. It also can substitute Methylene blue, thus intercalating into dsDNA and increasing its accessibility to redox reactions. This results in an increased sensitivity of electrochemical detection of DNA-epirubicin interactions [143].

Mitoxantrone is used in the clinical treatment of leukemia, ovarian cancer and breast cancer. It poses a lower risk of cardiotoxic effects when compared to doxorubicin and daunomycin. The reaction of mitoxantrone with dsDNA is stimulated by electrostatic interactions of side chains containing amino groups with phosphate residues of DNA. As other antracyclines, G-C base pairs are specifically involved in mitoxantron intercalation. The interaction of this drug with DNA, both in the solution and on the electrode surface, results in the lowering of the two irreversible anodic peak currents (at 0.51 V and 0.78 V) attributed to the oxidation of 5,8-dihydroxyl fragment and aminoalkyl substituents [144]. Amplification of mitoxantrone signal has been achieved by the incorporation of the DNA into chitosan carbon paste [145]. In optimal conditions, 0.030–3.50 mg/L of the drug could be detected using the anodic DPV peak at 0.46 V (LOD 1.3 nM) with a 7-minute incubation period.

Pirarubicin intercalates into DNA adsorbed on a graphite electrode [146]. Analytical signal of DNA sensors developed for detection of anthracycline drugs was monitored by own drug reduction peak using DPV in phosphate buffer. Incorporation of DNA increased the sensitivity of drug detection by about 3,000 times against bare glassy carbon, with an LOD of 4.3 pM and a dynamic range of 1 to 50 pM.

Analytical aspects of DNA sensors for the detection of anthracycline drugs have been summarized in Table 2, with emphasis on the application of nanomaterials.

Table 2. Analytical characteristics of detection of anthracycline drugs with electrochemical DNA sensors using nanomaterials.

Immobilization protocol	**Measurement principle**	**Detection technique**	**Dynamic range, LOD**	**Real sample**	**Ref.**
Doxorubicin (Adriamycin)					
GCE/MWCNT-COOH/PPAA/Ag_{nano}/ss- or ds-DNA	Changes in an analyte reduction current, adriamycin as indicator of hybridization event	DPV	LOD 1 μM	–	[147]
GCE/MWCNT-COOH/Ag_{nano} (DNA in solution)	Changes in the analyte reduction current, adriamycin as a DNA intercalator	DPV, CV (DNA detection)	8.2–19 nM, LOD 1.7 nM	–	[138]
Au/(MWCNT-SH/GNPs)$_n$	Changes in the doxorubicin oxidation current, doxorubicin as redox intercalator for hybridization event detection	DPV	LOD 1 μM (n = 6)	–	[141]
E-AB biosensor	MB as electroactive label	SQWV	1 μM–0.1 mM	Undiluted blood serum, whole blood	[148]
Pt/MWCNTs (DNA in solution)	Changes in the doxorubicin redox current, doxorubicin as redox intercalator for DNA detection	CV	0.05–4 μg/mL, 0.002 μg/mL	Plasma samples	[137]
GCE/PANI+DNA, polymerization media contains DNA	Changes in the cathodic current of the ferricyanide redox probe	CV	CV: 0.2 mM–10 nM; 10 nM–0.1 nM, LOD 0.1 pM	"Doxorubicin-LANS®", artificial samples with BSA and plasma electrolytes	[140]
GCE/poly(NR)/polycarboxylated thiacalix[4]arene/NR/DNA	Changes in the NR cathodic peak current and charge transfer resistance in the presence of $[Fe(CN)_6]^{3-/4-}$	CV, EIS	CV: 0.1–100 nM, 0.05 nM EIS:1 nM–0.1 mM, 0.1 nM	"Doxorubicin-LANS®" and "Rastocin", artificial plasma electrolytes	[112]
SPE/AuNPs/pTTBA/DNA/CL/AuNPs (Microfluidic device with FASS and FASI channels)	Temporal changes of the current	EKC-EC	7.5–50 pM, LOD 3.6 fM	Human urine samples	[149]

Table 2 contd. ...

...Table 2 contd.

Immobilization protocol	Measurement principle	Detection technique	Dynamic range, LOD	Real sample	Ref.
Au/MUA/(PAA/DNA)$_n$	SPR angle shift	SPR	1.0 pM–0.1 μM, LOD 0.7 pM (n = 8)	–	[108]
GCE/poly(NR)/PSS/DNA/octaaminothiacalix[4]arene	Changes in charge transfer resistance in presence of $[Fe(CN)_6]^{3-/4-}$	EIS	20 μM–1 nM, LOD 0.1 nM	–	[150]
GCE/poly(azure B)/DNA	Changes in intrinsic redox activity of the polymeric layer	CV	0.1 μM–0.1 nM, LOD 70 pM	Doxorubicin-LANS® and Doxorubicin-Teva®, blood serum, artificial blood serum	[113]
GCE/PANI/DNA/PANI	Changes in charge transfer resistance in presence of $[Fe(CN)_6]^{3-/4-}$	EIS	1.0 pM–0.1 mM, LOD 0.6 pM	Artificial urine samples	[127]
GCE/PANI/DNA(MB)	Changes in the PANI cathodic peak current, displacement protocol (MB is displaced with analyte)	CV	0.1 nM–1 μM, 10 pM	Doxorubicin-LANS®, blood serum, artificial blood plasma	[151]
GCE/thiacalix[4]arenes bearing oligolactic fragments/PEI/DNA	Changes in the MB redox probe current	CV	*cone*: 1 nM–30 pM, LOD 10 pM *paco*: 0.1 μM–1 nM, LOD 0.5 pM *1,3-alt*: 0.1 μM–1 μM 0.3 nM, LOD 0.1 nM	–	[152]
		Valrubicin			
Au/MWCNTs/en/AuNPs	Changes in valrubicin cathodic peak current	CV	0.5–80.0 μM, LOD 0.018 μM	Blood serum, urine samples	[45]
		Mitoxantrone			
SPE/AuNPs/pTTBA/DNA/CL/AuNPs Microfluidic device with FASS and FASI channels	Changes in current as a function of time in dependence on analyte concentration	EKC-EC	2–60 pM, LOD 1.2 fM	Human urine samples	[153]

Table 2 contd. ...

...Table 2 contd.

Immobilization protocol	Measurement principle	Detection technique	Dynamic range, LOD	Real sample	Ref.
GCE/MWCNT-Ag-PT/DNA	Changes in mitoxantrone anodic peak current	DPV	50 nM–0.1 mM, LOD 13 nM	Blood serum, urine samples	[154]
Epirubicin					
GSPE/FMWCNTs-IL-Chit/AuNPs/Apt/Dna	Changes in the cucurbituril (5 μM) peak current	DPV	0.007–0.1 mM, 0.3–7.0 mM, LOD 3 nM	–	[155]
PGE/DNA/NrG/PP	Changes in the guanine oxidation signal	DPV	0.004–55.0 μM, 1 nM	Epirubicin hydrochloride solutions, urine samples	[156]
Daunorubicin					
SPE/AuNPs/pTTBA/DNA/CL/AuNPs (Microfluidic device with FASS and FASI channels)	Changes in current as a function of time in dependence on analyte concentration	EKC-EC	8–50 pM, LOD 5.5 fM	Human urine samples	[153]
CPE/Ag-4-ATP-MWCNT/DNA	Changes in ferricyanide probe current as redox indicator	DPV	1 nM–10 μM, LOD 0.3 nM	Human urine, Human blood serum samples	[154]
PGE/rGO/DNA	Changes in guanine oxidation signal after daunorubicin intercalation (DNA detection)	DPV	LOD 6 μM	–	[157]
GCE/PANI+DNA, polymerization media contains DNA	Changes in the cathodic current of the ferricyanide redox probe	CV	10 μM–0.5 nM, LOD 0.1 nM	Artificial samples containing BSA and plasma electrolytes	[140]
GCE/poly(NR)/polycarboxylated thiacalix[4]arene/NR/DNA	Changes in the NR cathodic peak current and charge transfer resistance in presence of $[Fe(CN)_6]^{3-/4-}$	CV, EIS	CV: 0.1–10 nM, 0.1 nM EIS: 1 nM–10 mM, 0.1 nM	–	[112]
Al foil/PNIPAM-g-TiO2/PNIPAM-co-PMMA) (DNA in solution)	Changes in daunorubicin reduction current interaction with DNA	CV	0.33 mM	–	[158]
GCE/AuNPs/polyTTBA/PS-aptamer/AuNPs	Changes in reduction current of daunorubicin	DPV	0.1 and 60.0 nM LOD 52.3 pM	Human urine samples	[149]

Table 2 contd. ...

...Table 2 contd.

Immobilization protocol	Measurement principle	Detection technique	Dynamic range, LOD	Real sample	Ref.
Idarubicin					
GCE/PANI+DNA, polymerization media contains DNA	Changes in the cathodic current of the ferricyanide redox probe	CV	0.1 mM–1 nM, 0.2 nM	–	[140]
GCE/poly(NR)/polycarboxylated thiacalix[4]arene/NR/DNA	Changes in the NR cathodic peak current and charge transfer resistance in presence of $[Fe(CN)_6]^{3-/4-}$	CV, EIS	CV: 1 nM–1 mM, LOD 0.5 nM EIS: 1 nM–0.1 mM, LOD 1 nM	–	[112]
SPE/AuNPs/pTTBA/DNA/CL/AuNPs (Microfluidic device with FASS and FASI channels)	Changes in current as a function of time in dependence on analyte concentration	EKC-EC	5.0–55.0 pM, LOD 2.2 pM	Human urine samples	[149]

Abbreviations: Ag-4-ATP – Ag nanoparticles on 4-aminothiophenol; Ag_{nano} – silver nanoparticles; Apt – aptamer; AuNPs, GNPs – gold nanoparticles; Chit – chitosan; CL – Cardiolipin; E-AB biosensor – electrochemical aptamer-based biosensor; EKC-EC – electrokinetic chromatography with electrochemical detection; en – ethylenediamine; FASS – field amplified sample stacking; FASI – field amplified sample injection; GCE – glassy carbon electrode; GPE – graphite paste electrode; GSPE – graphite screen-printed electrode; IL – ionic liquid; MB – Methylene blue; MUA – 11-mercaptoundecanoic acid; MWCNT – multi-walled carbon nanotube; NR – Neutral Red; NrG – nitrogen-doped reduced graphene; PAA – poly(allylamine hydrochloride); PANI – polyaniline; PEI – poly(ethylene imine); PGE – pencil graphite electrode; PNIPAM-co-PMMA – poly(N-isopropylacrylamide)-co-poly(methyl methacrylate) nanofibers; PNIPAM-g-TiO_2 – titanium dioxide nanoparticles grafted with thermoresponsive poly(N-isopropylacrylamide); polyTTBA, pTTBA – poly-5,2':5',2"-terthiophene-3'carboxylic acid; PP – polypyrrole; PPAA-poly (*trans*-3-(3-pyridyl) acrylic acid); PS – phosphatidylserine; PSS – poly(sodium styrenesulfonate); PT – polythiophene; SPE – screen-printed electrode.

4.2 Other anti-tumor drugs

Chalcones (α,β-unsaturated ketones) exhibit a wide spectrum of anti-tumor, antibacterial and anti-flammatory properties. Their interaction with DNA has been monitored by the introduction of ferrocene moiety in the chalcone structure by a few researchers [159]. Therein, DNA from chicken blood was mixed with the drug in deaerated solution and cyclic voltammograms of ferrocene redox conversion were recorded with direct current voltammetry. Decrease in the peaks corresponded to the affinity constant equal to 5.2×10^3 M^{-1} and 1:1 stoichiometry of the interaction.

6-Mercaptopurine is a drug used in chemotherapy of different forms of leukemia. Its reaction with DNA introduced in carbon paste results in the accumulation of oxidation signal recorded by DPV [160]. The dynamic range of concentrations determined is 5.0 to 20 aM (LOD 2.0 aM) for an accumulation period of 6 minutes. This method has been tested on the tablets of this drug. The use of pencil graphite electrode modified with polypyrrol and multi-walled carbon nanotubes shifted the LOD to 80 nM (concentration range: 0.2 to 100 μM) [29]. Interaction of 6-mercaptopurine with dsDNA has been confirmed by monitoring the changes in guanine and adenine signals with drug concentration as well as by IR spectroscopy.

Another purine thioanalog, 6-thioguanine, is used in the treatment of systematic connective tissue diseases like leukemia and lymphomas. Belonging to the category of anti-metabolite drugs, 6-thioguanine is transferred in the cancer cell in the form of 6-thioguanylic acid; it interferes with the synthesis of guanine nucleotides and finally halts the cell cycle. 6-thioguanidine was first investigated with hanging mercury drop electrode where its reduction peak was observed in neutral media by direct current voltammetry [161]. In the presence of dsDNA, this peak disappeared, and another one at –1.27 V appeared and increased with the DNA concentration. This peak, attributed to adenine reduction, makes it possible to be detected by square-wave voltammetry in the ranges of 16–360 μM and 400–3000 μM (two linear ranges on the calibration curve), with an LOD of 1.1 μM of 6-thioguanine. The same drug is oxidized at 0.88 V at a glassy carbon electrode covered with single-walled carbon nanotubes and dsDNA [162]. The effect gradually increases with incubation time. The formation of the complex where drug molecules are attached to the minor groove area of the DNA helix has also been confirmed by EIS. An LOD of 0.25 μM has been reached with the DNA sensor in optimal conditions.

Mitomycin C is used in clinical chemotherapy against a broad range of solid tumors. As evidence suggests, reduction of mitomycin C on a hanging mercury drop electrode with adsorbed DNA results in the covalent binding of reactants with the guanine residue capable of reduction at –0.49 V [163]. To exclude the influences of the self-signal of the drug, the electrode is transferred in a working buffer after the accumulation step. Reductive activation follows the disappearance of the guanine reduction peak. It is interesting to note that a similar reaction in acidic conditions also forms adducts, but is of a different nature and capable of reduction at lower cathodic potentials.

Mitomycin C has also been investigated in oil/water systems in the presence of DNA from fish sperm [164]. In the referred study, pencil graphite electrode was first polarized to electrostatically accumulate DNA and then transferred in a microemulsion containing the drug. In DPV regime, specific DNA-drug interactions resulted in a decrease of the guanine signal observed in the range from 0.1 to 10 μg/mL of DNA.

A similar methodology has been reported for the investigation of the interactions of anti-tumor flavonoids (3-hydroxyflavone, hesperidin) with dsDNA adsorbed on a hanging mercury drop electrode [165]. Herein, the drugs were first converted into complexes with Cu(II) ions, which intercalate into DNA and retain their ability of cathodic reduction at –1.45 V; square-wave and direct current voltammetry were used for the quantification of dsDNA.

A DNA-sensor based on a glassy carbon electrode modified with reduced graphene oxide using adsorptive stripping DPV has been used for the detection of methotrexate, a broad-spectrum anti-cancer drug [76]. After its accumulation on the DNA helix, the guanine oxidation peak current was measured; a dynamic range from

0.055 to 2.2 μM (LOD 7.6 nM) was achieved. The biosensor was tested in diluted urine.

Topotecan, an effective drug against ovarian cancer, cervical cancer and small cell lung cancer, belongs to the family of camptothecin derivatives. For its interaction with dsDNA, a graphene paste electrode is used. The drug produces an irreversible peak at about 0.68 V, which dramatically increases in the presence of DNA and shifts to more anodic potentials [166]. The sensor makes it possible to detect the drug in a concentration range of 0.7 to 90 μM (LOD 0.37 μM). This biosensor has been tested in spiked serum and urine samples and shows recoveries in the range of 99% to 102%.

Ajmalin, a cytostatic drug that inhibits the replication and transcription of DNA, can be oxidized on a glassy carbon electrode, with the formation of an irreversible anodic peak at +0.6 V on cyclic voltammogram [167, 168]. The DNA adsorbed on the same electrode decreases the peak height due to the intercalation of the drug molecules. This phenomenon has been used for the accumulation of ajmalin on DNA-containing support, followed by the release of the drug in concentrated $LiClO_4$ solution. An LOD of 0.3 nM was achieved in optimal accumulation conditions.

Etoposide interacts with ssDNA and dsDNA from salmon sperm through combined intercalation and electrostatic interaction [169]. The binding contents determined for DNA adsorbed on a screen-printed electrode were found to be 4.1×10^5 M^{-1} (pH 4.5) and 5.2×10^5 M^{-1} (pH 7.4) for 1:1 stoichiometry of the complex. The affinity of the drug towards cation-radical was observed to be slightly higher. This investigation was performed using DPV and confirmed by UV-vis spectroscopy.

Furazolidone can be reduced on a glassy carbon electrode covered with multi-walled carbon nanotubes and dsDNA from calf thymus in a concentration range of 40 to 100 μg/L (LOD 25 μg/L) [170]. The presence of DNA enhances the peak of furazolidone reduction and shifts it to less cathodic values. The stoichiometry and binding constant of the DNA-drug interaction can be assessed using the Hill's model (5.9×10^3 M^{-1} and 1:1 complex formation).

4.3 Other drugs and non-specific interactions (DNA as adsorbent)

Efavirenz, an anti-HIV type 1 drug, can be detected using a DNA biosensor based on pencil graphite electrode [171]. The incubation of the drug with dsDNA solution results in the formation and regular increase of the guanine peak recorded by adsorptive stripping DPV at 1.0 V and 180 second accumulation, and decrease of the self Efavirenz peak at 1.3 V. Physical adsorption of the DNA on the same electrode does not affect the voltammograms. A similar transducer covered with polypyrrole/reduced graphene oxide was used for the subsequent accumulation of dsDNA, didanosine drug playing the same role of DNA transferase inhibition as Efavirenz described above [172]. The interaction of the drug and DNA decreased the oxidation peaks of guanine and adenine as measured by DPV. The dynamic range of concentration determination was 0.02–50 μM (LOD 8.0 nM). These DNA sensors were validated in spiked samples of urine using HPLC as the independent method of analysis.

Rutin, a well-known flavonoid with antioxidant properties, interacts with dsDNA and ssDNA in an aqueous solution and on the surface of a glassy carbon electrode covered with Langmuir-Blodgett film formed by octadecylamine [173]. This interaction results in a decreased rutin reduction peak current recorded in the presence of DNA. The associated calculations indicate the saturation of DNA molecules corresponding to 1 rutin molecule per two nucleobase pairs.

Khellin is a drug used as a therapeutic agent in the treatment of vitiligo and psoriasis. Its reaction with DNA can be used for its detection in a concentration range of 2 μM to 50 nM (LOD 10 nM); its oxidation peak as recorded by stripping DPV is observed as the analytical signal [174]. DNA from calf thymus is electrostatically accumulated on the carbon paste electrode prior to its contact with the drug. This method has been validated on spiked samples of blood serum and tablets.

Oxytetracycline can be detected through square-wave anodic stripping voltammetry on a carbon paste electrode contained of 40% multi-walled carbon nanotubes and 40% DNA from calf thymus [175]. The accumulation at –1.7 V results in the formation of a sharp peak at –0.18 V (attributed to the drug). The peak current is observed to be linear to the analyte concentration in the range of 1 to 10 ng/L (LOD 0.4 ng/L). In the referred study, however, the mechanism of signal generation has not been discussed. Surprisingly, doxycycline, chlorotetracycline and tetracycline did not interfere with most of the analyzed species. Among inorganic species, only Fe and Hg ions interfered with the measurements.

The irreversible reduction peak of nicotin decreases from –1.1 to –1.6 V on a glassy carbon electrode after physical adsorption of ssDNA [176]. No arguments regarding specific interactions of the reactants on the electrode surface have been found; however, the formal description of the concentration profiles of the signal shows a binding constant of 831 M^{-1}.

Colchicin gets accumulated on the dsDNA adsorbed on a boron-doped diamond electrode. The associated DPV shows two oxidation peaks, the first one (at 1.1 V) being proportional to the concentration of the drug in a range of 1 to 100 μM (LOD 0.26 μM) [177].

Methotrexate inhibits the functioning of dehydrofolate reductase (which plays an essential role in the synthesis of thymine nucleotide) and hence exhibits cytotoxicity towards cancer cells. Its interaction with DNA has been explored through DPV on glassy carbon and highly ordered pyrographite electrodes [178]. The oxidation peaks related to the damaged DNA fragments were recorded after the incubation of the DNA sensor obtained through dsDNA, poly(dA) and poly(dG) molecules physically adsorbed on the electrode surface. The reaction of dsDNA with the drug resulted in the appearance of two anodic peaks (related to oxo-ganine and oxo-adenine). The reaction was very fast and resulted in DNA condensation, as confirmed by atomic force microscopy and by the resultant lowering of the mentioned peaks. The preferred affinity of methotrexate towards adenine-rich DNA fragments was confirmed using homo-polynucleotide sequences.

The interaction of Ciprofloxacin with calf thymus DNA has been studied through changes in the irreversible oxidation peak of the drug recorded on a waxed graphite electrode at 0.88 V by differential pulse voltammetry [179]. Assuming that the drug fully loses its electrochemical activity after reacting with the DNA, a non-linear fitting of the concentration profile results in the determination of the binding constant as $1.36 \times 10^5\ M^{-1}$ and complexation stoichiometry as at most one molecule of the drug per two pairs of nucleobases. Increased signal of ciprofloxacin oxidation is obtained on graphene nanoflakes distributed on the surface of a glassy carbon electrode [180]. Co-adsorption of salmon sperm dsDNA increases the signal due to increased charge on the surface and electrostatic accumulation of positively charged DNA molecules. The graphene-based DNA sensor makes it possible to detect 0.1 to 100 μM of ciprofloxacin (LOD 0.1 μM). In later studies, the interaction of ciprofloxacin with dsDNA was recorded using an electrochemical sensor on a platform of pencil graphite electrode covered with electropolymerized polypyrrole, single-walled carbon nanotubes and DNA [181]. The electrode was used for the detection of the drug accumulated on DNA molecules, with a dynamic range of 0.008 to 30 μM (LOD 4.0 nM). The incubation of the electrode in ciprofloxacin solution decreased the oxidation currents of guanine (0.865 V) and adenine (1.169 V) on DPV voltammograms. The usability of this DNA sensor was substantiated by measurements in spiked urine samples against HPLC assay. Contrary to anthracycline intercalators, hydrophobic interactions are largely considered responsible for drug complexation in this case.

Similar results of ciprofloxacin oxidation have been obtained on boron-doped diamond electrodes [182]. Therein, a binding constant of $5.91 \times 10^5\ M^{-1}$ was obtained through square wave voltammetry for dsDNA intercalation; DPV made detection possible in the 0.50–60 μM range, with an LOD of 0.44 μM.

Domperidone, a drug used for treating gastrointestinal ailments, interacts with DNA along the minor groove of the DNA helix. Such an interaction affects the electrochemical behavior of the drug. To accumulate donperidone and characterize its reaction with dsDNA, electrochemical sensors consisting of carbon paste electrodes with oxidized multi-walled carbon nanotubes have been designed [183]. The associated cyclic voltammogram contains one pair of reversible peaks at 0.4–0.6 V and an irreversible peak at 0.9–1.1 V. All the peaks tend to decrease after increased addition of DNA to the drug solution. With changes in the peak currents, the binding constant is calculated to be $7.9 \times 10^4\ M^{-1}$.

Chlorpromazine, a psychotropic drug, intercalates into dsDNA immobilized on Au electrode through the terminal –SH group [184]. As a result, higher peak currents at 0.67 V and a lower LOD of 0.6 μM are achieved using DPV for signal recording.

Amitriptyline, a drug used for antipsychotic and sedative aggression, can be detected using an electrochemical DNA sensor based on composite materials consisting of graphite, SiO_2, Al_2O_3, Nb_2O_5 and dsDNA [185]. The initial material is obtained by sol-gel method; it is then mechanically mixed with DNA, graphite powder and mineral oil. Without the drug, two peaks are recorded on the voltammogram, attributed to the oxidation of guanine and adenine (0.8 V and 1.11 V). The addition of the drug increases the currents associated with the mediation of electron transfer. In optimal conditions, 10–80 μM of amitriptyline can be detected (LOD 0.12 μM).

Ketamine, an anesthesia drug, can interact with ssDNA and dsDNA adsorbed on electrochemically pre-oxidized carbon paste electrodes. The reaction results in the accumulation of the drug on ssDNA, resulting in a higher irreversible oxidation peak current at 1.14 V [186]. Parallel recording of the UV-vis spectra makes it possible to calculate the binding constant as 1.0×10^7 for dsDNA and $3.0 \times 10^7\ M^{-1}$ for ssDNA. Hybridization of the DNA sequence complementary to the DNA probe decreases the peak due to the competition of its binding at guanine residues. This disposable DNA sensor can detect ketamine in concentrations as low as 1.98 nM.

4.4 DNA-damaging factors

DNA-damaging factors have been studied for DNA sensors from the very beginning of the biosensor era [187–190]. Since 1990s, the electrochemistry of dsDNA and ssDNA adsorbed on polarized screen-printed carbon electrodes has come to be in use for both determination of DNA quantities and assessment of their interactions with potentially damaging factors. In these works, oxidation of guanine residues was mostly investigated through potentiometric stripping analysis at a constant current. Daunomycin, an anthracyline drug, was used as the intercalator. The potential decreased with DNA damage, whereas the guanine signal increased [191]. Additional information can be obtained from the comparison of signals recorded with ssDNA from the same source. Various anti-tumor drugs (cisplatin, acridine orange), aromatic amines (2-anthramine, 2-naphtylamine), mycotoxins, industrial pollutants (PCB, phthalates), as well as industrial waste waters, which exhibit genotoxicity as per standard texts (ToxAlert, umu-assay), have been tested to show possible areas of application of such biosensors. It has, however, come to be known that low selectivity and rather high concentration of the tested pollutants limit the performance of biosensors. Furthermore, the necessity for anodic polarization of the electrode during the entire duration of DNA accumulation complicates sensor preparation, especially in field conditions.

Interaction of antiproliferative metallodrugs with dsDNA has been investigated through changes in guanine oxidation currents recorded on glassy carbon electrodes [192, 193]. In one of the studies, the reactions of cis-platin, carboplatin, 2,2'-bypiridylbis(pyridine) Pt(II), titanocene dichloride and trans-tetrachlorodimethylsulfoxide imidazole ruthenate (III) with ssDNA and dsDNA from calf thymus electrostatically accumulated on screen-printed carbon electrode at 1.6 V were investigated. Therein, the relative decay of the guanine oxidation peak recorded by square-wave voltammetry at 0.95 V was used as a measure of specific interactions. In another study, the reaction of cis-platin and DNA was explored with an electrochemical sensor assembled on a glassy carbon electrode covered with electrochemically reduced graphene oxide and adsorbed dsDNA [194]. Guanine peaks (at 1.0 V on a bare electrode and 0.76 V on modified glassy carbon electrode) and adenine peaks (at 1.25 V on a bare electrode and 1.05 V on modified glassy carbon electrode) were recorded with adsorptive transfer stripping voltammetry and DPV. The peak heights were observed to decrease with increased drug concentration.

Cis-platin and oxaliplatin have been studied on disposable graphite electrodes modified with single-walled carbon nanotubes and adsorbed DNA [195]. The reaction results in reduced guanine signals and increased charge transfer resistance measured in a short time period (3–5 minutes).

Temozolomide is an antineoplastic alkylating agent which works against brain tumors. In solution phase, it undergoes spontaneous hydrolysis to yield methyldiazonium ions (Scheme 8).

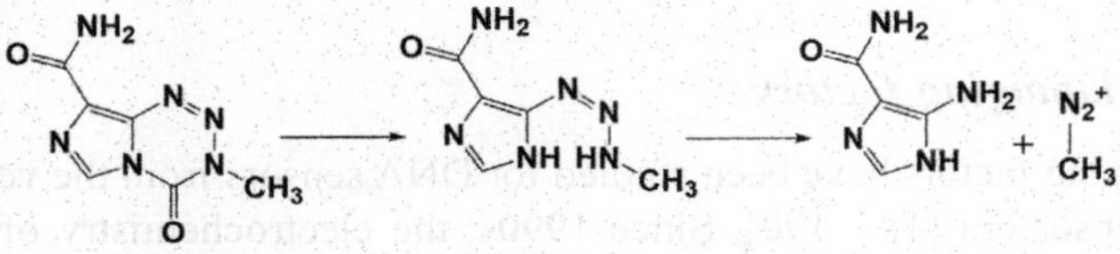

Scheme 8. The scheme of hydrolysis of temozolomide.

Intermediate and final products can also affect DNA conformation and damage. Electrochemical investigations of reactions performed in the presence of dsDNA from calf thymus on glassy carbon electrodes, through DPV, have confirmed DNA damage. In this case, the peaks attributed to the products of such a damage—8-oxo-guanine, 2,8-dioxoadenine and guanine—were recorded by DPV [196]. The reactions were performed in solution phase or with the DNA physically adsorbed on the electrode prior to its contact with the drug. In case of the latter, the DNA sensor was first moved to the buffer solution with no drug and then the oxidation peaks of the damaged DNA were recorded. Changes in the peak currents for differently incubated DNA sensors showed condensation of DNA into double-helical structures and specific interactions between the drug and its metabolites and guanine residue of the DNA chain [197]. A similar phenomenon has been observed with temozolomide accumulated onto dsDNA layer assembled on a pre-oxidized pencil graphite electrode [198]. The DNA damage caused by the products of drug conversion was observed within 7 days of their contact. The guanine oxidation peak currents recorded by DPV depended on the methylation degree of the DNA oligonucleotides.

Nitroimidazoles are also considered as a source of reactive oxygen species produced by the reduction of the nitro group in the drug molecule. The interaction of benznidazole and dsdna and ssDNA from calf thymus has been investigated through DPV on the interface of a glassy carbon electrode [199]. In the absence of DNA, the reduction is observed in one step in an alkaline medium, and in two separate steps with intermediate formation of hydroxylamine derivative in a weakly acidic medium. The addition of DNA results in the formation of two oxidation peaks—attributed to the guanosine (1.05 V) and adenosine (1.20 V) fragments.

Isoproterenol is an adrenoergic receptor agonist used in the treatment of cardiac arrest and shock. Its interaction with cardiomyocyte DNA has been investigated using glassy carbon electrodes modified with mutli-walled carbon nanotubes, polyaniline and gold nanoparticles assembled by drop casting of oppositely charged layers [52]. Therein, the assembling of the surface layer was monitored by EIS and fluorescence microscopy and the drug-DNA interaction was observed by means of DPV. The incubation of DNA with the drug resulted in the appearance and enhancement of the guanine signal at 0.84 V. Meanwhile, the self peak of isoproterenol oxidation at 0.38 V decreased synchronously. Furthermore, DNA damage (breakage in DNA double strands) was confirmed by protein expression and electrophoretic detection of appropriate protein markers in *in vivo* experiments. Top et al. [200] proposed impedimetric detection of DNA-isoproterenol complexation using screen-printed electrodes covered with electrostatically accumulated DNA. Significant difference in the behavior of cancer and normal DNA was shown.

Amlodipin is a third-generation calcium antagonist used against high blood pressure, hypertension and cardiac arrhythmia. Its reaction with DNA has been investigated using boron-doped diamond electrodes modified with physically adsorbed DNA [201], wherein the modification was monitored through changes in the cyclic voltammogram of ferricyanide redox pair. Increasing DNA quantities decreased the peak currents and increased the peak potential difference. Preliminary incubation of amlodipin with DNA prior to its adsorption changed the redox indicator voltammograms in the opposite direction, indicating partial shielding of DNA active sites from damage. This conclusion was confirmed by monitoring DPV of the guanine peaks in the absence of an external redox indicator.

Leuprolide interacts with dsDNA from fish sperm accumulated on an anodically activated pencil graphite electrode [202]. This interaction results in an increase in the guanine oxidation signal. In optimal working conditions, 0.20–6.00 ppm of the drug (LOD 0.04 ppm) can be detected. The biosensor in the referred study has been used for the analysis of peuprolide in pharmaceutical dosage form. Contrary to that, another anti-cancer drug, fulvestrand, decreases the guanine signal measured by DPV on the same electrode in a range of 1 to 20 ppb (LOD 0.31 ppb) [203].

Glivec, an anti-leukemia drug, causes the oxidation of adenine fragments of DNA with the formation of peaks attributed to 2,8-dehydroxyadenine at about 1.28 V, as is revealed DPV [204]. The reaction is found to be rather fast: the peak reaches its maximum after 3 minutes of incubation. In addition to the above, smaller peaks of guanine and oxoguanine are also observed on the voltammogram.

5. Conclusion

Recent advancements in the development of DNA sensors for drug detection encompas all modern trends in the progress of bioelectroanalysis, including broad application of nanomaterials, miniaturization tendency, application of disposable sensors or their transducers, and joint consideration of electrochemical techniques and physical methods of interfacial research. Electron exchange and interaction with DNA result in remarkable changes in the optical properties of the electrode interface, facilitating the assessment of binding constants and stoichiometry of the reaction. Comparing the results obtained in homogeneous conditions and those on the electrode surface offers new prospective avenues in the study of the mechanism of DNA-drug interactions, as well as for distinguishing between electrostatic and steric control of these interactions.

Unfortunately, there are no examples of DNA sensors used for screening anti-tumor drugs or for researching new types of biological activities of newly synthesized compounds. At present, anthracyclines are the only assurance of the future benefits that are associated with such an approach. However, interest in this class of cytostatics is rather related to their function as a redox probe in other DNA sensors. Rather challenging are applications of DNA sensors in a real sample assay. The spiked samples are mostly used for this purpose, and possible interferences with target species still remain an important topic for future investigation.

Acknowledgments

TH acknowledges the financial support of Science Agency VEGA, Project No. 1/0419/20.

References

[1] Thevenot, D.R., K. Toth, R.A. Durst and G.S. Wilson. 2001. Electrochemical biosensors: Recommended definitions and classification. Biosens. Bioelectron. 16: 121–131.
[2] Dunn, M.R., R.M. Jimenez and J.C. Chaput. 2017. Analysis of aptamer discovery and technology. Nat. Rev. Chem. 1: 0076.
[3] Gui, R., H. Jin, H. Guo and Z. Wang. 2018. Recent advances and future prospects in molecularly imprinted polymers-based electrochemical biosensors. Biosens. Bioelectron. 100: 56–70.
[4] Lee, S.T., D. Beaumont, X.D. Su, K. Muthoosamy and S.Y. New. 2018. Formulation of DNA chimera templates: Effects on emission behavior of silver nanoclusters and sensing. Anal. Chim. Acta 1010: 62–68.
[5] Miao, P., Y. Tang, B. Wang, J. Yin and L. Ning. 2015. Signal amplification by enzymatic tools for nucleic acids. TrAC Trends Anal. Chem. 67: 1–15.
[6] Lima, H.R.S., J.S. da Silva, E.A.O. Farias, P.R.S. Teixeira, C. Eiras and L.C.C. Nunes. 2018. Electrochemical sensors and biosensors for the analysis of antineoplastic drugs. Biosens. Bioelectron. 108: 27–37.
[7] Pourbasheer, E., Z. Azari and M.R. Ganjali. 2019. Recent advances in biosensors based nanostructure for pharmaceutical analysis. Curr. Anal. Chem. 15: 152–158.
[8] Erdem, A. 2007. Nanomaterial-based electrochemical DNA sensing strategies. Talanta 74: 318–325.
[9] Wang, J. 2005. Nanomaterial-based electrochemical biosensors. Analyst 130: 421–426.
[10] Chen, A. and S. Chatterjee. 2013. Nanomaterials based electrochemical sensors for biomedical applications. Chem. Soc. Rev. 42: 5425–5438.
[11] Iijima, S. 1991. Helical microtubules of graphitic carbon. Nature 354: 56–58.
[12] Zhao, Q., Z. Gan and Q. Zhuang. 2002. Electrochemical sensors based on carbon nanotubes. Electroanalysis 14: 1609–1613.
[13] Zhu, Z. 2017. An overview of carbon nanotubes and graphene for biosensing applications. Nano-Micro. Lett. 9: 25.
[14] Jung, H. and C.W. Bielawski. 2019. Asphaltene oxide promotes a broad range of synthetic transformations. Commun. Chem. 2: 113.
[15] Daniel, S., T.P. Rao, K.S. Rao, S.U. Rani, G.R.K. Naidu, H.-Y. Lee and T. Kawai. 2007. A review of DNA functionalized/grafted carbon nanotubes and their characterization. Sens. Actuat. B. 122: 672–682.
[16] Gao, L., C. Lian, Y. Zhou, L. Yan, Q. Li, C. Zhang, L. Chen and K. Chen. 2014. Graphene oxide-DNA based sensors. Biosens. Bioelectr. 60: 22–29.
[17] Shao, Y., J. Wang, H. Wu, J. Liu, I.A. Aksay and Y. Lin. 2009. Graphene based electrochemical sensors and biosensors: A review. Electroanalysis 22: 1027–1036.
[18] Zhou, M., Y. Zhai and S. Dong. 2009. Electrochemical sensing and biosensing platform based on chemically reduced graphene oxide. Anal. Chem. 81: 5603–5613.

[19] Mallesha, M., R. Manjunatha, C. Nethravathi, G.S. Suresh, M. Rajamathi, J.S. Melo and T.V. Venkatesha. 2011. Functionalized-graphene modified graphite electrode for the selective determination of dopamine in presence of uric acid and ascorbic acid. Bioelectrochemistry 81: 104–108.
[20] Fang, Y. and E. Wang. 2013. Electrochemical biosensors on platforms of graphene. Chem. Commun. 49: 9526–9539.
[21] Morales De la Cruz, K., G. Alarcón-Angeles and A. Merkoçi. 2019. Nanomaterial-based sensors for the study of DNA interaction with drugs. Electroanalysis 31: 1845–1867.
[22] Wei, L., J. Borowiec, L. Zhu and J. Zhang. 2012. Electrochemical investigation on the interaction of diclofenac with DNA and its application to the construction of a graphene-based biosensor. J. Solid State Electrochem. 16: 3817–3823.
[23] Erdem, A., M. Muti, P. Papakonstantinou, E. Canavar, H. Karadeniz, G. Congur and S. Sharma. 2012. Graphene oxide integrated sensor for electrochemical monitoring of mitomycin C–DNA interaction. Analyst 137: 2129–2135.
[24] Chen, J., B. Fu, T. Liu, Z. Yan and K. Li. 2018. A graphene oxide-DNA electrochemical sensor based on glassy carbon electrode for sensitive determination of methotrexate. Electroanalysis 30: 288–295.
[25] Erdem, A., H. Karadeniz and A. Caliskan. 2009. Single-walled carbon nanotubes modified graphite electrodes for electrochemical monitoring of nucleic acids and biomolecular interactions. Electroanalysis 21: 464–471.
[26] Evtugyn, G., A. Porfireva, R. Shamagsumova and T. Hianik. 2020. Advances in electrochemical aptasensors based on carbon nanomaterials. Chemosensors 8: 96.
[27] Tang, W., M. Zhang and X. Zeng. 2014. Establishment of dsDNA/GNs/chit/GCE biosensor and electrochemical study on interaction between 6-mercaptopurine and DNA. Biomed. Mater. Eng. 24: 1071–1077.
[28] Tang, W., W. Li, Y. Li, M. Zhang and X. Zeng. 2015. Electrochemical sensors based on multi-walled nanotubes for investigating the damage and action of 6-mercaptopurine on double-stranded DNA. New J. Chem. 39: 8454–8460.
[29] Karimi-Maleh, H., F. Tahernejad-Javazmi, N. Atar, M.L. Yola, V.K. Gupta and A.A. Ensafi. 2015. A novel DNA biosensor based on a pencil graphite electrode modified with polypyrrole/functionalized multiwalled carbon nanotubes for determination of 6-mercaptopurine anticancer drug. Ind. Eng. Chem. Res. 54: 3634–3639.
[30] Silva, T.A., F.C. Moraes, B.C. Janegitz and O. Fatibello-Filho. 2017. Electrochemical biosensors based on nanostructured carbon black: A review. J. Nanomaterials 2017: 14.
[31] Shuai, H.-L., K.-J. Huang and Y.-X. Chen. 2016. A layered tungsten disulfide/acetylene black composite based DNA biosensing platform coupled with hybridization chain reaction for signal amplification. J. Mater. Chem. B 4: 1186–1196.
[32] Kuzin, Yu., D. Kappo, A. Porfireva, D. Shurpik, I. Stoikov, G. Evtugyn and T. Hianik. 2018. Electrochemical DNA sensor based on carbon black–poly(Neutral red) composite for detection of oxidative DNA damage. Sensors 18: 3489.
[33] Turkevich, J., P.C. Stevenson and J. Hillier. 1951. A study of the nucleation and growth processes in the synthesis of colloidal gold. Discuss. Faraday Soc. 11: 55–75.
[34] Yeh, Y.-C., B. Creran and V.M. Rotello. 2012. Gold nanoparticles: Preparation, properties and application in bionanotechnology. Nanoscale 4: 1871–1880.
[35] Elahi, N., M. Kamali and M.H. Baghersad. 2018. Recent biomedical applications of gold nanoparticles: A review. Talanta 184: 537–556.
[36] Luo, X., A. Morrin, A.J. Killard and M.R. Smyth. 2006. Application of nanoparticles in electrochemical sensors and biosensors. Electroanalysis 18: 319–326.
[37] Chiang, H.C., Y. Wang, Q. Zhang and K. Levon. 2019. Optimization of the electrodeposition of gold nanoparticles for the application of highly sensitive, label-free biosensor. Biosensors 9: 50.
[38] Zabihollahpoor, A., M. Rahimnejad, G. Najafpour and A.A. Moghadamnia. 2019. Gold nanoparticle prepared by electrochemical deposition for electrochemical determination of gabapentin as an antiepileptic drug. J. Electroanal. Chem. 825: 281–286.

[39] Campuzano, S., M. Pedrero, C. Montemayor, E. Fatás and J.M. Pingarrón. 2006. Characterization of alkanethiol-self-assembled monolayers-modified gold electrodes by electrochemical impedance spectroscopy. J. Electroanal. Chem. 586: 112–121.

[40] Colangelo, E., J. Comenge, D. Paramelle, M. Volk and Q. Chen. 2017. Characterizing self-assembled monolayers on gold nanoparticles. Bioconjugate Chem. 28: 11–22.

[41] Fischer, L., M. Tenje, M. Heiskanen, A.R. Masuda, N. Castillo, J. Bentien, J. Emneus, M.H. Jakobsen and A. Boisen. 2009. Gold cleaning methods for electrochemical detection applications. Microelectronic Eng. 86: 1282–1285.

[42] Li, C.-Z., Y. Liu and J. H.T. Luong. 2005. Impendance sensing of DNA binding drugs using gold substrates modified with gold nanoparticles. Anal. Chem. 77: 478–485.

[43] Shen, Q., X. Wang and D. Fu. 2008. The amplification effect of functionalized gold nanoparticles on the binding of anticancer drug dacarbizane to DNA and DNA bases. Appl. Surf. Sci. 255: 577–580.

[44] He, X., W. Liu, X. Zhang, X. Zhang and J. Chen. 2014. Electrochemical determination of bleomycins based on dual-amplification of 4-mercaptophenyl boronic acid-capped gold nanoparticles and dopamine-capped gold nanoparticles. Anal. Methods 6: 6893–6899.

[45] Hajian, R., Z. Mehrayin, M. Mohagheghian, M. Zafari, P. Hosseini and N. Shams. 2015. Fabrication of an electrochemical sensor based on carbon nanotubes modified with gold nanoparticles for determination of valrubicin as a chemotherapy drug: Valrubicin-DNA interaction. Mater. Sci. Eng. C 49: 769–775.

[46] Wang, J., K. Ma, H. Yin, Y. Zhou and S. Ai. 2018. Aptamer based voltammetric determination of ampicillin using a single-stranded DNA binding protein and DNA functionalized gold nanoparticles. Microchim. Acta 185: 68–74.

[47] Jahandari, S., M.A. Taher, H. Karimi-Maleh, A. Khodadadi and E. Faghih-Mirzaei. 2019. A powerful DNA-based voltammetric biosensor modified with Au nanoparticles, for the determination of Temodal; an electrochemical and docking investigation. J. Electroanal. Chem. 840: 313–318.

[48] Gholivand, M.B. and M. Torkashvand. 2016. The fabrication of a new electrochemical sensor based on electropolymerization of nanocomposite gold nanoparticle-molecularly imprinted polymer for determination of valganciclovir. Materials Sci. Eng. C 59: 594–603.

[49] Najari, S., H. Bagheri, Z. Monsef-Khoshhesab, A. Hajian and A. Afkhami. 2018. Electrochemical sensor based on gold nanoparticle-multiwall carbon nanotube nanocomposite for the sensitive determination of docetaxel as an anticancer drug. Ionics 24: 3209–3219.

[50] Kuralay, F. and A. Erdem. 2015. Gold nanoparticles/polymer nanocomposite for highly sensitive drug- DNA interaction. Analyst 140: 2876–2880.

[51] Wang, W., Y. Cheng, L. Yan, H. Zhu, G. Li, J. Li and W. Sun. 2015. Highly sensitive electrochemical sensor for dopamine with a double-stranded deoxyribonucleic acid/gold nanoparticle/graphene modified electrode. Analyt. Meth. 7: 1878–1883.

[52] Wang, J., Y. Li, C. Li, X. Zeng, W. Tang and X. Chen. 2017. A voltammetric study on the interaction between isoproterenol and cardiomyocyte DNA by using a glassy carbon electrode modified with carbon nanotubes, polyaniline and gold nanoparticles. Microchim. Acta 184: 2999–3006.

[53] Jia, Z., Y. Ma, L. Yang, C. Guo, N. Zhou, M. Wang, L. He and Z. Zhang. 2019. $NiCo_2O_4$ spinel embedded with carbon nanotubes derived from bimetallic NiCo metal-organic framework for the ultrasensitive detection of human immune deficiency virus-1 gene. Biosens. Bioelectr. 133: 55–63.

[54] Cheng, D., X. Xiao, X. Li, C. Wang, Y. Liang, Z. Yu, C. Jin, N. Zhou, M. Chen, Y. Dong, Y. Lin, Z. Xie and C. Zhang. 2018. A non-enzymatic electrochemical sensing platform based on hemin@MOF composites for detecting hydrogen peroxide and DNA. J. Electrochem. Soc. 165: B885–B892.

[55] Liao, X., H. Fu, T. Yan and J. Lei. 2019. Electroactive metal–organic framework composites: Design and biosensing application. Biosens. Bioelectron. 146: 111743.

[56] Shao, H., J. Lu, Q. Zhang, Y. Hu, S. Wang and Z. Guo. 2018. Ruthenium-based metal organic framework (Ru-MOF)-derived novel Faraday-cage electrochemiluminescence biosensor for ultrasensitive detection of miRNA-141. Sens. Act. B. 268: 39–46.

[57] Lu, J., G. Getz, E.A. Miska, E.A. Saavedra, J. Lamb, D. Peck, A.S. Cordero, B.L. Ebert, R.H. Mak, A.A. Ferrando, J.R. Downing, T. Jacks, H.R. Horvitz and T.R. Golub. 2005. MicroRNA expression profiles classify human cancers. Nature 435: 834–838.

[58] Jou, A.F., C.H. Lu, Y.C. Ou, S.S. Wang, S.L. Hsu, I. Willner and J.A. Ho. 2015. Diagnosing the miR-141 prostate cancer biomarker using nucleic acid-functionalized CdSe/ZnS QDs and telomerase. Chem. Sci. 6: 659–665.

[59] Forster, R.J., P. Bertoncello and T.E. Keyes. 2009. Electrogenerated chemiluminescence. Ann. Review of Anal. Chem. 2: 359–385.

[60] Qin, F., S. Jia, F. Wang, S. Wu, J. Song and Y. Liu. 2013. Hemin@metal–organic framework with peroxidase-like activity and its application to glucose detection. Catalysis Sci. Technol. 3: 2761–2768.

[61] Gong, Q., Y. Wang and H. Yang. 2017. A sensitive impedimetric DNA biosensor for the determination of the HIV gene based on graphene-Nafion composite film. Biosens. Bioelectron. 89: 565–569.

[62] Huang, Y., Z. Gao, H. Luo and N.B. Li. 2017. Sensitive detection of HIV gene by coupling exonuclease III-assisted target recycling and guanine nanowire amplification. Sens. Act. B 238: 1017–1023.

[63] Deng, K., C. Li, J. Huang and X. Li. 2017. Rolling circle amplification based on signal-enhanced electrochemical DNA sensor for ultrasensitive transcription factor detection. Sens. Actuat. B 238: 1302–1308.

[64] Li, Q., W. Cheng, D. Zhang, T. Yu, Y. Yin, H. Ju and S. Dinget. 2012. Rapid and sensitive strategy for salmonella detection using an *InvA* gene-based electrochemical DNA sensor. Int. J. Electrochem. Sci. 7: 844–856.

[65] Zhang, Y., X. Geng, J. Ai, Q. Gao, H. Qi and C. Zhang. 2015. Signal amplification detection of DNA using a sensor fabricated by one-step covalent immobilization of amino-terminated probe DNA onto the polydopamine-modified screen-printed carbon electrode. Sens. Actuat. B 221: 1535–1541.

[66] Wu, S., Y. Tang, L. Chen, X. Ma, S. Tian and J. Sun. 2015. Amplified electrochemical hydrogen peroxide reduction based on hemin/G-quadruplex DNAzyme as electrocatalyst at gold particles modified heated copper disk electrode. Biosens. Bioelectron. 73: 41–46.

[67] Shariati, M., M. Ghorbani, P. Sasanpour and A. Karimizefreh. 2019. An ultrasensitive label free human papilloma virus DNA biosensor using gold nanotubes based on nanoporous polycarbonate in electrical alignment. Anal. Chim. Acta 1048: 31–41.

[68] Soni, A., C.M. Pandey, S. Solanki and G. Sumana. 2019. Synthesis of 3D-coral like polyaniline nanostructures using reactive oxide templates and their high performance for ultrasensitive detection of blood cancer. Sens. Act. B 281: 634–642.

[69] Gholivand, M.-B. and A. Akbari. 2019. A sensitive electrochemical genosensor for highly specific detection of thalassemia gene. Biosens. Bioelectron. 129: 182–188.

[70] Soni, A., C.M. Pandey, M.K. Pandey and G. Sumana. 2019. Highly efficient polyaniline-MoS_2 hybrid nanostructures based biosensor for cancer biomarker detection. Anal. Chim. Acta 1055: 26–35.

[71] Senel, M., M. Dervisevic and F. Kokkokoglu. 2019. Electrochemical DNA biosensors for label-free breast cancer gene marker detection. Anal. Bioanal. Chem. 411: 2925–2935.

[72] Khosravi-Nejad, F., M. Teimouri, S. Jafari Marandi and M. Shariati. 2019. The highly sensitive impedimetric biosensor in label free approach for hepatitis B virus DNA detection based on tellurium doped ZnO anowires. Appl. Phys. A 125: 616.

[73] Divsar, F. 2019. A label-free photoelectrochemical DNA biosensor using a quantum dot–dendrimer nanocomposite. Anal. Bioanal. Chem. 411: 6867–6875.

[74] Lv, M.-M., S.-F. Fan, Q.-L. Wang, Q-Y. Lv, X. Song and H.-F. Cui. 2020. An enzyme-free electrochemical sandwich DNA assay based on the use of hybridization chain reaction and gold nanoparticles: application to the determination of the DNA of Helicobacter pylori. Microchim. Acta 187: 73.

[75] Flynn, N.T., T.N.T. Tran, M.J. Cima and R. Langer. 2003. Long-term stability of self-assembled monolayers in biological media. Langmuir 19: 10909–10915.

[76] Chen, Y., P. Li, J.A. Modica, R.J. Drout and O.K. Farha. 2018. Acid-resistant mesoporous metal–organic framework toward oral insulin delivery: Protein encapsulation, protection, and release. J. Am. Chem. Soc. 140: 5678–5681.

[77] Ma, J., W. Chai, J. Lu, T. Tian, S. Wu, Y. Yang, J. Yang and C. Li. 2019. Coating a DNA self-assembled monolayer with a metal organic framework-based exoskeleton for improved sensing performance. Analyst 144: 3539–3545.

[78] Evtugyn, G., S. Belyakova, A. Porfireva and T. Hianik. 2020. Electrochemical aptasensors based on hybrid metal-organic frameworks. Sensors 20: 6943.

[79] Wallace, G.G., G.M. Spinks, I.A.P. Kane-Maguire and P.R. Teasdale. 2003. Conductive Electroactive Polymers. Intelligent Materials Systems. CRC Press, Boca Raton.

[80] Evtugyn, G. and T. Hianik. 2016. Electrochemical DNA sensors and aptasensors based on electropolymerized materials and polyelectrolyte complexes. TrAC Trends Anal. Chem. 79: 168–178.

[81] Malhotra, B.D., A. Chaubey and S.P. Singh. 2006. Prospects of conducting polymers in biosensor. Anal. Chim. Acta 578: 59–74.

[82] Ates, M. 2013. A review study of (bio)sensor systems based on conducting polymers. Mater. Sci. Eng. C 33: 1853–1859.

[83] Inzelt, G. 2011. Rise and rise of conducting polymers. J. Solid State Electrochem. 15: 1711–1718.

[84] Puskás, Z. and G. Inzelt. 2005. Formation and redox transformations of polyphenazine. Electrochim. Acta 50: 1481–1490.

[85] Barsan, M.M., M.E. Ghica and C.M.A. Brett. 2015. Electrochemical sensors and biosensors based on redox polymer/carbon nanotube modified electrodes: A review. Anal. Chim. Acta 881: 1–23.

[86] Xu, Y., X. Ye, Y.L. Yang, P. He and Y. Fang. 2006. Impedance DNA biosensor using electropolymerized polypyrrole/multiwalled carbon nanotubes modified electrode. Electroanalysis 18: 1471–1478.

[87] Rezaei, B., M.Kh. Boroujeni and A.A. Ensafi. 2016. Development of Sudan II sensor based on modified treated pencil graphite electrode with DNA, o-phenylenediamine, and gold nanoparticle bioimprinted polymer. Sens. Actuators B 222: 849–856.

[88] Osaka, T., T. Momma, S. Komaba, H. Kanagawa and S. Nakamura. 1994. Electrochemical process of formation of an insulating polypyrrole film. J. Electroanal. Chem. 372: 201–207.

[89] Liu, K., W.-Z. Wei, J.-X. Zeng, X.-Y. Liu and Y.-P. Gao. 2006. Application of a novel electrosynthesized polydopamine-imprinted film to the capacitive sensing of nicotine. Anal. Bioanal. Chem. 385: 724–729.

[90] Cosnier, S. and M. Holzinger. 2011. Electrosynthesized polymers for biosensing. Chem. Soc. Rev. 40: 2146–2156.

[91] Okamoto, H. and T. Kotaka. 1999. Effect of counter ions in electrochemical polymerization media on the structure and responses of the product polyaniline films. III. Structure and properties of polyaniline films prepared via electrochemical polymerization. Polymer 40: 407–417.

[92] Prakash, R. 2002. Electrochemistry of polyaniline: Study of the pH effect and electrochromism. J. Appl. Polymer Sci. 83: 378–385.

[93] Shao, Y., Y. Jin and S. Dong. 2002. DNA-templated assembly and electropolymerization of aniline on gold surface. Electrochem. Commun. 4: 773–779.

[94] Malinauskas, A. 2004. Self-doped polyanilines. J. Power Sources 126: 214–220.

[95] Sadki, S., P. Schottland, N. Brodie and G. Sabouraud. 2000. The mechanisms of pyrrole electropolymerization. Chem. Soc. Rev. 29: 283–293.

[96] Krische, B. and M. Zagorska. 1989. The polythiophene paradox. Synthetic Metals 28: C263–C268.

[97] Jadamiec, M., M. Lapkowski, M. Matlengiewicz, A. Brembilla, B. Henry and L. Rodehüser. 2007. Electrochemical and spectroelectrochemical evidence of dimerization and oligomerization during the polymerization of terthiophenes. Electrochim. Acta 52: 6146–6154.

[98] Patra, A., M. Bendikov and S. Chand. 2014. Poly(3,4-ethylenedioxyselenophene) and its derivatives: Novel organic electronic materials. Acc. Chem. Res. 47: 1465–1474.

[99] Huynh, T.-P., P.S. Sharma, M. Sosnowska, F. D'Souza and W. Kutner. 2015. Functionalized polythiophenes: Recognition materials for chemosensors and biosensors of superior sensitivity, selectivity, and detectability. Progress Polym. Sci. 47: 1–25.

[100] Palomera, N., J.L. Vera, E. Meléndez, J.E. Ramirez-Vick, M.S. Tomar, S.K. Arya and S.P. Singh. 2011. Redox active poly(pyrrole-N-ferrocene-pyrrole) copolymer based mediator-less biosensors. J. Electroanal. Chem. 658: 33–37.
[101] Pauliukaite, R., M.E. Ghica, M. Barsan and C.M.A. Brett. 2007. Characterisation of poly(neutral red) modified carbon film electrodes; application as a redox mediator for biosensors. J. Solid State Electrochem. 11: 899–908.
[102] Kertesz, V. and G.J. Van Berkel. 2001. Electropolymerization of methylene blue investigated using on-line electrochemistry/electrospray mass spectrometry. Electroanalysis 13: 1425–1430.
[103] Schlereth, D.D. and A.A. Karyakin. 1995. Electropolymerization of phenothiazine, phenoxazine and phenazine derivatives—characterization of the polymers by UV-visible difference spectroelectrochemistry and Fourier-transform IR spectroscopy. J. Electroanal. Chem. 395: 221–232.
[104] Karyakin, A.A., E.E. Karyakina and H.-L. Schmidt. 1999. Electropolymerized azines: A new group of electroactive polymers. Electroanalysis 11: 149–155.
[105] Yang, C., J. Yi, X. Tang, G. Zhou and Y. Zeng. 2006. Studies on the spectroscopic properties of poly(neutral red) synthesized by electropolymerization. Reactive Functional Polym. 66: 1336–1341.
[106] Ge, D., S. Huang, R. Qi, J. Mu, Y. Shen and W. Shi. 2009. Nanowire based polypyrrole hierarchical structures synthesized by a two-step electrochemical method. ChemPhysChem 10: 1916–1921.
[107] Kuzin, Yu., A. Ivanov, G. Evtugyn and T. Hianik. 2016. Voltammetric detection of oxidative DNA damage based on interactions between polymeric dyes and DNA. Electroanalysis 28: 2956–2964.
[108] Ivanov, A., Yu. Kuzin and G. Evtugyn. 2018. SPR sensor based on polyelectrolyte complexes with DNA inclusion. Sens. Actuators B 281: 574–581.
[109] Liu, C., D. Jiang, G. Xiang, L. Liu, F. Liu and X. Pu. 2014. An electrochemical DNA biosensor for the detection of *Mycobacterium tuberculosis*, based on signal amplification of graphene and a gold nanoparticle–polyaniline nanocomposite. Analyst 139: 5460–5465.
[110] Muthusankar, G., R. Sasikumar, S.-M. Chen, G. Gopu, N. Sengottuvelan and S.-P. Rwei. 2018. Electrochemical synthesis of nitrogen-doped carbon quantum dots decorated copper oxide for the sensitive and selective detection of non-steroidal anti-inflammatory drug in berries. J. Colloid Interface Sci. 523: 191–200.
[111] Li, X. and X. Kan. 2019. A boronic acid carbon nanodots/poly(thionine) sensing platform for the accurate and reliable detection of NADH. Bioelectrochem. 130: 107344.
[112] Evtugyn, G., A. Porfireva, V. Stepanova and H. Budnikov. 2015. Electrochemical biosensors based on native DNA and nanosized mediator for the detection of anthracycline preparations. Electroanalysis 27: 629–637.
[113] Porfireva, A., V. Vorobev, S. Babkina and G. Evutgyn. 2019. Electrochemical sensor based on poly(Azure B)-DNA composite for doxorubicin determination. Sensors 19: 2085.
[114] Stoikov, D.I., A.V. Porfireva, D.N. Shurpik, I.I. Stoikov and G.A. Evtyugin. 2019. Electrochemical DNA sensors on the basis of electropolymerized thionine and Azure B with addition of pillar[5] arene as an electron transfer mediator. Russ. Chem. Bull. 68: 431–437.
[115] Evtugyn, G.A., V.B. Stepanova, A.V. Porfireva, A.I. Zamaleeva and R.R. Fakhrullin. 2014. Electrochemical DNA sensors based on nanostructured organic dyes/DNA/polyelectrolyte complexes. J. Nanosci. Nanotechnol. 14: 6738–6747.
[116] Ricci, F. and K.W. Plaxco. 2008. E-DNA sensors for convenient, label-free electrochemical detection of hybridization. Microchim. Acta 163: 149–155.
[117] Shen, Q., M. Fan, Y. Yang and H. Zhang. 2016. Electrochemical DNA sensor-based strategy for sensitive detection of DNA demethylation and DNA demethylase activity. Anal. Chem. Acta 934: 66–71.
[118] Davies, J., D. Burke, J.R. Olliver, L.J. Hardie, C.P. Wild and M.N. Routledge. 2007. Methylene blue but not indigo carmine causes DNA damage to colonocytes *in vitro* and *in vivo* at concentrations used in clinical chromoendoscopy. Gut 56: 155–156.
[119] Hajian, R., N. Shams and M. Mohagheghian. 2009. Study on the interaction between doxorubicin and deoxyribonucleic acid with the use of Methylene blue as a probe. J. Braz. Chem. Soc. 20: 1399–1405.

[120] Kara, P., K. Kerman, D. Ozkan, B. Meric, A. Erdem, Z. Ozkan and M. Ozsoz. 2002. Electrochemical genosensor for the detection of interaction between methylene blue and DNA. Electrochem. Commun. 4: 705–709.

[121] Vardevanyan, P.O., A.P. Antonyan, M.A. Parsadanyan, M.A. Shahinyan and L.A. Hambardzumyan. 2013. Mechanisms for binding between Methylene blue and DNA. J. Appl. Spectr. 80: 595–599.

[122] Rohs, R. and H. Sklenar. 2004. Methylene blue binding to DNA with alternating AT base sequence: Minor groove binding is favored over intercalation. J. Biomol. Struct. Dynam. 21: 699–711.

[123] Erdem, A., K. Kerman, B. Meric, U.S. Akarka and M. Ozsoz. 2000. Novel hybridization indicator methylene blue for the electrochemical detection of short DNA sequences related to hepatitus B virus. Anal. Chim. Acta 422: 139–149.

[124] Kelley, S.O., J.K. Barton, N.M. Jackson and M.G. Hill. 1997. Electrochemistry of Methylene blue bound to a DNA-modified electrode. Bioconjugate Chem. 8: 31–37.

[125] Farjami, E., L. Clima, K.V. Gothelf and E.E. Ferapontova. 2010. DNA interactions with a Methylene Blue redox indicator depend on the DNAlength and are sequence specific. Analyst 135: 1443–1448.

[126] Kekegy-Nagy, L. and E. Ferapontova. 2019. Directional preference of DNA-mediated electron transfer in gold-tethered DNA duplexes: Is DNA a molecular rectifier? Angew. Chem. 58: 3048–3052.

[127] Kekedy-Nagy, L., S. Shipovskov and E.E. Ferapontova. 2019. Electrocatalysis of ferricyanide reduction mediated by electrontransfer through the DNA duplex: Kinetic analysis by thin layer voltammetry. Electrochim. Acta 318: 703–710.

[128] Kulikova, T., A. Porfireva, G. Evtugyn and T. Hianik. 2019. Electrochemical DNA sensors with layered polyaniline—DNA coating for detection of specific DNA interactions. Sensors 19: 469.

[129] Graves, D.E. and L.M. Velea. 2000. Intercalative binding of small molecules to nucleic acids. 2000. Curr. Org. Chem. 4: 915–929.

[130] Pérez-Arnaiz, C., N. Busto, J.M. Leal and B. García. 2014. New insights into the mechanism of the DNA/doxorubicin interaction. J. Phys. Chem. B 118: 1288−1295.

[131] Kurbanoglu, S., B. Dogan-Topal, E.P. Rodriguez, B. Bozal-Palabiyik, S.A. Ozkan and B. Uslu. 2016. Advances in electrochemical DNA biosensors and their interaction mechanism with pharmaceuticals. J. Electroanal. Chem. 775: 8–26.

[132] Hasanzadeh, M. and N. Shadjou. 2016. Pharmacogenomic study using bio- and nanobioelectrochemistry. Drug-DNA interaction. Mater. Sci. C 61: 1002–1007.

[133] Piedale, J.A.P., I.R. Fernandez and A.M. Oliveira-Brett. 2002. Electrochemical sensing of DNA–adriamycin interactions. Bioelectrochem. 56: 81–83.

[134] Oliveira-Brett, A.M., M. Vivan, I.R. Fernandes and J.A.P. Piedade. 2002. Electrochemical detection of *in situ* adriamycin oxidative damage to DNA. Talanta 56: 959–970.

[135] Kostoryz, E.L. and D.M. Yourtee. 2001. Oxidative mutagenesis of doxorubicin-Fe(III) complex. Mutation Res. 490: 131–139.

[136] Cheng, G., H. Qu, D. Zhang, J. Zhang, P. He and Y. Fang. 2002. Spectroelectrochemical study of the interaction between antitumor drug daunomycin and DNA in the presence of antioxidants. J. Pharm. Biomed. Anal. 29: 361–369.

[137] Hajian, R., Z. Tayebi and N. Shams. 2017. Fabrication of an electrochemical sensor for determination of doxorubicin in human plasma and its interaction with DNA. J. Pharm. Anal. 7: 27–33.

[138] Zhang, K. and Y. Zhang. 2010. Electrochemical behavior of adriamycin at an electrode modified with silver nanoparticles and multi-walled carbon nanotubes, and its application. Microchim. Acta 169: 161–165.

[139] Soleymani, J., M. Hasanzadeh, N. Shadjou, M.K. Jafari, J.V. Gharamaleki, M. Yadollahi and A. Jouyban. 2016. A new kinetic-mechanistic approach to elucidate electrooxidation of doxorubicin hydrochloride in unprocessed human fluids using magnetic graphene based nanocomposite modified glassy carbon electrode. Mater. Sci. Eng. C 61: 638–650.

[140] Shamagsumova, R., A. Porfireva, V. Stepanova, Y. Osin, G. Evtugyn, and T. Hianik. 2015. Polyaniline–DNA based sensor for the detection of anthracycline drugs. Sens. Actuators, B 220: 573–582.

[141] Zhang, Y., K. Zhang and H. Ma. 2009. Electrochemical DNA biosensor based on silver nanoparticles/poly(3-(3-pyridyl)acrylic acid)/carbon nanotubes modified electrode. Anal. Biochem. 387: 13–19.

[142] Cheng, G.F., J. Zhao, Y.-H. Tu, P.-G. He and Y.-Z. Fang. 2005. Study on the interaction between antitumor drug daunomycin and DNA. Chinese J. Chem. 23: 576–580.

[143] Hajian, R., E. Eknhlasi and R. Daneshvar. 2012. Spectroscopic and electrochemical studies on the interaction of epirubicin with fish sperm DNA. E-Journal Chem. 9: 1587–1598.

[144] Li, N., Y. Ma, C. Yang, L. Guo and X. Yang. 2005. Interaction of anticancer drug mitoxantrone with DNA analyzed by electrochemical and spectroscopic methods. Biophys. Chem. 116: 199–205.

[145] Torkzadeh-Mahani, A., A. Mohammadi, M. Torkzadeh-Mahani and M. Mohamadi. 2017. A label-free electrochemical DNA biosensor for the determination of low concentrations of mitoxantrone in serum samples. Int. J. Electrochem. Sci. 12: 6031–6044.

[146] Paziewska-Nowak, A., J. Jankowska-Śliwińska, M. Dawgul and D.G. Pijanowska. 2017. Selective electrochemical detection of pirarubicin by means of DNA-modified graphite biosensor. Electroanalysis 29: 1810–1819.

[147] Zhang, Y., H. Ma, K. Zhang, S. Zhang and J. Wang. 2009. An improved DNA biosensor built by layer-by-layer covalent attachment of multi-walled carbon nanotubes and gold nanoparticles. Electrochim. Acta 54: 2385–2391.

[148] Li, H., P. Dauphin-Ducharme, G. Ortega and K.W. Plaxco. 2017. Calibration-free electrochemical biosensors supporting accurate molecular measurements directly in undiluted whole blood. J. Am. Chem. Soc. 139: 11207–11213.

[149] Chandra, P., S.A. Zaidi, H.-B. Noh and Y.-B. Shim. 2011. Separation and simultaneous detection of anticancer drugs in a microfluidic device with an amperometric biosensor. Biosens. Bioelectron. 28: 326–332.

[150] Porfireva, A.V., K.S. Shibaeva, V.G. Evtyugin, L.S. Yakimova, I.I. Stoikov and G.A. Evtyugin. 2019. An electrochemical DNA sensor for doxorubicin based on a polyelectrolyte complex and aminated thiacalix[4]arene. J. Anal. Chem. 74: 542–550.

[151] Kulikova, T.N., A.V. Porfireva, R.V. Shamagsumova and G.A. Evtugyn. 2018. Voltammetric sensor with replaceable polyaniline-DNA layer for doxorubicin determination. Electroanalysis 30: 2284–2292.

[152] Stepanova, V., V. Smolko, V. Gorbatchuk, I. Stoikov, G. Evtugyn and T. Hianik. 2019. DNA-Polylactide modified biosensor for electrochemical determination of the DNA-drugs and aptamer-aflatoxin M1 interactions. Sensors 19: 4962.

[153] Chandra, P., H.B. Noh, M.S. Won and Y.B. Shim. 2011. Detection of daunomycin using phosphatidylserine and aptamer co-immobilized on Au nanoparticles deposited conducting polymer. Biosens. Bioelectron. 26: 4442–4449.

[154] Saljooqi, A., T. Shamspur and A. Mostafavi. 2019. The MWCNT-Ag-PT GCE electrochemical sensor functionalized with dsDNA for mitoxantrone sensing in biological media. IEEE Sens. J. 19: 4364–4368.

[155] Hashkavayi, A.B., J.B. Raoof and R. Ojani. 2017. Preparation of epirubicin aptasensor using curcumin as hybridization indicator: Competitive binding assay between complementary strand of aptamer and epirubicin. Electroanalysis 29: 1–9.

[156] Khodadadi, A., E. Faghih-Mirzaei, H. Karimi-Maleh, A. Abbaspourrad, S. Agarwal and V.K. Gupta. 2019. A new epirubicin biosensor based on amplifying DNA interactions with polypyrrole and nitrogen-doped reduced graphene: Experimental and docking theoretical investigations. Sens. Actuat. B 284: 568–574.

[157] Eksin, E., E. Zor, A. Erdem and H. Bingol. 2017. Electrochemical monitoring of biointeraction by graphene-based material modified pencil graphite electrode. Biosens. Bioelectron. 92: 207–214.

[158] Gong, Z.L., D.Y. Tang, X.D. Zhang, J. Ma and Y. Mao. 2014. Self-assembly of thermoresponsive nanocomposites and their applications for sensing daunorubicin with DNA. Appl. Surf. Sci. 316: 194–201.

[159] Shah, A., R. Qureshi, A.M. Khan, F.L. Ansari and S. Ahmad. 2009. Determination of binding parameters and mode of ferrocenyl chalconeDNA interaction. Bull. Chem. Soc. Jpn. 82: 453–457.

[160] Zhu, J.-J., K. Gu, J.-Z. Xu and H.-Y. Chen. 2001. DNA modified carbon paste electrode for the detection of 6-mercaptopurine. Anal. Lett. 34: 329–337.
[161] Mirmomtaz, E., A. Zirakbash and A.A. Ensafi. 2016. An electrochemical DNA sensor for determination of 6-thioguanine using adsorptive stripping voltammetry at HMDE: An anticancer drug DNA interaction study. Russ. J. Electrochem. 52: 320–329.
[162] Unal, D.N., E. Eksin and A. Erdem. 2017. Carbon nanotubes modified graphite electrodes for monitoring of biointeraction between 6-thioguanine and DNA. Electroanalysis 29: 2292–2299.
[163] Perez, P., C. Teijeiro and D. Marin. 1999. Interactions of surface-confined DNA with electroreduced mitomycin C comparison with acid-activated mitomycin C. Chem.-Biol. Interact. 117: 65–81.
[164] Karadeniz, H., L. Alparslan, A. Erdem and E. Karasulu. 2007. Electrochemical investigation of interaction between mitomycin C and DNA in a novel drug-delivery system. J. Pharm. Biomed. Anal. 45: 322–326.
[165] Temerk, Y.M., M.S. Ibrahim, M. Kot and W. Schuhmann. 2013. Interaction of antitumor flavonoids with dsDNA in the absence and presence of Cu (II). Anal. Bioanal. Chem. 405: 3839–3846.
[166] Beitollahi, H., G. Dehghannoudeh, H.M. Moghaddam and H. Forootanfar. 2017. A sensitive electrochemical DNA biosensor for anticancer drug Topotecan based on graphene carbon paste electrode. J. Electrochem. Soc. 164: H812–H817.
[167] Babkina, S.S., G.K. Budnikov and N.A. Ulakhovich. 2009. Bioaffine methods for determining ajmaline using an amperometric DNA-sensor and an immunoenzyme epectrophotometric test system. J. Anal. Chem. 64: 958–963.
[168] Babkina, S.S. and N.A. Ulakhovich. 2011. Determination of pharmaceuticals based on indole alkaloids with amperometric dna-sensors and enzyme immunoassay test-system. Anal. Lett. 44: 837–849.
[169] Radi, A.-E., H.M. Nassef and A. Eissa. 2013. Voltammetric and ultraviolet-visible spectroscopic studies on the interaction of etoposide with deoxyribonucleic acid. Electrochim. Acta 113: 164–169.
[170] Fotouhi, L. and F. Bahmani. 2013. MWCNT modified glassy carbon electrode: Probing furazolidone-DNA interactions and DNA determination. Electroanalysis 25: 757–764.
[171] Dogan-Topal, B., B. Uslu and S.A. Ozkan. 2009. Voltammetric studies on the HIV-1 inhibitory drug Efavirenz: The interaction between dsDNA and drug using electrochemical DNA biosensor and adsorptive stripping voltammetric determination on disposable pencil graphite electrode. Biosens. Bioelectron. 24: 2358–2364.
[172] Karimi-Maleh, H., A. Bananezhad, M.R. Ganjali, P. Norouzi and A. Sadrnia. 2018. Surface amplification of pencil graphite electrode with polypyrrole and reduced graphene oxide for fabrication of a guanine/adenine DNA based electrochemical biosensors for determination of didanosine anticancer drug. Appl. Surf. Sci. 441: 55–60.
[173] Tian, X., F. Li, L. Zhu and B. Ye. 2008. Study on the electrochemical behavior of anticancer herbal drug rutin and its interaction with DNA. J. Electroanal. Chem. 621: 1–6.
[174] Radi, A. 1999. Voltammetric study of khellin at a DNA-coated carbon paste electrode. Anal. Chim. Acta 386: 63–68.
[175] Ly, S.Y., C.H. Lee and Y.S. Jung. 2007. Measuring oxytetracycline using a simple prepared DNA immobilized on a carbon nanotube paste electrode in fish tissue. J. Korean Chem. Soc. 51: 412–417.
[176] Wang, L., H. Xiong, X. Zhang and S. Wang. 2009. Electrochemical behaviors of nicotine and its interaction with DNA. Electrochem. Commun. 11: 2129–2132.
[177] Stanković, D.M., L. Švorc, J.M.L. Mariano, A. Ortner and K. Kalcher. 2017. Electrochemical determination of natural drug colchicine in pharmaceuticals and human serum sample and its interaction with DNA. Electroanalysis 29: 2276–2281.
[178] Ponthina, A.D.R., S.M.A. Jorge, A.-M.C. Paquim, V.C. Diculescu and A.M. Oliveira-Brett. 2011. Phys. Chem. Chem. Phys. 13: 5227–5234.
[179] Shi, Q., S. Wang, B. Zhu and M. Ji. 2008. Electrochemical studies of reaction of ciprofloxacin and DNA. Front. Chem. China 3: 52–56.
[180] Lim, S.A. and M.U. Ahmed. 2016. A simple DNA-based electrochemical biosensor for highly sensitive detection of ciprofloxacin using disposable graphene. Anal. Sci. 32: 687–693.

[181] Cheraghi, S., M.A. Taher, H. Karimi-Maleh and E. Faghih-Mirzaeide. 2017. A nanostructure label-free DNA biosensor for ciprofloxacin analysis as a chemotherapeutic agent: an experimental and theoretical investigation. New J. Chem. 41: 4985–4989.

[182] Garbellini, G.S., R.C. Rocha-Filho and O. Fatibello-Filho. 2015. Voltammetric determination of ciprofloxacin in urine samples and its interaction with dsDNA on a cathodically pretreated boron-doped diamond electrode. Anal. Methods 7: 3411–3418.

[183] El-Desoky, H., M. Ghoneim, M. Abdel-Galeil and A. Khalifa. 2017. Electrochemical sensor based on functionalized multiwalled carbon nanotubes, domperidone determination, DNA binding and molecular docking. J. Electrochem. Soc. 164: H1133–H1147.

[184] Jankowska-Śliwińska, J., M. Dawgul and D.G. Pijanowska. 2014. DNA intercalation-based amperometric biosensor for chlorpromazine detection. Proc. Eng. 87: 747–750.

[185] Marco, J.P., K.B. Borges, C.R.T. Tarley, E.S. Ribeiro and A.C. Pereira. 2013. Development and application of an electrochemical biosensor based on carbon paste and silica modified with niobium oxide, alumina and DNA ($SiO_2/Al_2O_3/Nb_2O_5$/DNA) for amitriptyline determination. J. Electroanal. Chem. 704: 159–168.

[186] Asghary, M., J.B. Raoof, R. Ojani and E. Hamidi-Asl. 2015. A genosensor based on CPE for study the interaction between ketamine as anesthesia drug with DNA. Intern. J. Biol. Macromol. 80: 512–519.

[187] Mascini, M., I. Palchetti and G. Marazza. 2001. DNA electrochemical biosensors. Fresenius J. Anal. Chem. 369: 15–22.

[188] Rauf, S., J.J. Gooding, K. Akhtar, M.A. Ghauri, M. Rahman, M.A. Anwar and A.M. Khalid. 2005. Electrochemical approach of anticancer drugs-DNA interaction. J. Pharm. Biomed. Anal. 37: 205–217.

[189] Svitková, V. and J. Labuda. 2017. Construction of electrochemical DNA biosensors for investigation of potential risk chemical and physical agents. Monatsh. Chem. 148: 1569–1579.

[190] Fojta, M., A. Daňhel, L. Havran and V. Vyskočil. 2016. Recent progress in electrochemical sensors and assays for DNA damage and repair. Trends in Anal. Chem. 79: 160–167.

[191] Marazza, G., I. Chanella and M. Mascini. 1999. Disposable DNA electrochemical biosensors for environmental monitoring. Anal. Chim. Acta 387: 297–307.

[192] Mascini, M., G. Bagni, M.L. Di Pietro, M. Ravera, S. Baracco and D. Osella. 2006. Electrochemical biosensor evaluation of the interaction between DNA and metallo-drugs. BioMetals 19: 409–418.

[193] Ravera, M., G. Bagni, M. Mascini, J.C. Dabrowiak and D. Osella. 2007. The activation of platinum(II) antiproliferative drug in carbonate medium evaluated by means of a DNA-biosensor. J. Inorg. Chem. 101: 1023–1027.

[194] Yardim Y., M. Vandeput, M. Çelebi, Z. Şentürk and J.-M. Lauffmann. 2017. A reduced graphene oxide-based electrochemical DNA biosensor for the detection of interaction between cisplatin and DNA based on guanine and adenine oxidation signals. Electroanalysis 29: 1451–1458.

[195] Yapasan, E., A. Caliskan, H. Karadeniz and A. Erdem. 2010. Electrochemical investigation of biomolecular interactions between platinum derivatives and DNA by carbon nanotubes modified sensors. Mater. Sci. Eng. 169: 169–173.

[196] Lopes, I.C., S.C.B. Oliveira and A.M. Oliveira-Brett. 2013. *In situ* electrochemical evaluation of anticancer drug temozolomide and its metabolites–DNA interaction. Anal. Bioanal. Chem. 405: 3783–3790.

[197] Tiğ, G.A., G. Günendi, T.E. Bolelli, İ. Yalçın and Ş. Pekyardımcı. 2016. Study on interaction between the 2-(2-phenylethyl)-5-methylbenzimidazole and dsDNA using glassy carbon electrode modified with poly-3-amino-1,2,4-triazole-5-thiol. J. Electroanal. Chem. 776: 9–17.

[198] Topkaya, S.N., G. Serindere and M. Ozder. 2016. Determination of DNA hypermethylation using anticancer drug-temozolomide. Electroanalysis 28: 1052–1059.

[199] La-Scalea, M.A., S.H.P. Serrano, E.I. Ferreira and A.M. Oliveira Brett. 2002. Voltammetric behavior of benznidazole at a DNA-electrochemical biosensor. J. Pharm. Biomed. Anal. 29: 561–568.

[200] Top, M., O. Er, G. Congur, A. Erdem and F.Y. Lambrecht. 2016. Intracellular uptake study of radiolabeled anticancer drug and impedimetric detection of its interaction with DNA. Talanta 160: 157–163.

[201] Svítková, V., S. Švikruhová, M. Vojs, M. Marton and L. Švorc. 2016. DNA-modified boron-doped diamond electrode as a simple electrochemical platform for detection of damage to DNA by antihypertensive amlodipine. Monatsh. Chem. 147: 1365–1373.

[202] Dogan-Topal, B. and S.A. Ozkan. 2011. A novel sensitive electrochemical DNA biosensor for assaying of anticancer frug leuprolide and its adsorptive stripping voltammetric determination. Talanta 83: 780–788.

[203] Dogan-Topal, B. and S.A. Ozkan. 2011. Electrochemical determination of anticancer drug fulvestrant as dsDNA modified pencil graphite electrode. Electrochim. Acta 56: 4433–4438.

[204] Diculescu, V.C., M. Vivan and AM. Oliveira Brett. 2006. Voltammetric behavior of antileukemia drug Glivec. Part III: *In situ* DNA oxidative damage by the Glivec electrochemical metabolite. Electroanalysis 18: 1963–1970.

3

Nanomaterials in Matrix X-ray Sensors for Computed Tomography

Alexander N. Yakunin,[1,*] *Sergey V. Zarkov,*[2] *Yuri A. Avetisyan,*[2] *Garif G. Akchurin*[2,3] and *Valery V. Tuchin*[2,3,4]

1. Introduction: A brief history of development of X-Ray: Innovations to increase the depth and resolution of images

As of today, more than 125 years have passed since the discovery of X-rays by Wilhelm Conrad Röntgen while studying the nature of cathode rays in a Crookes gas-discharge vacuum tube [1]. He had recorded the penetration of X-rays through his palm with a metal ring on his finger, using a photographic plate. The image showed a significant contrast of soft tissues, bones and metal through variation in grayscale. Thereafter, the diagnostic use of X-rays in medicine began almost immediately. However, the physical mechanism for the generation of X-rays was only discovered in 1912—a result of research by Max von Laue, William Henry Bragg and William Lawrence Bragg.

Experimental observations of X-ray diffraction in crystals allows one to establish the electromagnetic nature of the radiation and determine its wavelength [2, 3]. It is known that the spectrum of X-ray radiation lies in the wavelength range from 0.005 to 10 nm [4]. The upper limit corresponds to extreme ultraviolet radiation and

[1] Laboratory of System Problems in Control and Automation in Mechanical Engineering, Institute of Precision Mechanics and Control, FRC "Saratov Scientific Centre of the Russian Academy of Sciences," 410028 Saratov, Russia.
[2] Laboratory of Laser Diagnostics of Technical and Living Systems, Institute of Precision Mechanics and Control, FRC "Saratov Scientific Centre of the Russian Academy of Sciences," 410028 Saratov, Russia.
[3] Scientific Medical Center, Saratov State University, 410012 Saratov, Russia.
[4] Laboratory of Laser Molecular Imaging and Machine Learning, Tomsk State University, 634050 Tomsk, Russia.
Emails: szarcov@gmail.com; yuaavetisyan@mail.ru; akchuringg@mail.ru; tuchinvv@mail.ru
* Corresponding author: anyakunin@mail.ru

the lower limit corresponds to gamma radiation. The spectrum of X-ray waves is determined by the energy eV_a (eV) of electrons accelerated by an electrostatic field (V_a being the potential difference between the cathode and the anode) and colliding with a metal anode (anti-cathode).

The efficiency of converting the energy of accelerated electrons into X-ray electromagnetic radiation (which has a continuous spectrum and is called bremsstrahlung) is determined by the empirical relation $\eta = a \cdot Z \cdot V_a$, where Z is the atomic number of the target material and a is a constant equal to 1.1×10^{-9} V_a^{-1} [5]. For a tungsten target ($Z = 74$) and an accelerating potential of 100 kV, the efficiency is 0.0081, i.e., it is less than 1%; most of the electron energy in this case is converted into heat. The minimum wavelength of X-ray bremsstrahlung is λ_{min}(nm) = $1240/eV_a$ (eV), while maximum radiation is realized at $\lambda_{max} = (3/2)$ λ_{min} [4]. The spectral dependence of the intensity of bremsstrahlung radiation on wavelength is determined from the relation $I_\lambda = \beta \cdot Z \cdot (\lambda - \lambda_{min})/(\lambda^3 \cdot \lambda_{min})$, where β is the proportionality coefficient [4].

At a certain anode voltage V_a, against a background of broadband bremsstrahlung radiation, narrow resonance spectral peaks appear due to X-ray radiation from the atoms of the anode material. This so-called characteristic radiation is a line spectrum—a set of narrow doublet spectral lines arising from the ionization of an atom of the anode material, as a result of a collision with an accelerated electron. These spectral lines are due to radiative transition from the upper energy levels of the filled electron shells of the atom (L (n = 2) and M (n = 3)) to the innermost shell of the atom (the so-called K (n = 1) series) in accordance with the quantum-mechanical selection rules. The mechanism of occurrence of the L and M spectral series is similar. The wavelengths of the characteristic spectrum are determined by the theory of Moseley (1913) [6, 7]. An important consequence of this theory is that the inverse of the square root of the wavelength of the resonant radiation is proportional to the charge number Z of the chemical element of the anode material. This regularity is used in practice for the unambiguous identification of the chemical composition of a material.

In 1913, an American researcher, Wiiliam David Coolidge, proposed and experimentally investigated a high-vacuum X-ray tube with an electron source in the form of an incandescent tungsten cathode. Therein, the cathode temperature was found to significantly affect the emission current, and the voltage between the cathode and the anode determined the electron energy [8]. This technology for obtaining X-ray radiation is the one mainly used at present.

The main advantages of probing biological objects of sizes in the range of the wavelength of X-ray radiation are determined by the two factors, the first being that the refractive index of any biological tissue in the X-ray region is close to 0.99. Also, for $Z > 20$, the effects of X-ray radiation scattering can be neglected, and for $Z < 20$, the scattering coefficient of biological tissues is much less than the absorption coefficient. Therefore, the propagation of the rays is close to straightforward (ballistic). Second, the low absorption of X-ray radiation (in the order of 0.2 cm^{-1}) at $V_a > 100$ kV [9] provides an increase in the transmission thickness to 50–70 cm, with the possibility of precise measurement of absorption coefficient. In the X-ray spectral range used in medical fluorography and tomography (at an anode voltage of about 50 kV), the typical absorption coefficient of human bone tissue does not exceed 0.48 cm^{-1}; the corresponding numbers are 0.178 cm^{-1} for muscles and 0.18 cm^{-1} for blood [9]. The absorption coefficient μ_a of X-ray radiation by atoms is determined from the relation $\mu_a = C \cdot \lambda^3 \cdot Z^4$ (C is the proportionality coefficient) [4]. The cubic dependence of μ_a on λ allows one to significantly reduce absorption when using short-wavelength X-rays, the generation of which in X-ray tubes for computer tomography (CT) is provided at high anode voltages (up to 150 kV).

The sensitivity of photographic film or solid-state detectors in digital systems, which allows for the detection of three to four orders of intensity gradation, turns out to be sufficient to effectively solve the tomography problem—reconstructing the spatial configuration of structural inhomogeneities of the studied object from the results of two-dimensional projections—using the Radon mathematical transform [10]. The physical basis of the method is the exponential law of extinction of X-ray beams propagating in a straight line. The first mathematical algorithms for X-ray CT were developed in 1917 by the Austrian mathematician I. Radon. Further, in 1971, the American physicist A. Cormack, who discovered a new solution to the problem of tomographic reconstruction, and the English engineer-physicist G. Hounsfield, who designed and tested the first computerized X-ray tomograph, were awarded the Nobel Prize in Physiology and Medicine "for the development of computed tomography" [11, 12].

The advantages and potentialities of X-ray CT applied to biological objects of considerable size (humans, animals) become obvious when compared with the CT parameters of biological tissues in the traditional optical wavelength range [13]. It is known that in the spectral region of transparency of biological tissues (NIR range: 700–1100 nm), the absorption coefficient of various types of biological tissues is comparable to the typical absorption coefficient in the X-ray range and is less than 0.5 cm^{-1}. However, in the same spectral region, the scattering coefficient of optical radiation of almost all biological tissues, excluding eye tissues (sclera and retina), exceeds the absorption coefficient by one or two orders of magnitude. Therefore, the developed projection methods of X-ray tomography cannot be automatically applied in the optical range. For these purposes, optical low-coherence or laser confocal methods have been developed; they use the features of the spatial and temporal coherence of probing optical beams and underlie the corresponding tomographs. The main area of medical application of 3D OCT is ophthalmology, namely, the study of pathologies of the retina with a transparent lens and cornea with a maximum probing depth of about 1 mm, or a catheter light guide study of the mucous walls of internal hollow organs with a thickness of no more than 1 mm and a depth resolution of 3–6 μm [13].

The progress of CT tomographs is directly related to the increase in the number of detectors, that is, to the increase in the number of simultaneously collected projections. Hence, in the first tomograph (developed in 1973), one X-ray emitter and one receiver were used. To create two-dimensional projections, they were simultaneously rotated around the patient in one-degree increments. It took about 1 hour to obtain a complete tomogram. The succeeding generations of tomographs included one X-ray emitter and an array of detectors with circular rotation with a frequency of up to three hertz and a simultaneous translational displacement of the patient's position along the axis of rotation. The maximum spatial resolution of such 3D CT systems is currently determined by the voxel size (approximately 0.5 mm × 0.5 mm × 0.5 mm). The advantage of an array of detectors is that the number of detectors in a row can be easily increased to obtain more slices per revolution of the X-ray tube.

2. Advantages of matrix X-Ray sensors and the role of nanostructured field emitters in radiation dose reduction and resolution increase

A further step in the direction of development of CT systems with increased scanning speed and spatial resolution was the creation of an array of miniature sources of electron beams based on nanostructured cathodes with field emission [14–23]. The advantages of such systems include the simplification of the device by eliminating the cathode heating circuit and, in turn, the auxiliary heater unit. A significant increase in scan speed is achieved by reducing the switching time (low inertia) compared to systems based on hot cathodes. An increase in the number of elementary sources of electron beams and X-ray detectors (an increase in the size of source and detector matrices) provided a significant improvement in the spatial resolution of X-ray CT systems. Each X-ray source is activated to generate one view. The entire scan is performed electronically by triggering all X-ray sources, which can be performed at a high speed. In addition to improved temporal resolution, the absence of any moving parts in stationary CT arguably offers improved reliability [14].

The last decade was marked by the development, research and practical application of the technology for creating arrays of field emitters based on carbon nanotubes (CNTs) for CT systems [15, 16, 24–29]. X-ray tubes with CNT emitters are used in micro-CT systems [26] and for high resolution stationary digital breast tomosynthesis [27]. The magnitude of the electron current in the field emission of CNT emitters is determined by the potential difference applied to create a strong external electric field. Direct monitoring of electron emission makes it easy to obtain both nanosecond X-ray pulses and a time-constant X-ray flux. In micro-CT applications, this allows for precise timing of short X-ray pulses to provide better transient cardiac or respiratory strobing for obtaining better images and reducing stress on patients [26]. Electron emission in CNT emitters does not require heating; therefore, several emitters can be densely located in one vacuum tube. In tomosynthesis systems, one X-ray tube with CNT emitters and a linear array of fixed focal spots replaces a moving X-ray source [27]. Each projection is obtained by including different focal points. This array of focal spots speeds up tomosynthesis scans and removes image blur caused by a moving source. CNT emitters can also be organized in the form of a two-dimensional (2D) matrix, which is effective in creating X-ray sources for diagnostic and radiotherapy applications [28].

The possibility of successful implementation of matrix current sources is due to the progress in technologies ensuring the required durability of emitters based on

CNTs with an acceptable emissivity. CNTs exhibit excellent characteristics such as high aspect ratio, high thermal conductivity and low chemical reactivity. However, the degradation of the field emission of CNTs at elevated current densities is a major obstacle in designing durable X-ray tubes. Therefore, the technological aspects of ensuring thermal stability are constantly in focus for research. For instance, in one piece of research, an increase in the thermal stability of CNTs was achieved using the technology of their treatment with boron and phosphorus compounds [29]. As other examples of successful technological solutions, one can cite the results of developments as presented in Figure 1 and Figure 2 [24, 25].

To form the isotropic focal spot at anode target, the emitter has an elliptical area of 2.0 mm × 0.5 mm as shown in Figure 1a [24]. The initial CNTs without post-treatment are vertically aligned; after post-treatments they form the array of bundles (Figure 1b). One CNT bundle (Figure 1c) has a diameter of 3 μm. CNT bundles are precisely located with an island-to-island distance of 15 μm.

Researchers have also obtained CNT matrices with high current density, stable current and high adhesion strength to the substrate by designing a unique

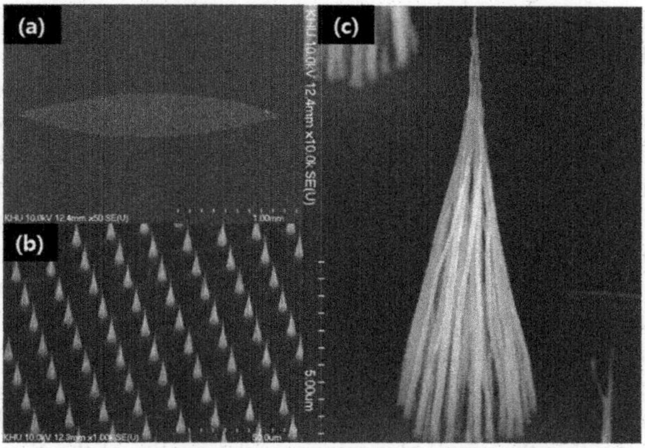

Figure 1. SEM images of the elliptical emitter with dimensions 2.0 mm × 0.5 mm, formed by resist-assisted patterning process (a); magnified image of CNT array emitters with 3 μm islands and 15 μm pitch (b); a magnified image of one CNT-bundle (c). Reproduced from Ref. [24] with permission of MDPI.

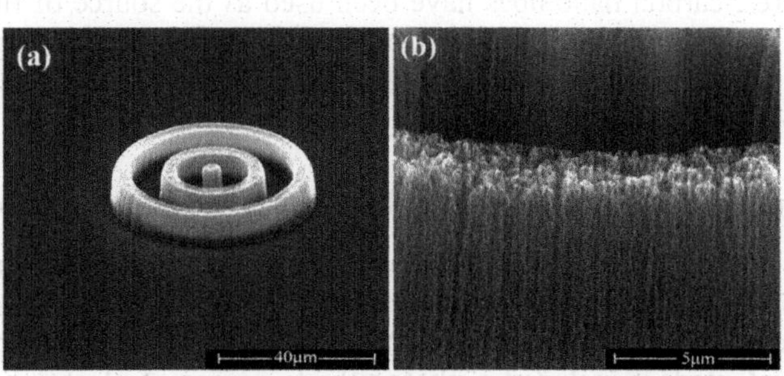

Figure 2. SEM image of an as-grown CNT array: overall shape of the CNT array (a); higher resolution image of the CNT array (b). Reproduced from Ref. [25] with the permission of the American Vacuum Society, The Authors.

multilayer metal structure for CNT growth [25]. The multilayer structure provides good adhesion of CNTs and a reduction in thermal resistance at the "substrate-CNT" interface. Eventually, this radically increases the thermal stability of the field emitter.

Improvement of digital CT (DCT) systems for mammography and brain neuropathology, aimed at the monitoring and early diagnosis of oncological diseases, requires the urgent development of knowledge-intensive technologies in the interest of modern medicine [30]. The experience of developing such systems with X-ray sources based on emitters with field emission of electrons [31, 32] shows that the emitter current must be sufficiently high for high-quality signal detection and image reconstruction. Estimates based on the results of experimental data [32] indicate that the average emission current density should not be less than 80 mA/cm^2 during the entire service life of the device.

It should be noted that the problem of ensuring the reliability and durability of field emission emitters is fundamental because of the inherent exponential dependence of emission current density on the strength of the local electrostatic field. The result of such a criticality may be spontaneous thermal breakdown at small fluctuations of accelerating voltage, leading to the irreversible degradation of emission properties and, ultimately, the destruction of the emitter [33].

The two main technical limitations of modern CT technology are low spatial resolution and long scan times [34]. Therefore, a large number of researchers have aimed at replacing devices consisting of a rotating or moving X-ray tube with matrix radiation sources [15–23, 35]. The use of such arrays yields two advantages. First of all, the source matrix is activated without any mechanical movement, but by switching control potentials. Therefore, the speed of scanning and generation of projection images, which are necessary for the synthesis of a volumetric image of the object under study, increases. An important role to this end is played by the efficiency of the algorithms for digital processing of large data sets [36]. The second advantage is that the elimination of mechanical movement of the X-ray source decreases image blur and increases spatial resolution. Furthermore, reducing the size of the elementary current source (within the pixel) also improves the quality of the X-ray image. This aspect has led many researchers to orient themselves towards the development of matrix emitters with field emission of electrons [15, 16, 24–29]. In almost all of these structures, carbon nanotubes have been used as the source of field emission. In one such study, an array of emitters based on tungsten trioxide nanowires was synthesized, with artificially formed defects to increase the field emission current density to about 14 mA/cm^2 [37].

3. Prospects for optically controlled matrix X-Ray sensors

The idea of further increasing the scanning speed and spatial resolution of DCT systems was developed during the implementation of optical (instead of electric) control of the processes of generation of X-ray beams. Relatively recently, an X-ray source was patented; it included a photocathode source of electrons, wherein the photocathode was illuminated by a spatially scanning laser beam [38]. An acousto-optic modulator was used to control the intensity of the laser beam and the exposure time. This photocathode source of electrons can include halides of alkali metals (CsBr, CsI) with a low work function A; then, if photon energy $h\nu > A$, photoemission of electrons into vacuum occurs. The emitted electrons are accelerated by the applied electrostatic field and generate X-rays by bombarding the metal anode (material options—Tungsten, Copper, Rhodium, Molybdenum). Another solution, which makes it possible to initiate a sufficiently high emission current from CNT cathodes at a relatively low accelerating voltage, was described in [39]. For this purpose, the CNT array was irradiated with a 150-mW laser beam with wavelength $\lambda = 650$ nm, focused onto a spot, the average intensity of which was 50 W/cm^2. Despite the fact that the photon energy of laser radiation is approximately two times lower than the electron work function for CNTs, the CNT array was heated to a temperature of 2000°C within a short time duration due to the effective absorption of laser radiation. The high temperature yielded thermally stimulated field emission.

The composite structure of a blade type with field emission has been described in literature [40]. Here, low-voltage field emission resulted due to the lowering of the potential barrier at the "metal blade – vacuum" interface in the zone of localization of the electrostatic field. The results of an experimental study showed its durability of 8700 hours at an average emission current density of at least 100 mA/cm^2. The achieved emission levels provide the prospect of using such structures in current sources of X-ray systems. High photosensitivity of the structure was also witnessed in a wide wavelength range [41, 42].

As a development in the construction of a pixel current source with photoinduced electron emission, a new hybrid material (HM) based on gold nanostars (GNS) embedded in a nanoscale film of diamond-like carbon (DLC) has been described in a study [43].

Below in Sections from 4 to 6 presented results of a study of the influence of plasmon effects in GNS on the optical properties in the wavelength range from UV to NIR, the features of localization of field and heat sources, the generation of high-energy hot electrons and the mechanisms of their transport into vacuum. An assessment of the advantages of the material in comparison with the original composite and the prospects for using an X-ray CT system in a matrix source has also been provided.

4. Features of tunnel photoemission into vacuum

In this section of the chapter, the processes associated with the generation and tunneling of hot electrons at the "nanoparticle–vacuum" or "semiconductor film–vacuum" interface in a strong external electrostatic field will be discussed. This differs them from most of the works [44–46] devoted to the problems of solar energy conversion or catalysis. The motivation for the work was the previously obtained encouraging results [47, 48], as well as experimental data [33, 40–42, 49–51] on tunneling emission of hot electrons from metal (Mo) and composite structures such as a "metal–DLC (α-C) film" in a vacuum under conditions of combined exposure to a strong electrostatic field and laser radiation. It has been found that irradiation of interdigital structures such as "metal–DLC film" of a vacuum microcathode with CW laser radiation of milliwatt power (power density of approximately 100 W/cm^2)

at discrete wavelengths from 405 to 1550 nm produces a tunnel photocurrent which varies linearly with the light intensity [50]. With an increase in the voltage at the anode, an exponential increase in the photocurrent is observed, which is characteristic for tunneling emission. Similar experiments on the emission of hot electrons from a metal and the above-mentioned composite, into a vacuum, have also been carried out under irradiation with nanosecond laser pulses of mJ level (10 mJ/cm^2 fluence) in the 400–1200 nm range [51]. It is typical of the red threshold of the photoelectric effect for Mo to exceed the photon energy of the laser beam by a factor of 4–6, indicating precisely the tunneling mechanism of emission.

In an HM, which is GNS embedded in a DLC film, two electron transport mechanisms are implemented in the presence of a strong electrostatic field, as shown in Figure 3. For "metal–DLC film" structures, an additional potential barrier appears at the metal–semiconductor interface, along with the drift transfer of nonequilibrium charges in the DLC film and their subsequent tunneling through the potential barrier at the "semiconductor–vacuum" interface, as shown in Figure 3b.

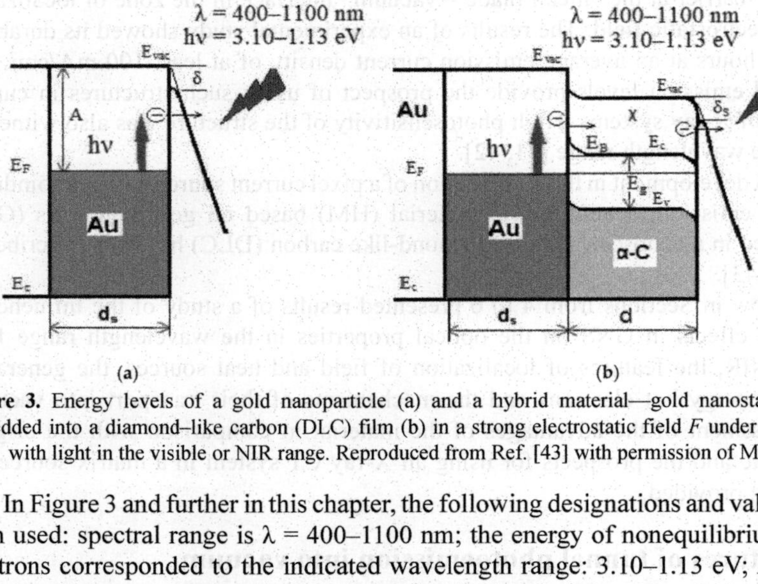

Figure 3. Energy levels of a gold nanoparticle (a) and a hybrid material—gold nanostars (GNS), embedded into a diamond–like carbon (DLC) film (b) in a strong electrostatic field F under irradiation with light in the visible or NIR range. Reproduced from Ref. [43] with permission of MDPI.

In Figure 3 and further in this chapter, the following designations and values have been used: spectral range is λ = 400–1100 nm; the energy of nonequilibrium "hot" electrons corresponded to the indicated wavelength range: 3.10–1.13 eV; E_F is the Fermi level; E_{VAC} is the vacuum level; E_g is the width of the band gap in the DLC; E_C and E_V are the energy levels corresponding to the bottom of the conduction band and the top of the valence band in DLC; χ is the electronic affinity; E_B is the height of the potential barrier at the metal–semiconductor interface; $\delta = (e^3\beta F)^{1/2} \approx 1.2\,(F(\text{V/nm}))^{1/2}$ is the decrease in the height of the potential barrier in a strong electrostatic field at the metal–vacuum interface (see Figure 3a), taking into account the Schottky effect and δ_S for a semiconductor film with reduced field strength F in the film by a factor of ε; ε is the dielectric constant of the film material and β is the form factor of strengthening the electrostatic field due to the curvature of the surface at the semiconductor–vacuum interface; work function $A = E_{VAC} - E_F$; the shape of the potential barrier at the metal–vacuum interface is determined by the ratio $(-eF\zeta - e^2/4\zeta)$, where F is the strength of the electrostatic field, e is the absolute value of the electron charge and ζ is the distance from the interface; d_s is the "skin depth" (in the order of several tens of nanometers for noble metals); and d is the thickness of the diamond-like nanofilm (10–30 nm).

Modification of the emission structure (see Figure 4a) by replacing the existing emitter with an HM structure and manufacturing electrodes from a light-transmitting indium tin oxide (ITO) alloy (see Figure 4b) makes it possible to implement the array X-ray emitter design (see Figure 4c) for DCT systems [49]. The performance of the device is provided by the use of a high-speed laser scanner.

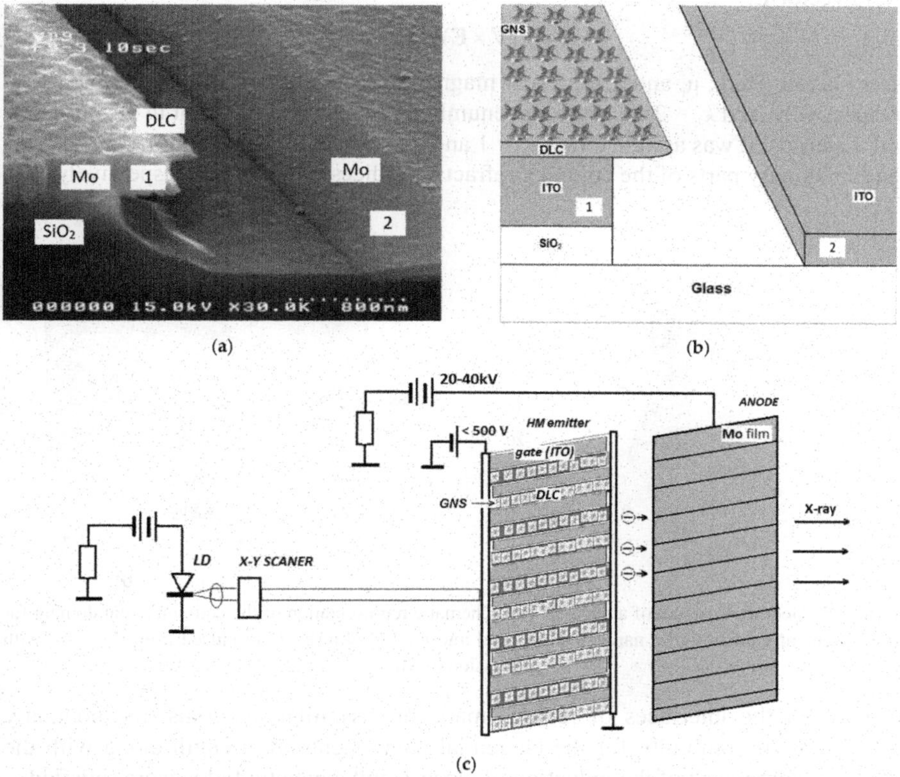

Figure 4. SEM image of a previously developed emitter in the form of a Mo–DLC blade (1–emitter, 2–gate) (a); schematic representation of the proposed hybrid nanostructured cathode cell based on hybrid material (HM) (1–emitter, 2–gate) (b); block diagram of an optically controlled X–ray emitter array (c). Reproduced from Ref. [43] with permission of MDPI.

5. Statement of the problem of modeling a hybrid nanomaterial based on GNS-DLC

Plasmon-resonant GNS are of particular interest as a new type of metal nanoparticles used in various technical and biomedical applications [44–46, 52–57]. Compared to nanospheres, the advantage of their use is high absorption of radiation in the NIR region—which is important, in particular, for laser medicine and biophotonics, including biosensing, optoporation, and transfection of cells [52, 53]. Compared to gold nanorods, GNS are characterized by a lower orientational and polarization sensitivity towards induced optical [57–59] and temperature [60] fields. In the succeeding paragraphs, we consider the possibility of intensifying the photoemission of electrons from composite photocathodes based on GNS embedded in a nano-scale film of diamond-like carbon.

For the electrodynamic calculation of the optical characteristics of GNSs (see Figure 5a) in the medium, a 3D finite element model (shown in Figure 5b) was used. In the schematically shown computational domain, the wave equation for the electric field vector **E**:

$$\nabla \times \mu_r^{-1}(\nabla \times \mathbf{E}) - k_\lambda^2 \varepsilon_r \mathbf{E} = 0 \tag{1}$$

was solved. Here, μ_r and ε_r represent magnetic permeability and dielectric function respectively, and $k_\lambda = 2\pi/\lambda$ is the wavenumber of the irradiating light in vacuum. For all materials, it was assumed that $\mu_r = 1$ and $\varepsilon_r = (n - ik)^2$, where n and k are the real and imaginary parts of the complex refractive indices of materials, respectively.

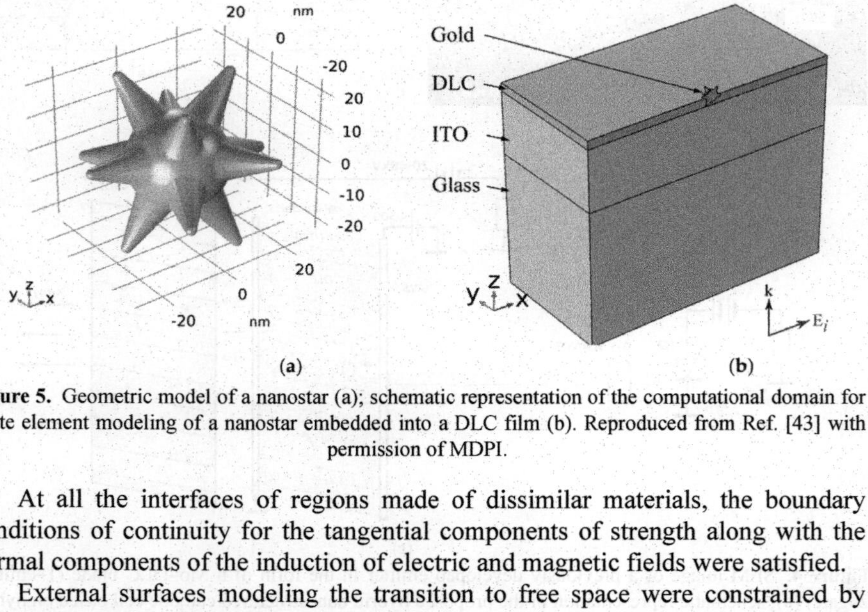

Figure 5. Geometric model of a nanostar (a); schematic representation of the computational domain for finite element modeling of a nanostar embedded into a DLC film (b). Reproduced from Ref. [43] with permission of MDPI.

At all the interfaces of regions made of dissimilar materials, the boundary conditions of continuity for the tangential components of strength along with the normal components of the induction of electric and magnetic fields were satisfied.

External surfaces modeling the transition to free space were constrained by "perfectly matched layers" [61, 62]. The case of normal incidence of a plane wave in the positive direction of the z-axis with an electric vector directed along the x-axis was considered. For the magnetic field strength **H** in the symmetry plane x–z with the normal vector $\mathbf{y}_0$, the "perfect magnetic conductor" boundary condition [61, 62] was used:

$$\mathbf{y}_0 \times \mathbf{H} = 0. \tag{2}$$

A symmetric GNS model with spikes uniformly oriented along the angle was considered. The thermo-optical parameters of the hybrid material were simulated, taking into account the effect of limiting the electron path length in the spikes of the nanostar. This noticeably improves the agreement between theory and experiment [55].

The specific absorption of radiation was calculated by the formula [45]:

$$Q = \nu \mathrm{Im}(\varepsilon_r) |\mathbf{E}|^2, \tag{3}$$

where ν and **E** are the frequency and the vector of electric field strength, respectively.

To verify the model, the results of the calculation of absorption and scattering cross-sections were compared with the experimentally determined spectral dependences of the extinction of a suspension of GNS in water and GNS on a glass plate [57]. This showed a good quantitative agreement between the theoretical and experimental results. In particular, when using the two-fraction approximation of the GNS ensemble model in water, the coincidence of the first resonance peak of the normalized extinction spectrum near 900 nm was obtained in theory and experiment. Additionally, the difference in the positions of the second resonance peak near 1900 nm did not exceed 100 nm. The theoretical and experimental peaks coincided within a few percent. Therefore, the use of the described mathematical model seems reasonable.

Photocathodes of complex structures and shapes are characterized by the localization of the concentration of photoexcited electrons in separate spatial regions called "hot spots". The estimation of the spatial distribution of the concentration of "hot" electrons in the CW irradiation mode has been carried out within the framework of simplified formalism [44–46] based on the following equation:

$$\frac{d^2 N(\mathbf{r})}{ds dE} = |E_{norm}(\mathbf{r})|^2 \frac{2e^2 E_F^2}{\pi^2 \gamma (h\nu)^4}, \tag{4}$$

where $d^2N(\mathbf{r})/dsdE$ is the number of hot electrons for unit area s and energy E in the vicinity of the considered point **r** on the surface, $E_{\mathrm{norm}}(\mathbf{r})$ is the component of the complex amplitude of the electric field normal to the surface at point **r**, e is the absolute value of electron charge, E_{F} is the Fermi energy, γ is the rate of energy relaxation of single electrons, and $h\nu$ is the energy of the irradiating photons.

In accordance with [45], the estimate of the influence of the spectral dependence of the function multiplied by $|E_{\mathrm{norm}}(\mathbf{r})|^2$ in Equation (4) leads to the conclusion that this function can be approximated by a constant with an error of no more than 25%. Thus, for the analysis of the concentration distribution of hot electrons, the key characteristic is $|E_{\mathrm{norm}}(\mathbf{r})|^2$, which is investigated in detail below.

Photoemission current density is also significantly influenced by the electrostatic field, which lowers the height of the potential Schottky barrier at the metal-vacuum and DLC-vacuum interfaces. Calculations show that the spatial distribution of the electrostatic field in structures with field emission is significantly heterogeneous [51, 63]. Furthermore, on the basis of numerical calculations, an analysis of the localization of optical field (to identify areas of effective generation of hot electrons) and the spatial distribution of electrostatic field (to intensify photoelectron emission) was carried out. The influence of the spatial separation of the regions of localization of the electrostatic field and hot electrons on the photoemission process is discussed in the succeeding section.

6. Modelling of plasmon effects, hot electrons and field emission in hybrid nanomaterial

To describe the bulk dielectric functions of materials (gold [64], DLC [65], ITO [66]), the interpolated tabular data given in the above works were used. The dielectric constant of the glass substrate was taken to be $\varepsilon_r = 2.25$. The geometrical parameters of GNS (used in the design of the computational model) were considered in accordance with the GNS synthesized and experimentally investigated in Ref. [43]. Thermo-optical properties of such GNS have been studied theoretically earlier [34]. The results obtained made it possible to explain the experimentally observed threshold nature of their photomodification when irradiated with nanosecond laser pulses. Therefore, the geometrical model of GNS was a sphere of 22 nm diameter, with twelve S-cone spikes with a height and base diameter of 10 nm each and a tip curvature radius of 1.5 nm.

From the calculations of the optical parameters of a GNS presented in Figure 6, it can be seen that in the case of GNS in a vacuum, the absorption and scattering cross-section spectra have one narrow resonance peak near $\lambda = 600$ nm. For GNS embedded in the HM structure, the shape of the spectra becomes much more complicated—the spectra noticeably broaden and an additional pronounced resonance peak appears in the NIR region near $\lambda = 1000$ nm.

It can be seen that halfway from the maxima, the width of the resonance curve of the absorption cross section increases by a factor of 5, and, taking into account the second peak, by a factor of nine. The appearance of this peak in the NIR region was expected and is associated with an increase in the real part of the dielectric function of the medium surrounding the part of the GNS immersed in the DLC film

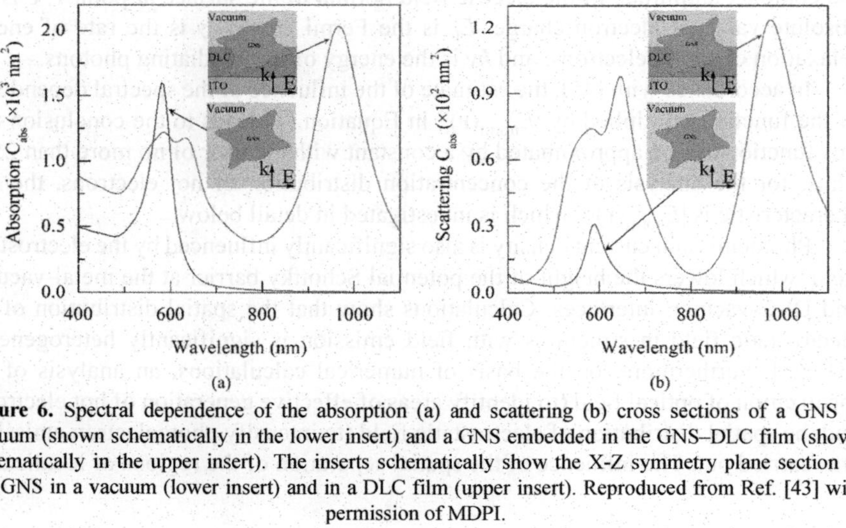

Figure 6. Spectral dependence of the absorption (a) and scattering (b) cross sections of a GNS in vacuum (shown schematically in the lower insert) and a GNS embedded in the GNS–DLC film (shown schematically in the upper insert). The inserts schematically show the X-Z symmetry plane section of the GNS in a vacuum (lower insert) and in a DLC film (upper insert). Reproduced from Ref. [43] with permission of MDPI.

[53, 54]. Hence, it can be expected that the main contribution to this shift is made by precisely those rays that are immersed in the DLC film. The non-trivial multi-peak shapes of the spectra are due to the fact that the GNS is only partially immersed in the DLC film. Various spikes, partially or completely immersed in a vacuum or in a film, mainly contribute to the short-wavelength and NIR region of the spectral curves respectively.

Slices in Figure 7 (upper panel) demonstrate how the localization zones of |E| are redistributed along different spikes of the GNS with changing laser wavelength. The relative change in the field amplitude, accounting for size correction, is from 5% at 808 nm to 19% at 970 nm. In addition to the spike tips, localization zones are also observed on the exit/entry lines of the spikes in the DLC film. Therefore, in addition to the tips on the four spikes of the GNS, the field characteristics are analyzed at two points on the exit/entry lines of the spikes in the DLC film. The positions of these six points are shown in the inserts in the following figures. The regularities of localization of absorption zones are practically identical to those given for |E| above (see Figure 7 (lower panel)).

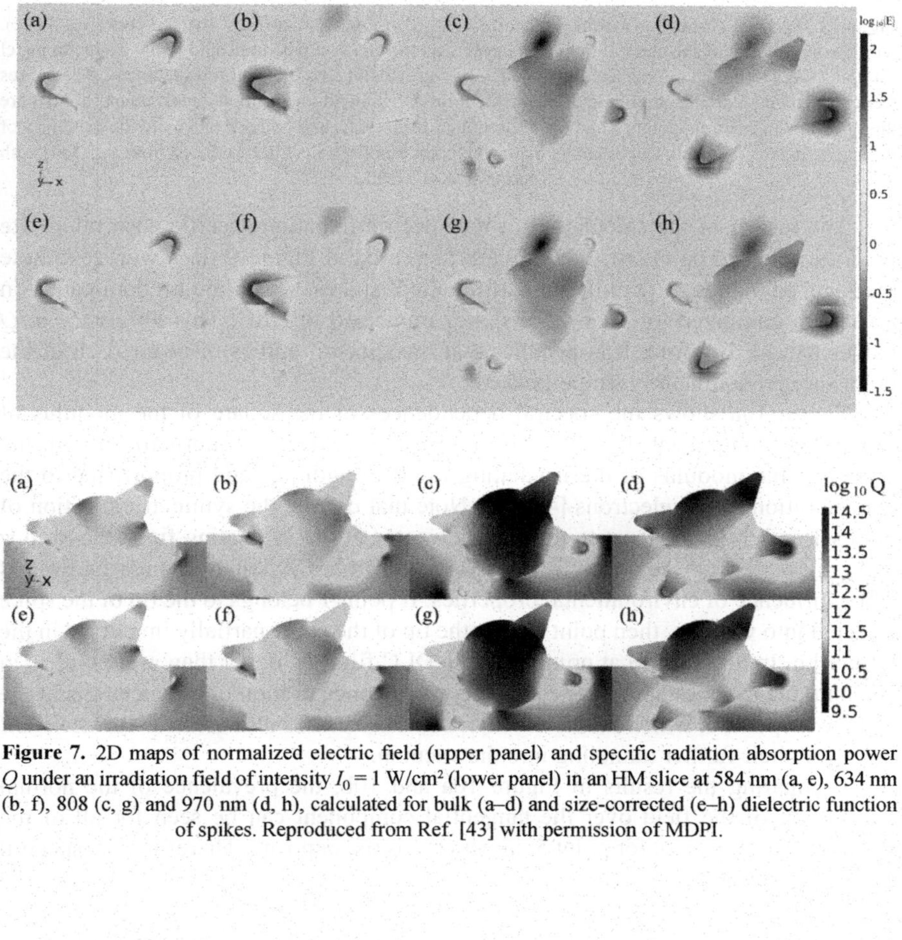

Figure 7. 2D maps of normalized electric field (upper panel) and specific radiation absorption power Q under an irradiation field of intensity $I_0 = 1$ W/cm^2 (lower panel) in an HM slice at 584 nm (a, e), 634 nm (b, f), 808 (c, g) and 970 nm (d, h), calculated for bulk (a–d) and size-corrected (e–h) dielectric function of spikes. Reproduced from Ref. [43] with permission of MDPI.

The results in Figure 6 confirm the previously stated assumption about the cause of the multi-peak shape of the absorption and scattering cross-sections—various spikes, partially or completely immersed in a vacuum or in a film, make the main contribution to the short-wave and NIR regions of the spectral curves respectively. The difference between the solid and dashed curves in the fragment in Figure 8c, which reached 23% in the NIR, indicates the need for considering size correction in the dielectric function of the GNS spikes.

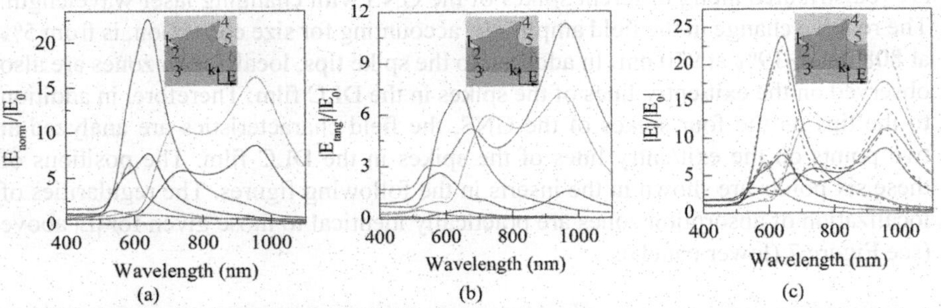

Figure 8. Spectral curves for modules of normal $|E_{norm}|$ (a), tangential $|E_{tang}|$ (b) and complete $|E|$ (c), with the amplitudes of the electric field at six points on the GNS surface indicated in the insets. In panel (c), the solid and dotted curves were obtained with and without size correction respectively. The curves were normalized with the amplitude of the incident field $|\boldsymbol{E}_0|$. Field localization zones: points 1 and 4 are the spike tips in vacuum, points 3 and 6 are the spike tips in DLC, and points 2 and 5 are the exit/entry of the spikes in the DLC film. Curve colors match the point number colors. Reproduced from Ref. [43] with permission of MDPI.

The results of the calculation of the specific radiation specific absorption rate Q at the selected points of the GNS are presented in Figure 9. It is worthy to note the reduced values of Q at the tips of the GNS spikes, which are predominantly in a vacuum compared to other GNS spikes immersed in DLC. This difference in Q values can be 3–9-fold. It is peculiar that the spectral curves of Q for each of the points are predominantly single-peaked.

Figure 10a shows the spectral dependences of the square of the modulus of the normal component of the electric/irradiating field $|E_{norm}|$, normalized with the square of the modulus of the irradiating field strength $|E_0|$ and proportional to the concentration of hot electrons [44–46]. Note that despite the symmetric position of points 1 and 6 with respect to the electric vector of the irradiating field, the ratio $\psi = |E_{norm}|^2/|E_0|^2$ for these points differs by a factor of 5. Obviously, the reason for this is the influence of environmental properties. If point 1 belongs to the tip of the spike directed into vacuum, then point 6 is on the tip of the spike partially immersed in the DLC film. In this case, ψ at point 6 (in the DLC film) is higher than ψ at point 1 (in vacuum). Thus, one should expect an increased concentration of hot electrons in the spike with point 6, which favors the injection of electrons into the near-surface DLC layer and their further tunneling into the vacuum.

Comparing the results in Figure 10a and 10b, the prevalence of the normal component of the field over the tangential component can be seen for all of the studied points – except for point 5, in which it is the opposite. Therefore, the spectral

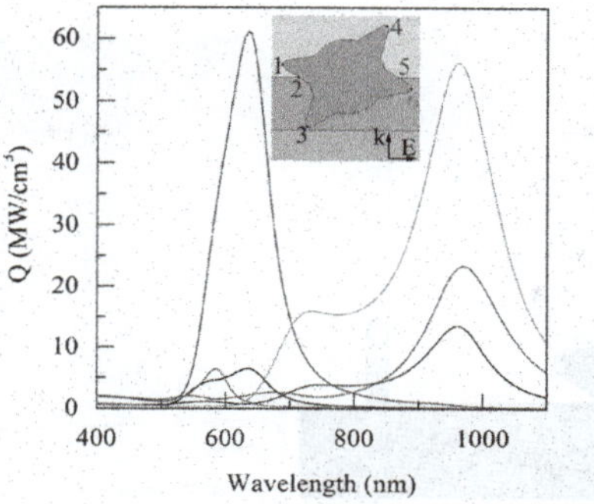

Figure 9. Spectral curves of the specific radiation specific absorption rate Q at the indicated points on the GNS surface (Intensity of the irradiating field $I_0 = 1$ W/cm²). The designations of points and curves are the same as in Figure 8. Reproduced from Ref. [43] with permission of MDPI.

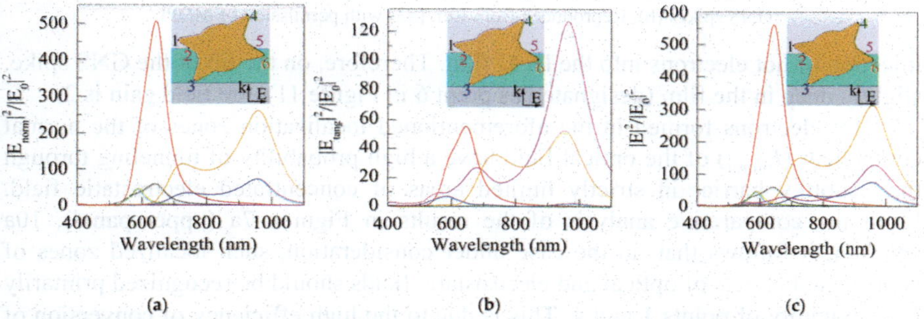

Figure 10. Spectral curves for the squares of the modulus of normal $|E_{norm}|$ (a) and tangential $|E_{tang}|$ (b) field components and total field (c), normalized with the square of the modulus of the irradiating field $|E_0|$, at six specified points on the GNS surface. The designations of the points and the curves are the same as in Figure 8. Reproduced from Ref. [43] with permission of MDPI.

curves of the total field in Figure 10c are also very close for all points, except for point 5.

To assess the role of the electrostatic field in stimulating the photoemission of an HM based on GNS-DLC, it is important to know the features of localization of field under the influence of a high external voltage. In this respect, the information on the concentration (amplification) of the external field strength F_0 on the heterogeneities of the GNS form embedded in DLC is of the greatest interest. The results of modeling the distribution of this gain $\xi_F = |F(\mathbf{r})|/|F_0|$ (where $F(\mathbf{r})$ is the vector of the electrostatic field strength at the current point with the radius vector $\mathbf{r}$) are shown in Figure 11a. It can be seen that the orientation of the rays relative to the external field vector F_0 significantly affects the distribution of ξ_F. Local maxima ξ_F are observed at the points of the GNS rays, but the value of each of the maxima increases with a decrease in the angle between the spike axis and the direction of the vector F_0. Thus, for a spike with a minimum deviation from the direction of the vector F_0, the value of ξ_F is maximum and equal to 6.7. For a spike with a maximum deviation from the direction of the vector F_0, the value of ξ_F is halved to 3.3. It should be noted that, according to the calculation, the external field F_0 also has a positive effect on the conditions of

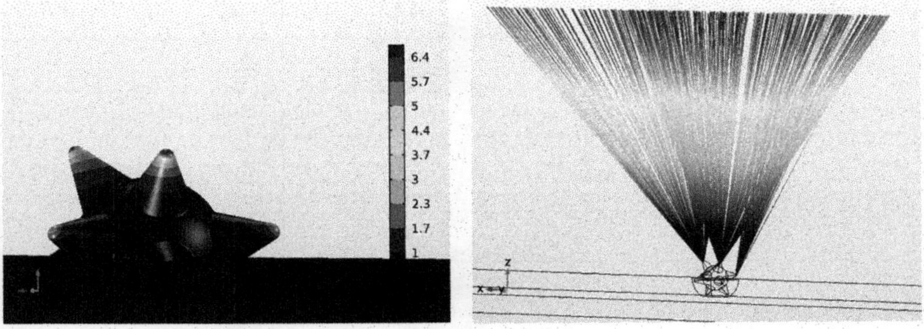

(a) (b)

Figure 11. A map of the gain distribution ξ_F of the electrostatic field on the surface of GNS embedded in the DLC film and its surroundings (a), the vector of an external uniform field with strength F_0 is directed along the *Z*-axis; results of modeling the trajectories of electrons emitted from the particular points of GNS spikes (b). Reproduced from Ref. [43] with permission of MDPI.

injection of hot electrons into the DLC film. Therefore, on the tip of the GNS spike, slightly deep in the film (designated as point 6 in Figure 11), the field gain is 2.

Hot electrons formed in the aforementioned localization zones of the normal component (E_{norm}) of the optical field have a high probability of tunneling through the Schottky barrier in strictly limited areas of concentrated electrostatic field. From the comparative analysis of the results in Figures 7a (upper panel), 10a and 11a, it follows that, in the case under consideration, such localized zones of coordinated effects of optical and electrostatic fields should be recognized primarily in the vicinity of points 1 and 4. This is due to the high efficiency of conversion of the laser beam energy into hot electrons (see Figure 10a) and the relatively high electrostatic field gain (see Figure 11a). Additional consideration of the influence of the plasmon effect as well (see Figure 10c) allows us to conclude that the conditions for the generation of hot electrons are approximately equivalent in the vicinity of the tips of both spikes (points 1 and 4). However, there is a difference between them in terms of the magnitude of amplification of the electrostatic field; at point 4, ξ_F is twice as high as that at point 1. This affects the magnitude of the lowering of the Schottky barrier and will consequently lead to more intense tunneling of hot electrons into vacuum. The results of modeling the trajectories of photoelectrons from the tips of the GNS spikes are shown in Figure 11b. Under conditions of uniformly accelerated motion in a uniform electrostatic field, this beam will form a spot with a diameter of no more than 60 μm on the anode located at a distance of 20 mm from the emitter.

The high efficiency of generating hot electrons in the vicinity of points 2 and 6 (see Figures 7a (upper panel), 10a, 11a and 11b) triggers an alternative mechanism—injecting electrons into the DLC film, transporting them to the film surface and further emission into the vacuum [40, 42]. The lowest degree of matching of the optical and electrostatic fields in the vicinity of points 3 and 5 lowers the probability of both generation and successful transport of hot electrons into the vacuum.

It should also be noted that the scope of the described matrix sources is not limited exclusively to DCT systems based on the analysis of changes in the intensity of transmitted hard X-ray photons. They can be successfully implemented in a number of existing promising imaging systems that use the coherent properties of ultraviolet and soft X-ray radiation, giving an increased resolution [67, 68].

7. Conclusion

Based on the analysis of the development of X-ray CT, the significance of the appearance of distributed matrix sources of an electron beam generating X-ray beams has been discussed. Due to the increase in speed, such systems provide a high degree of detail in the X-ray image and decrease the radiation dose. The transition from thermionic cathodes to current sources with field emission from nanostructured emitters provides compactness, energy efficiency and durability to X-ray CTs. Further improvement in the characteristics of such systems is associated with the implementation of the idea of ultrafast optical control of the current of each elementary emitter of the cathode array.

A new hybrid material based on GNS embedded in a nanoscale DLC film has also been theoretically investigated in this chapter. The results obtained therein make it possible to assess the prospects of using the effect of an increased concentration of hot electrons in nanostructures with plasmonic materials for photostimulated field emission of electrons and prospective applications in the field of biomedical instrumentation and sensing. In particular, the material is promising for the development of a matrix source with photoemission for X-ray computed tomography sensors, providing an increase in the resolution and a decrease in the criticality of local heating during tunneling electron emission of an X-ray gun with a matrix cathode.

The broadband effect of optical excitation of the hybrid material in the visible and NIR wavelength ranges has been noted. The areas of localization of electrostatic and optical fields in GNS, in which their effective matching is ensured, have also been determined.

Acknowledgment

This research was partially funded by the Russian Foundation for Basic Research (Project No. 19-54-06008). The results of Sections 1–5 were obtained within the framework of the State Assignment of Ministry of Science and Higher Education of the Russian Federation (Theme No. FFNM-2021-0002). VVT was supported by the Tomsk State University Development Programme (Priority-2030).

References

[1] Kaye, G. 1934. Wilhelm Conrad Röntgen: And the early history of the Roentgen Rays. Nature 133: 511–513. https://doi.org/10.1038/133511a0.

[2] Eckert, M. 2012. Max von Laue and the discovery of X-ray diffraction in 1912. Annalen der Physik. 524: A83–A85. DOI: 10.1002/andp.201200724.

[3] Bragg, W.H. and W.L. Bragg. 1913. The reflection of X-rays by crystals. Proc. R. Soc. Lond. A 88: 428–438. http://doi.org/10.1098/rspa.1913.0040.

[4] Handbook of X-ray Spectrometry, 2nd ed., Van Grieken, R. and A. Markowicz (eds.). Marcel Dekker, Inc: New York and Basel, 2002, 984 pp., ISBN 0-8247-0600-5.
[5] Michael, G. 2001. X-ray computed tomography. Phys. Educ. 36(6): 442–451. https://doi.org/10.1088/0031-9120/36/6/301.
[6] Moseley, H.G.J. 1913. The high-frequency spectra of the elements. The London, Edinburgh, and Dublin Philosophical Magazine and Journal of Science 26(156): 1024–1034. DOI: 10.1080/14786441308635052.
[7] Soltis, T., L. Folan and W. Eltareb. 2017. One hundred years of Moseley's law: An undergraduate experiment with relativistic effects. American Journal of Physics 85: 352–358. DOI: 10.1119/1.4977793.
[8] Coolidge, W.D. 1913. A powerful Röntgen Ray tube with a pure electron discharge. Phys. Rev. 2: 409. DOI: 10.1103/PhysRev.2.409.
[9] Elbakri, I.A. and J.A. Fessler. 2002. Statistical image reconstruction for polyenergetic X-Ray computed tomography. IEEE Trans. on Medical Imaging 21(2): 89–99. Doi: 10.1109/42.993128.
[10] Ólafsson, G. and E.T. Quinto. 2006. The radon transform, inverse problems, and tomography. Proc. of Symposia in Applied Mathematics 63. https://doi.org/10.1090/psapm/063.
[11] Cormack, A.M. 1980. Nobel Award address. Early two-dimensional reconstruction and recent topics stemming from it. Med. Phys. 7(4): 277–282. Doi: 10.1118/1.594708.
[12] Hounsfield, G.N. 1980. Nobel Award address. Computed medical imaging. Med. Phys. 7(4): 283–290. Doi: 10.1118/1.594709. PMID: 6993911.
[13] Tuchin, V. 2007. Tissue Optics: Light Scattering Methods and Instruments for Medical Diagnosis (2nd ed.). SPIE Press, Bellingham, WA, USA, 841 pp., ISBN: 0-8194-6433-3.
[14] Neculaes, V.B., P.M. Edic, M.A. Frontera Caiafa, G. Wang and B. De Man. 2014. Multisource X-Ray and CT: Lessons learned and future outlook. IEEE Access 2: 1568–1585. Doi: 10.1109/ACCESS.2014.2363949.
[15] Hadsell, M., G. Cao, J. Zhang, L. Burk, T. Schreiber, E. Schreiber, S. Chang, J. Lu and O. Zhou. 2014. Pilot study for compact microbeam radiation therapy using a carbon nanotube field emission micro-CT scanner. Med. Phys. 41: 061710. Doi: 10.1118/1.4873683.
[16] Li, X., J. Zhou, Q. Wu, M. Liu, R. Zhou and Z. Chen. 2019. Fast microfocus X-ray tube based on carbon nanotube array. J. Vac. Sci. Technol. B 37: 051203. Doi: 10.1116/1.5099697.
[17] Lin, Z., P. Xie, R. Zhan, D. Chen, J. She, S. Deng, N. Xu and J. Chen. 2019. Defect-enhanced field emission from WO_3 nanowires for flat-panel X-Ray sources. ACS Appl. Nano Mater. 2: 5206–5213. Doi: 10.1021/acsanm.9b01074.
[18] Gan, H., T. Zhang, Z. Guo, H. Lin, Z. Li, H. Chen, J. Chen and F. Liu. 2019. The growth methods and field emission studies of low-dimensional boron-based nanostructures. Appl. Sci. 9: 1019. Doi: 10.3390/app9051019.
[19] Koh, A.L., E. Gidcumb, O. Zhou and R. Sinclair. 2017. *In situ* field emission of carbon nanotubes in oxygen using environmental TEM and the influence of the imaging electron beam. Microsc. Microanal. 23: 910–911. Doi: 10.1017/S1431927617005219.
[20] Burk, L.M., Y.Z. Lee, J. Lu and O. Zhou. 2016. Carbon nanotube field-emission x-ray-based micro-computed tomography for biomedical imaging. pp. 201–226. *In*: Zhang, M., R.R. Naik and L. Dai (eds.). Carbon Nanomaterials for Biomedical Applications; Springer: Heidelberg, Germany; New York, NY, USA; Dordrecht, The Netherlands; London, UK; Cham, Switzerland. Doi: 10.1007/978-3-319-22861-7_6.
[21] Kang, J.-T., H.-R. Lee, J.-W. Jeong, J.-W. Kim, S. Park, M.-S. Shin, J.-H. Yeon, H. Jeon, S.-H. Kim, Y.C. Choi and Y.-H. Song. 2015. Fast and stable operation of carbon nanotube field-emission X-Ray tubes achieved using an advanced active-current control. IEEE Electron Device Lett. 36: 1209–1211. Doi: 10.1109/LED.2015.2478157.
[22] Shan, J., A.W. Tucker, Y.Z. Lee, M.D. Heath, X. Wang, D.H. Foos, J. Lu and O. Zhou. 2014. Stationary chest tomosynthesis using a carbon nanotube x-ray source array: A feasibility study. Phys. Med. Biol. 60: 81–100. Doi: 10.1088/0031-9155/60/1/81.
[23] Cole, M.T., R.J. Parmee and W.I. Milne. 2016. Nanomaterial-based X-ray sources. Nanotechnology 27: 082501. Doi: 10.1088/0957-4484/27/8/082501.
[24] Ryu, J.H., J.S. Kang and K.C. Park. 2012. Carbon nanotube electron emitter for X-ray imaging. Materials 5: 2353–2359. https://doi.org/10.3390/ma5112353.

[25] Li, X., J. Zhou, Q. Wu, M. Liu, R. Zhou and Z. Chen. 2019. Fast microfocus X-Ray tube based on carbon nanotube array. Journal of Vacuum Science & Technology B 37(5): 051203. Doi: 10.1116/1.5099697.

[26] Cao, G., L.M. Burk, Y.Z. Lee, X. Calderon-Colon, S. Sultana, J. Lu and O. Zhou. 2010. Prospective-gated cardiac micro-CT imaging of free-breathing mice using carbon nanotube field emission X-Ray. Med. Phys. 37(10): 5306–5312. Doi: 10.1118/1.3491806.

[27] Qian, X., A. Tucker, E. Gidcumb, J. Shan, G. Yang, X. Calderon-Colon, S. Sultana, J. Lu, O. Zhou, D. Spronk, F. Sprenger, Y. Zhang, D. Kennedy, T. Farbizio and Z. Jing. 2012. High resolution stationary digital breast tomosynthesis using distributed carbon nanotube X-Ray source array. Med. Phys. 39(4): 2090–2099. Doi: 10.1118/1.3694667.

[28] Wang, S., X. Calderon, R. Peng, E.C. Schreiber, O. Zhou and S. Chang. 2011. A carbon nanotube field emission multipixel x-ray array source for microradiotherapy application. Appl. Phys. Lett. 98(21): 213701. Doi: 10.1063/1.3595268.

[29] Floweri, O., H. Jo, Y. Seo and N. Lee. 2018. Improving the thermal stability of carbon nanotubes and their field emission characteristics by adding boron and phosphorus compounds. Carbon 139: 404–414. Doi: 10.1016/j.carbon.2018.06.072.

[30] Hu, Z., Z. Chen, C. Zhou, X. Hong, J. Chen, Q. Zhang, C. Jiang, Y. Ge, Y. Yang, X. Liu, H. Zheng, Z. Li and D. Liang. 2020. Evaluation of reconstruction algorithms for a stationary digital breast tomosynthesis system using a carbon nanotube X-ray source array. J. Xray Sci. Technol. 28: 1157–1169. Doi: 10.3233/XST-200668.

[31] Gonzales, B., D. Spronk, Y. Cheng, Z. Zhang, X. Pan, M. Beckmann, O. Zhou and J. Lu. 2013. Rectangular computed tomography using a stationary array of CNT emitters: Initial experimental results. Proc. SPIE 8668: 86685K. https://doi.org/10.1117/12.2008030.

[32] Gidcumb, E., B. Gao, J. Shan, C. Inscoe, J. Lu and O. Zhou. 2014. Carbon nanotube electron field emitters for x-ray imaging of human breast cancer. Nanotechnology 25(24): 245704. https://doi.org/10.1088/0957-4484/25/24/245704.

[33] Aban'shin, N.P., Y.A. Avetisyan, G.G. Akchurin, A.P. Loginov, S.P. Morev, D.S. Mosiyash and A.N. Yakunin. 2016. A planar diamond-like carbon nanostructure for a low-voltage field emission cathode with a developed surface. Tech. Phys. Lett. 42: 509–512. Doi: 10.1134/S1063785016050175.

[34] Qian, X., A. Tucker, E. Gidcumb, J. Shan, G. Yang, X. Calderon-Colon, S. Sultana, J. Lu, O. Zhou, D. Spronk, F. Sprenger, Y. Zhang, D. Kennedy, T. Farbizio and Z. Jing. 2012. High resolution stationary digital breast tomosynthesis using distributed carbon nanotube X-ray source array. Med. Phys. 39: 42090–42099. Doi: 10.1118/1.3694667.

[35] Cole, M.T., R.J. Parmee and W.I. Milne. 2016. Nanomaterial-based X-ray sources. Nanotechnology 27: 082501. Doi: 10.1088/0957-4484/27/8/082501.

[36] Hu, Z., Z. Chen, C. Zhou, X. Hong, J. Chen, Q. Zhang, C. Jiang, Y. Ge, Y. Yang, X. Liu, H. Zheng, Z. Li and D. Liang. 2020. Evaluation of reconstruction algorithms for a stationary digital breast tomosynthesis system using a carbon nanotube X-ray source array. J. Xray Sci. Technol. 28: 1157–1169. Doi: 10.3233/XST-200668.

[37] Lin, Z., P. Xie, R. Zhan, D. Chen, J. She, S. Deng, N. Xu and J. Chen. 2019. Defect-enhanced field emission from WO_3 nanowires for flat-panel X-Ray sources. ACS Appl. Nano Mater. 2: 5206–5213. Doi: 10.1021/acsanm.9b01074.

[38] Patent US 9,520.260 B2.

[39] Li, Z., B. Bai, C. Li and Q. Dai. 2015. Efficient photo-thermionic emission from carbon nanotube arrays. Carbon 96: 641–646. Doi: 10.1016/j.carbon.2015.09.074.

[40] Aban'shin, N.P., A.P. Loginov, D.S. Mosiyash and A.N. Yakunin. 2016. Theoretical and experimental study of characteristics of the planar tetrode with field emission of diamond-like carbon film. pp. 1–2. *In*: Proceedings of the 2016 29th International Vacuum Nanoelectronics Conference (IVNC), Vancouver, BC, Canada, 11–15 July. Doi: 10.1109/IVNC.2016.7551494.

[41] Yakunin, A.N., N.P. Aban'shin, D.S. Mosiyash, Y.A. Avetisyan and G.G. Akchurin. 2018. Features of field emission and formation of electron beam in a multi-electrode planar cathode. pp. 249–252. *In*: Proceedings of the International Conference on Actual Problems of Electron Devices Engineering (APEDE), Saratov, Russia, 27–28 September. Doi: 10.1109/apede.2018.8542332.

[42] Zimnyakov, D.A., N.P. Aban'shin, G.G. Akchurin, Y.A. Avetisyan, A.P. Loginov, S.A. Yuvchenko and A.N. Yakunin. 2019. Experimental study of a broadband vacuum photosensor with the tunnel emission from a metal nanoscale blade. Proc. SPIE 11022: 110220A. Doi: 10.1117/12.2521749.
[43] Yakunin, A.N., S.V. Zarkov, Y.A. Avetisyan, G.G. Akchurin, N.P. Aban'shin and V.V. Tuchin. 2021. Modeling of laser-induced plasmon effects in GNS-DLC-based material for application in X-ray source array sensors. Sensors 21: 1248. https://doi.org/10.3390/s21041248.
[44] Santiago, E.Y., L.V. Besteiro, X.-T. Kong, M.A. Correa-Duarte, Z. Wang and A.O. Govorov. 2020. Efficiency of hot-electron generation in plasmonic nanocrystals with complex shapes: Surface-induced scattering, hot spots, and interband transitions. ACS Photonics 7: 2807–2824. Doi: 10.1021/acsphotonics.0c01065.
[45] Kong, X.-T., Z. Wang and A.O. Govorov. 2017. Plasmonic nanostars with hot spots for efficient generation of hot electrons under solar illumination. Adv. Opt. Mater. 5: 1600594. Doi: 10.1002/adom.201600594.
[46] Besteiro, L.V., X.-T. Kong, Z. Wang, G.V. Hartland and A.O. Govorov. 2017. Understanding hot-electron generation and plasmon relaxation in metal nanocrystals: Quantum and classical mechanisms. ACS Photonics 4: 2759–2781. Doi: 10.1021/acsphotonics.7b00751.
[47] Forati, E., T.J. Dill, A.R. Tao and D. Sievenpiper. 2016. Photoemission-based microelectronic devices. Nat. Commun. 7: 13399. Doi: 10.1038/ncomms13399.
[48] Piltan, S. and D. Sievenpiper. 2018. Plasmonic nano-arrays for enhanced photoemission and photodetection. JOSA B 35: 208–213. Doi: 10.1364/JOSAB.35.000208.
[49] Yakunin, A.N., N.P. Aban'shin, Y.A. Avetisyan, G.G. Akchurin, G.G. Jr. Akchurin, A.P. Loginov, S.P. Morev and D.S. Mosiyash. 2019. Stabilization of field- and photoemission of a planar structure with a nanosized diamond-like carbon film. J. Commun. Technol. Electron. 64: 83–88. Doi: 10.1134/S1064226919010133.
[50] Akchurin, G.G., A.N. Yakunin, N.P. Aban'shin, B.I. Gorfinkel, G.G. Jr. Akchurin. 2013. Controlling the Red boundary of the tunneling photoeffect in nanodimensional carbon structures in a broad (UV–IR) wavelength range. Tech. Phys. Lett. 39: 544–547. Doi: 10.1134/S1063785013060151.
[51] Yakunin, A.N., N.P. Aban'shin, G.G. Akchurin, Y.A. Avetisyan, A.P. Loginov, S.A. Yuvchenko and D.A. Zimnyakov. 2019. A visible and near-IR tunnel photosensor with a nanoscale metal emitter: The effect of matching of hot electrons localization zones and a strong electrostatic field. Appl. Sci. 9: 5356. Doi: 10.3390/app924535.
[52] Fabris, L. 2020. Gold nanostars in biology and medicine: Understanding physicochemical properties to broaden applicability. J. Phys. Chem. C 124: 26540–26553. Doi: 10.1021/acs.jpcc.0c08460.
[53] Pylaev, T., E. Vanzha, E. Avdeeva, B. Khlebtsov and N. Khlebtsov. 2019. A novel cell transfection platform based on laser optoporation mediated by Au nanostar layers. J. Biophotonics 12: e201800166. Doi: 10.1002/jbio.201800166.
[54] Cristiano, M.N., T.V. Tsoulos and L. Fabris. 2020. Quantifying and optimizing photocurrent via optical modeling of gold nanostar-, nanorod-, and dimer-decorated MoS_2 and $MoTe_2$. J. Chem. Phys. 152: 014705. Doi: 10.1063/1.5127279.
[55] Pettine, J., P. Choo, F. Medeghini, T.W. Odom and D.J. Nesbitt. 2020. Plasmonic nanostar photocathodes for optically controlled directional currents. Nat. Commun. 11: 1–10. Doi: 10.1038/s41467-020-15115-0.
[56] Tsoulos, T.V., S. Atta, M.J. Lagos, M. Beetz, P.E. Batson, G. Tsilomelekis and L. Fabris. 2019. Colloidal plasmonic nanostar antennas with wide range resonance tunability. Nanoscale 11: 18662–18671. Doi: 10.1039/C9NR06533D.
[57] Khlebtsov, N.G., S.V. Zarkov, V.A. Khanadeev and Y.A. Avetisyan. 2020. A novel concept of two-component dielectric function for gold nanostars: Theoretical modeling and experimental verification. Nanoscale 12: 19963–19981. Doi: 10.1039/d0nr02531c.
[58] Zarkov, S., Y. Avetisyan, G. Akchurin, G. Jr. Akchurin, O. Bibikova, V. Tuchin and A. Yakunin. 2020. Numerical modeling of plasmonic properties of gold nanostars to prove the threshold nature of their modification under laser pulse. Opt. Eng. 59: 061628. Doi: 10.1117/1.OE.59.6.061628.
[59] Vigderman, L., B.P. Khanal and E.R. Zubarev. 2012. Functional gold nanorods: Synthesis, self-assembly, and sensing applications. Adv. Mater. 24: 4811–4841. Doi: 10.1002/adma.201201690.

[60] Zarkov, S.V., A.N. Yakunin, Y.A. Avetisyan, G.G. Akchurin, G.G. Jr. Akchurin and V.V. Tuchin. 2019. The peculiarities of localized laser heating of a tissue doped by gold nanostars. Proc. SPIE. Doi: 10.1117/12.2530519.

[61] Wave Optics Module User's Guide. Available online: https://doc.comsol.com/5.3/doc/com.comsol.help.woptics/WaveOpticsModuleUsersGuide.pdf (accessed on 10.03.2021).

[62] Jin, J.-M. 2010. Theory and Computation of Electromagnetic Fields; John Wiley & Sons, Inc.: Hoboken, NJ, USA, pp. 3–560, ISBN 978-0-470-53359-8.

[63] Aban'shin, N.P., B.I. Gorfinkel, S.P. Morev, D.S. Mosiyash and A.N. Yakunin. 2014. Autoemission structures of nanosized carbon with ionic protection. Studying the prospects of reliable control in forming structures. Tech. Phys. Lett. 40: 404–407. Doi: 10.1134/S1063785014050022.

[64] Johnson, P.B. and R.W. Christy. 1972. Optical constants of the noble metals. Phys. Rev. B 6: 4370. Doi: 10.1103/PhysRevB.6.4370.

[65] Zhou, X.L., T. Suzuki, H. Nakajima, K. Komatsu, K. Kanda, H. Ito and H. Saitoh. 2017. Structural analysis of amorphous carbon films by spectroscopic ellipsometry, RBS/ERDA, and NEXAFS. Appl. Phys. Lett. 110: 201902. Doi: 10.1063/1.4983643.

[66] König, T.A.F., P.A. Ledin, J. Kerszulis, M.A. Mahmoud, M.A. El-Sayed, J.R. Reynolds and V.V. Tsukruk. 2014. Electrically tunable plasmonic behavior of nanocube-polymer nanomaterials induced by a redox-active electrochromic polymer. ACS Nano 8: 6182–6192. Doi: 10.1021/nn501601e.

[67] Yakunin, A.N., Y.A. Avetisyan, G.G. Akchurin, S.V. Zarkov, N.P. Aban'shin, V.A. Khanadeev and V.V. Tuchin. 2022. Photoemission of plasmonic gold nanostars in laser-controlled electron current devices for technical and biomedical applications. Sensors 22: 4127. https://doi.org/10.3390/s22114127.

[68] Skruszewicz, S., S. Fuchs, J.J. Abel, J. Nathanael, J. Reinhard, C. Rödel, F. Wiesner, M. Wüjnsche, P. Wachulak, A. Bartnik, K. Janulewicz, H. Fiedorowicz and G.G. Paulus. 2021. Coherence tomography with broad bandwidth extreme ultraviolet and soft X-ray radiation. Appl. Phys. B 127: 55. https://doi.org/10.1007/s00340-021-07586-w.

4

Recent Advances and Prospects in Nanomaterials-Based Electrochemical Affinity Biosensors for Autoimmune Disease Biomarkers

Paloma Yáñez-Sedeño, Araceli González-Cortés, Susana Campuzano* and *José M. Pingarrón*

1. Introduction

The use of nanomaterials has been instrumental in improving the performance and demonstrating the potential of electrochemical affinity biosensors (mainly immunosensors, peptide-based or DNA biosensors) for early detection and monitoring of autoimmune diseases.

Nanomaterials can be used as carrier tags or for nanostructuring electrode surfaces and their use can prove to be instrumental in improving electron transfer to the electrode surface, in increasing the active surface area of the electrode and/or for enhancing its catalytic capability, thus resulting in biodevices with improved sensitivity and selectivity. Nanomaterials may also exert positive effects on the signal-to-noise ratio, hence allowing for low limits of detection and enhancing the calibration dynamic ranges, reproducibility, robustness and storage stability.

This chapter delves into the latest advances in electrochemical affinity-based biosensors using nanomaterials for the determination of biomarkers related to autoimmune diseases. They have been classified according to the role played by the nanomaterial in the biosensor, its type, and the disease to which the developed biosensor has been applied.

Departamento de Química Analítica, Facultad de CC. Químicas, Universidad Complutense de Madrid, E-28040 Madrid (Spain).

* Corresponding author: yseo@quim.ucm.es

2. Nanomaterials as electrode modifiers

The use of nanomaterials as modifiers of electrode surfaces has been widely used in electroanalytical chemistry, yielding three main achievements: (i) the excellent electrical conductivity of nanomaterials such as CNTs or graphene aids in the improvement of the responses of conventional electrodes for providing larger currents and higher signal-to-background ratios, ultimately resulting in a larger sensitivity; (ii) the possibility of nanomaterials functionalization by simple oxidation, grafting of organic molecules, or self-assembling monolayers, among other strategies, gives rise to a wide variety of suitable platforms for further immobilization of (bio)molecules; (iii) the ability of nanomaterials such as gold nanoparticles (AuNPs) to maintain the bioactivity of biomolecules involved in the preparation of biosensors, creating suitable environments for bioreagents immobilization, has contributed towards extending the lifetime of these bioplatforms. These improvements are illustrated in the following sections with some relevant examples classified according to the monitored disease.

2.1 Rheumatoid arthritis

The need for early prediction, diagnosis and monitoring of rheumatoid arthritis (RA) has led to the designing of a variety of electrochemical biosensors capable of detecting specific biomarkers. Many of these biosensors involve the use of nanomaterials. RA is a progressive and chronic inflammatory autoimmune condition that causes deformities in joints and pain in aged populations [1]. A few reports have recognized certain causal pathways and target proteins that lead to the generation of autoantibodies in RA [2]. Rheumatoid factors (RF) primarily consist of IgM, IgA and IgG antibodies directed against the Fc fragment within the patient's own IgG antibody molecules. The IgM isotype (IgM RF) exhibits the highest prevalence in RA diagnosis by virtue of its polyvalence; therefore, the sensitive detection of IgM-RF is of great importance. Recently, label-free and direct detection of IgM-RF using an impedimetric-interdigitated wave type microelectrode array (IDWμE) was reported. The IDWμE was fabricated on a glass slide substrate by deposition of nanolayers of Ti and Au functionalized with a self-assembled monolayer (SAM) of thioctic acid (TA) (Figure 1). The sensor design involved immobilization of IgG-Fc through the TA SAM onto the nanometric metal surface of IDWμE, thus providing a biocompatible environment preserving the biological activity of immunoreagents. Impedance measurements were target specific and linear with IgM-RF concentrations between 1 and 200 IU mL^{-1}. An LOD value of 0.22 IU mL^{-1} was achieved in human serum [3].

Anti-citrullinated peptide antibodies (ACPAs) are detected in most of RA patients [4, 5]. These antibodies recognize different self-proteins, such as filaggrin or vimentin, the arginyl residues of which have been post-translationally transformed into citrullyl residues appearing during the early stages of the disease and in healthy individuals who later develop RA. Despite the importance of ACPAs in diagnosis and prediction of response to biological therapy, only one electrochemical immunosensor for the detection of ACPAs has been found in the literature [6]. The biosensor

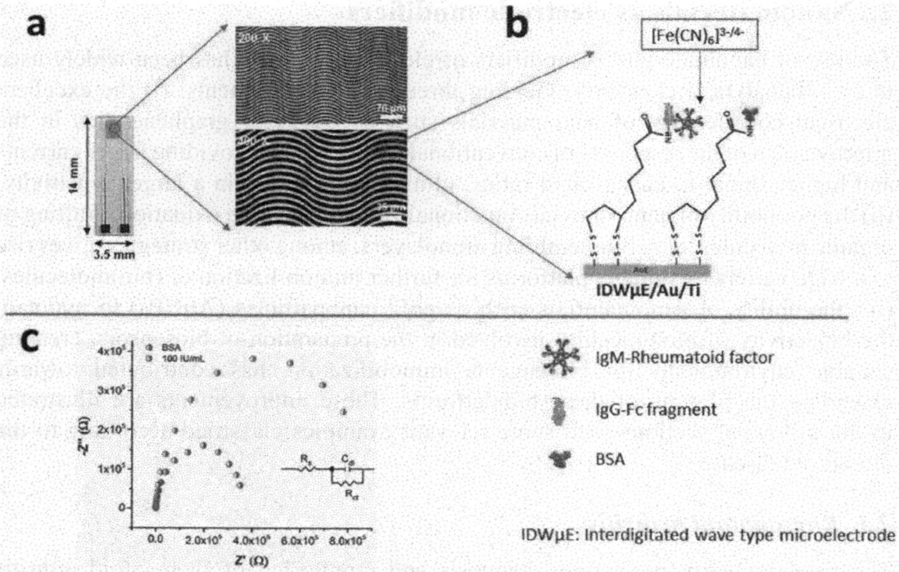

Figure 1. Photograph and microscopic images of the IDWμE array: (a) scheme of SAM functionalization of IDWμE and crosslinking of IgG-Fc fragments, (b) EIS measurements in the absence and in the presence of 100 IU mL^{-1} IgM-RF. Reproduced from [3] with permission.

consisted of an amperometric device where a chimeric fibrin–filaggrin synthetic peptide was used as the recognition element anchored to the surface of a multiwalled carbon nanotube–polystyrene electrochemical transducer. The determination of ACPAs was performed using a secondary antibody labeled with peroxidase (HRP-IgG), and hydrogen peroxide as the enzyme substrate. The presence of the nanocarbon composite as a modifier of the metallic thin film electrodes assured not only stable immobilization of the synthetic antigen through covalent bonding, but also the high conductivity and well-defined reversible faradaic response provided by 3,3′,5,5′-tetramethylbenzidine (TMB) used as the redox probe. The biosensor was tested in the sera of rabbits previously inoculated with the synthetic peptide as well as for the detection of ACPAs in human sera.

The pathogenic activity of ACPAs has been explained by considering the formation of immune complexes with the subsequent release of pro-inflammatory cytokines, including tumor necrosis factor alpha (TNF-α) from monocytes/macrophages [7]. Because of this, several electrochemical biosensors have been developed for the determination of TNF-α and interleukins such as IL-6, which are also biomarkers commonly used for monitoring RA disease activity. An illustrative example is a label-free impedimetric sensor for the determination of TNF-α using reduced graphene oxide (rGO) with AuNPs on an indium tin oxide (ITO) microdisk electrode (Figure 2A). In this case, the detection mechanism relied on the measurements of resistance changes due to $[Fe(CN)_6]^{3-/4-}$ redox probe movement towards the conductive channels of the AuNP-rGO films by the recognition of the target biomarker by its anti-TNF-α antibody. The antibody adsorbed electrode showed an increase in resistance (ΔR) with increasing antigen concentration. The sensor exhibited a linear range of 1–1,000 pg mL^{-1} with an LOD value of 0.78 pg mL^{-1} in human serum [8]. Mazloum-Ardakani et al. [9] reported a label-free electrochemical immunosensor for the determination of TNF-α, which involved entrapping the captured antibody into a nanocomposite containing fullerene-functionalized MWCNTs and ionic liquid (1-butyl-3-methylimidazolium bis (trifluoromethyl sulfonyl)imide) (Figure 2B). The determination of TNF-α was based on the obstruction of the electrocatalytic oxidation of catechol after the TNF-α binding to the surface through interaction with the anti-TNF-α, the protein layers on the modified electrode acting as a barrier for the electron-transfer. Under optimal conditions, the electrochemical immunosensor showed a dynamic range between 5.0 and 75 pg mL^{-1} TNF-α, with an LOD value of 2.0 pg mL^{-1}. The immunosensor was successfully applied to the determination of TNF-α in serum samples.

Interleukins not only provide valuable information about the evolution of RA but also aid in the early detection of other autoimmune diseases such as multiple sclerosis, type 1 diabetes, and psoriasis. Hence, numerous biosensors have been reported for

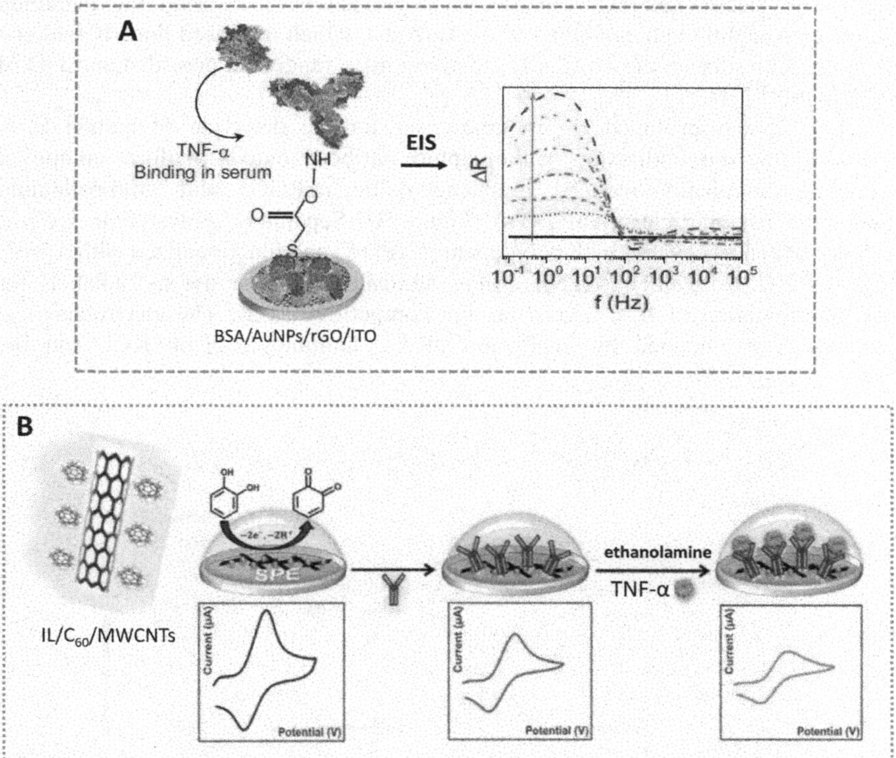

Figure 2. (A) Schematic diagrams showing the binding mechanism of TNF-α with the receptor antibodies on AuNPs-rGO/ITO electrode and EIS detection, and (B) the preparation of a TNF-α immunosensor using a C_{60}-CNTs/IL modified SPE. Adapted from [8] and [9].

the determination of interleukins. For instance, Chen and co-workers developed a method for the determination of IL-6 levels (involved in the pathogenesis of RA), using a liquid-gated field-effect transistor (FET) sensor with horizontally aligned single-walled carbon nanotubes (SWCNTs). The method relied on the transistor drain current responses upon interactions of IL-6 with the corresponding antibody (IL-6R) immobilized on the nanotubes' surface. The immunosensor exhibited an LOD of 1.37 pg mL^{-1} and a linear response in the 1–100 pg mL^{-1} IL-6 range. Furthermore, no significant degradation in the electronic performance was observed after storage under ambient conditions for up to three months, a possible explanation for the same being the strong adhesion and good horizontal alignment of CNTs on the quartz substrate [10]. A dual electrochemical and electrochemiluminescent (ECL) sandwich-type immunosensor has also been reported for the detection of IL-6; this involves the use of two kinds of TiO_2 mesocrystal nanoarchitectures. Herein, glassy carbon electrodes modified with a composite of TiO_2 (anatase) mesocages and a carboxy-terminated ionic liquid (CTIL) were used for the immobilization of anti-IL-6, whereas octahedral anatase TiO_2 mesocrystals were used as the matrix for immobilizing acid phosphatase, along with a detection antibody labelled with HRP. The electrochemical response was obtained by measuring the oxidation signal of 1-naphthyl phenol at 0.4 V vs. Ag/AgCl, which increased linearly with the IL-6 concentration in the 10 fg mL^{-1} to 90 ng mL^{-1} range and provided an LOD of 0.32 fg mL^{-1} [11].

Lou et al. developed an immunosensor for the detection of human IL-6, involving the immobilization of the capture antibody onto a platform composed of electrochemically reduced graphene oxide (ERGO) and gold/palladium bimetallic nanoparticles (AuPdNPs) (Figure 3). Separately, polystyrene spheres (PS, $\phi = 200$ nm) covered with polydopamine (PDA) and functionalized with AgNPs and IL-6 (IL-6-PS@PDA-AgNPs) were fabricated for their use as biolabels for the determination of IL-6 levels through competitive assay. The electrochemical response was obtained by incubation of the immunosensor in KCl, and the

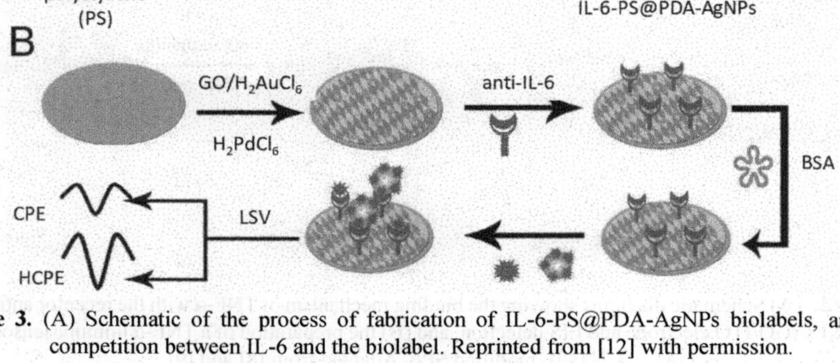

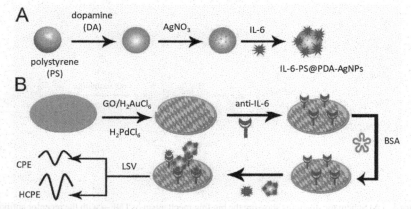

Figure 3. (A) Schematic of the process of fabrication of IL-6-PS@PDA-AgNPs biolabels, and (B) competition between IL-6 and the biolabel. Reprinted from [12] with permission.

silver component, acting as the target compound, was measured by linear sweep voltammetry (LSV). As an uncommon practice, it should be noted, an electrically heated carbon paste electrode (HCPE) was used for sensitivity enhancement. The immunosensor exhibited a wide linear response to IL-6 concentrations ranging from 0.1 to 10^5 pg mL^{-1}, with an LOD of 0.059 pg mL^{-1} [12].

Interleukin-17A (IL-17A or simply IL-17) has received substantial attention as a valuable biomarker associated with numerous autoimmune diseases such as RA and multiple sclerosis (MS), among others [13, 14]. Over-expression of IL-17 is associated with the pathogenesis of such processes and strongly correlated with the cell surface expression level of its receptor, IL-17RA, which is also considered as a valuable biomarker for the diagnosis and prognosis of these disorders. Jeong et al. constructed an electrochemical aptasensor for the detection of IL-17RA, using a specific aptamer [15]. Figure 4 shows that AuNPs were electrodeposited onto a gold electrode and the 5′-thiol modified aptamer immobilized. Quantitative detection of IL-17R was made by impedimetric measurements. The sensor exhibited a wide dynamic range of 10–10^4 pg mL^{-1} and an LOD of 2.13 pg mL^{-1}—which is lower than that of commercially available ELISA kits. The clinical applicability of the sensor was demonstrated using neutrophils isolated from asthma patients [15].

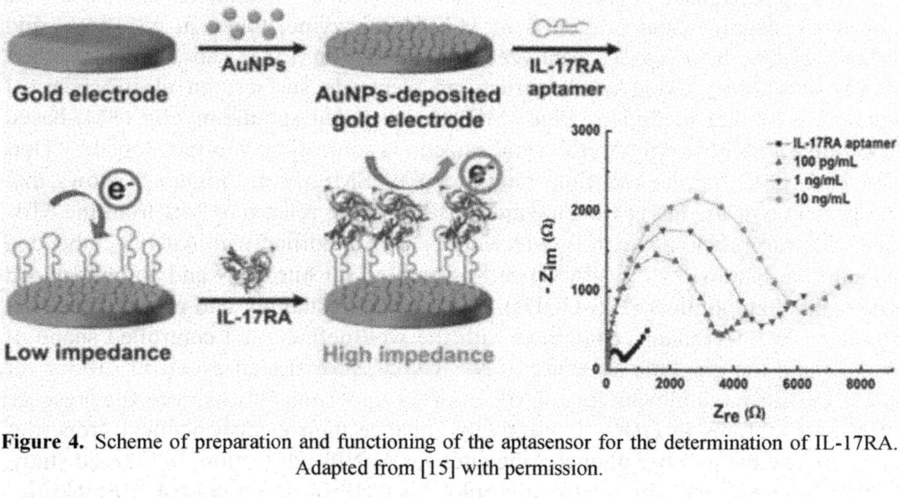

Figure 4. Scheme of preparation and functioning of the aptasensor for the determination of IL-17RA. Adapted from [15] with permission.

2.2 Multiple sclerosis and related diseases

Multiple sclerosis (MS) is an autoimmune disease of the central nervous system (CNS), characterized by progressive demyelination of neural axons. The disease causes slow damage to the nervous system, causing inhibited muscle coordination, loss of visual sense, and other effects [16]. Among the recognized biomarkers, the specific antibody of myelin basic protein (anti-MBP) is commonly used for MS monitoring. Derkus et al. [17] combined alginate (a polysaccharide forming hydrogels with high porosity) and TiO_2 nanoparticles (characterized by a high surface area and electron transfer rate promoting properties) for the fabrication of a biointerface acting as an efficient matrix for immobilization of MBP and its further interaction with the anti-MBP antibodies from the samples. Platinum electrodes modified with this biocomposite (MBP-alg/TiO_2) were prepared and the resulting label-free immunosensor was able to determine anti-MBP antibodies levels, with an LOD value of 0.18 ng mL^{-1}, by impedimetric measurements using ferro-ferricyanide as the redox probe. In addition, anti-MBP was determined in real samples of cerebrospinal fluid (CSF) and serum of MS patients. More recently, the same authors prepared a nanoimmunosensor for the simultaneous quantification of MPB and Tau proteins [18]. Both biomolecules are recognized as clinical biomarkers of MS, although the latter is also involved in other neurological disorders such Alzheimer's disease [19]. For the preparation of the immunosensor, the respective antibodies were immobilized onto SPCEs modified with GO and first-generation dendrimers of trimethylolpropane tris[poly(propyleneglycol)] (pPG) functionalized with amide groups. Differential pulse voltammetry (DPV) was used for the determination of proteins using pPG/CdS/anti-MBP and pPG/PbS/anti-Tau electrochemical signal probes in which the presence of CdS and PbS nanocrystals resulted in two distinct peaks—at –0.6 and –0.8 V vs. Ag/AgCl. The LOD values were 0.30 nM for MBP and 0.15 nM for Tau protein, which was sufficient to analyse the CSF and serum of MS patients.

Sphingomyelinase (SMEnzyme) has emerged as a reliable biomarker for improved diagnosis and prognosis of MS. This enzyme, found in lysosomes and in the extracellular space, catalyzes sphingomyelin (SM), an important class of phospholipids. Using a nanostructured electrode surface, an electrochemical method involving methylene blue (MB)-encapsulated sphingomyelin (SM)-based liposomes with 50% cholesterol, synthesized via sonication, was developed by Dutt and Park [20], for the real-time monitoring of SMEnzyme. Figure 5 shows that the presence of the target SMEnzyme causes the free release of MB from the MB-liposome formulation, which is detected at a GCE modified with AuNPs-embedded polyaniline nanowires (Au-PANI) self-assembled on nitrogen- and sulphur-doped graphene quantum dots (N,S-GQDs). The long-term stability and nanodimensional structure of PANI chain, combined with the well-defined and controlled shape of AuNPs and the graphitic structure of N,S-GQDs, provides an excellent surface for capturing the maximum amount of MB from the liposome. Furthermore, the presence of N and S atoms in GDQs respectively enhance their electrochemical properties and increase the number of anchoring sites for AuNPs adsorption. In the said study, DPV was used to obtain a calibration plot for SMEnzyme in the 0.1–10 mU mL^{-1} range with an LOD value of 0.0072 mU mL^{-1}. The developed biosensor was used for detecting SMEnzyme in spiked human serum and plasma and in cell culture supernatant [20].

Other diseases related to MS, which involve inflammatory symptoms of the central nervous system should also be duly considered. Among these, neuromyelitis optica (NMO), also known as Devic's disease, causes blindness and paralysis and is commonly confused with MS due to the similar symptoms. However, NMO is characterized by the presence of autoantibodies against the aquaporin-4 protein (AQP4)—recognized as specific biomarkers by [21]. Son et al. reported a CNTs-field-effect transistor (CNTs-FET) with immobilized AQP4 extracellular loop peptides

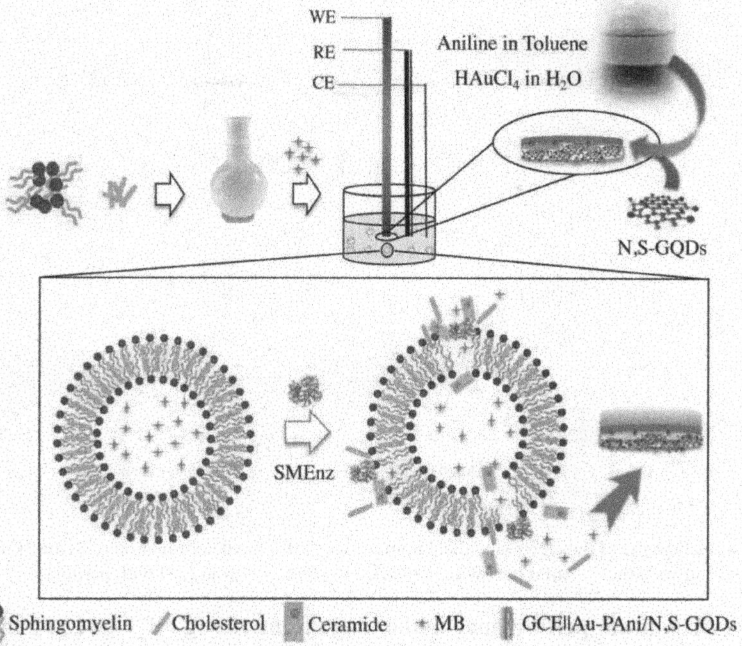

Figure 5. Preparation of methylene blue-encapsulated sphingomyelin liposome (MB-liposome) and fundamentals of detection of SMEnzyme at a GCE modified with Au-PANI and N,S-GQDs. Reproduced from [20] with permission.

for the rapid detection of the concerned antibodies in human serum without any sample treatment [22]. As is known, the unique one-dimensional nanoscale structure of CNTs and their large surface-to-volume ratio allow their electrical properties to be easily changed by surface adsorption of biomolecules. After the synthesis of peptides, phenylalanine (Phe) was added to the C-terminus to immobilize them onto CNTs via π-π interactions with the aromatic rings. The constructed biosensor showed p-type FET characteristics and was able to detect AQP4 antibodies present in human serum through a rapid decrease in conductance, with an LOD value of 1 ng L^{-1} [22].

Semaphorins and their receptors (plexins) are soluble cell-surface proteins that contribute to the process of morphogenesis in multiple organs [23]. A member of this family, Semaphorin 3E (Sema 3E), plays a crucial role in immune regulation, among other actions. The serum level of soluble Sema 3E shows high correlation with immunological diseases like systemic sclerosis [24]. Therefore, Sema 3E is considered as a potential biomarker for these disorders. Yuan et al. proposed a metal nanowire-based biosensor for the quantification of Sema 3E levels [25]. To achieve this, trimetallic CuAuPd nanowire networks (NNWs) with high specific surface area, good electron transport and catalytic properties and biocompatibility were synthesized and used as a multifunctional substrate for electron transfer, antibody immobilization and signal amplification via amperometric detection of H_2O_2

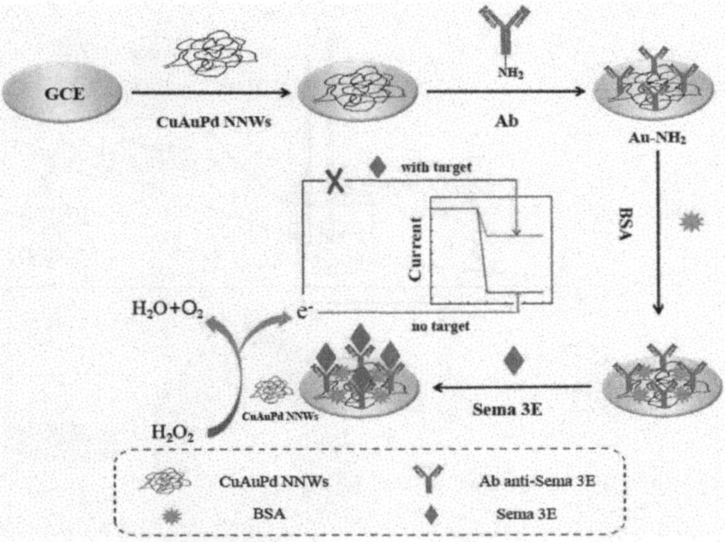

Figure 6. Schematic of the electrochemical immunosensor for the determination of Sema 3E levels using trimetallic CuAuPd nanowire networks. Reproduced from [25] with permission.

(Figure 6). A linear performance in the range from 100 fg mL^{-1} to 10 ng mL^{-1}, with a low LOD of approximately 1.5 fg mL^{-1}, was achieved.

2.3 Celiac disease

As a common knowledge, celiac disease (CD) is a gluten-dependent autoimmune disorder affecting a significant percentage of the population. One of the effects of this disease is the production of autoantibodies that cause the destruction of the intestinal mucosa [26]. At the same time, an essential route for non-invasive detection of the disease is the identification and quantification of antibodies produced upon gluten exposure, appearing in blood [27]. Among these, anti-gliadin antibody (AGA) and anti-tissue transglutaminase (anti-tTG) are the most frequently detected targets. Reliable analytical methods for the serological diagnosis of the disorder are still required for performing both an early diagnosis and the follow-up of a patient adhering to a gluten-free diet. For this purpose, Habtamu et al. prepared an amperometric immunosensor for the diagnosis of CD, using MWCNTs/SPCEs for covalent linking with the antigen (gliadin) and determination of the concentration of AGA by measuring current changes, using cyclic voltammetry with ferricyanide as the redox probe [28]. The antibody could be detected over a 2.7 to 650 μg mL^{-1} concentration range. The LOD value attained with this method was 0.13 μg mL^{-1}. For anti-tTG detection, an electrochemical immunosensor was recently developed with the use of gold nanoelectrodes (NEEs) prepared through the deposition of gold within the pores of polycarbonate track-etched membranes [29]. NEEs were functionalized with tissue transglutaminase and their incubation in the samples resulted in the capture of anti-tTG. Quantification was accomplished by incubation with an HRP-labelled secondary antibody, followed by measuring the electrochemical response of H_2O_2 in the presence of hydroquinone as the redox mediator. The immunosensor resulted in an LOD of 1.8 ng mL^{-1} and was successfully deployed for the analysis of human serum. Giannetto et al. [30] developed an amperometric immunosensor for the detection of both IgA and IgG serotypes of anti-tTG in human serum, involving covalent immobilization of tissue transglutaminase enzyme in its open conformation (open-tTG), which seems to be more specifically involved in the pathogenesis of CD [31]. A GCE was electrochemically functionalized with AuNPs and subsequently derivatized with an 11-mercaptoundecanoic acid SAM for covalent anchoring of the enzyme. This step was performed under carefully controlled conditions to keep the open conformation of the tTG. The immunosensor showed LOD values of 1.7 AUmL^{-1} for IgA and 2.7 AU mL^{-1} for Ig and was validated in sera from pediatric patients, against the results obtained with two specific ELISA kits.

2.4 Type 1 Diabetes

Type 1 Diabetes (T1D) is a chronic immune disorder that results from the destruction of β-cell function in the islets of Langerhans, resulting in deficient insulin levels which in turn raise blood glucose levels due to impaired glucose metabolism (hyperglycemia). The etiology of T1D is largely unknown, but a combination of genetic predisposition, environmental factors and dysregulated immune system is believed to be the cause of the disorder [32]. The action of various autoantibodies such as GADA (glutamic acid decarboxylase 65 autoantibody) has been reported as triggering factors for type 1 diabetes (T1D) [33] and the serum concentration of GADA has been recognized as a highly valuable biomarker for the prediction of T1D. Premaratne et al. developed an electrochemical immunosensor for the detection of GADA by modifying a gold electrode array with carboxylated graphene, followed by covalent attachment of GAD-65 antigen and complexation with GADA standard solutions; alternatively, the sample was carried onto protein A/G-functionalized magnetic microparticles (MAG) (Figure 7) [34]. A dynamic range obtained between 0.02 and 2 ng mL^{-1} and an LOD of 48 pg mL^{-1} were obtained by impedimetric measurements with ferro-ferricyanide redox probe. In the same experimental conditions, DPV decreased with increasing GADA concentration, providing a lower LOD value than that achieved with impedimetric transduction (34 pg mL^{-1}). In both the cases, these results were better than those obtained by surface plasmon resonance (SPR) using an immunoassay scheme similar to that used in the electrochemical approach. Real serum from patients with T1D and spiked serum samples were analyzed, with results in coherence with those provided by an ELISA kit [34].

2.5 Graves' disease and other autoimmune disorders

The immune system of Grave's disease (GD) patients produces antibodies which stimulate the thyroid gland to release excess amounts of thyroid hormone into the blood (hyperthyroidism). Production of substantial CXCL10 chemokines under interferon-γ (IFN-γ)-induced stimulation has been observed in such cases and, therefore, IFN-γ detection is of high significance for the diagnosis of GD. Various electrochemical biosensors involving nanomaterials have been constructed for

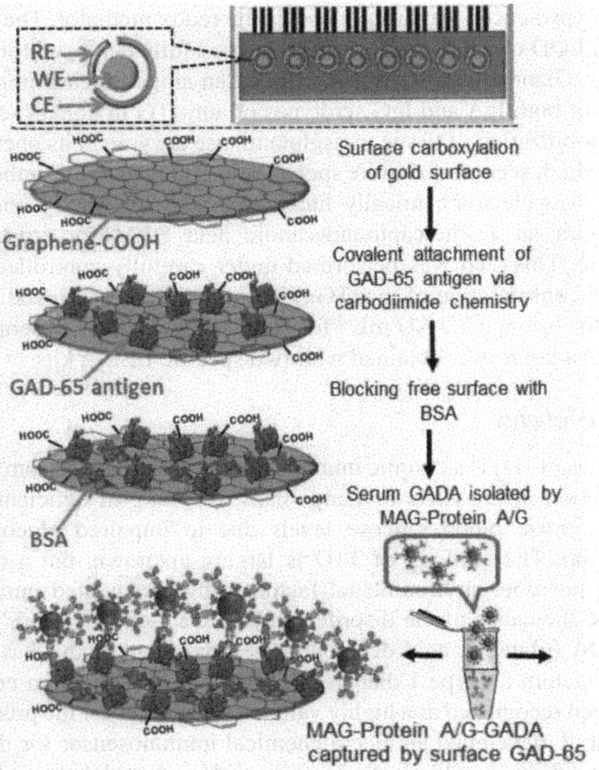

Figure 7. Scheme of preparation of MAG-Protein-A/G-GADA electrochemical immunosensor. Reprinted from [34] with permission.

this purpose. A relevant example here is a label-free immunosensor involving the immobilization of anti-IFN-γ on a TiO_2-modified electrode and the measurement of impedimetric changes after loading of proteins. A dynamic range obtained between 0.0001 and 0.1 ng mL^{-1}, and an LOD of 0.74 pg mL^{-1} IFN-γ were obtained [35].

IFN-γ is also involved in other autoimmune diseases (e.g., psoriasis). In this context, Bao et al. reported an electrochemical biosensor providing a low LOD of 0.6 fM IFN-γ [36]. The biosensor used a zeolitic imidazolate framework-8 (ZIF-8) as the immobilization substrate. This nanomaterial shows advantages such as ease of synthesis, large surface area and pore size, good chemical stability and negligible cytotoxicity. However, its low conductivity hinders electron-transfer, thus affecting the biosensor sensitivity negatively. To minimize this drawback, the authors prepared graphene@ZIF-8 hybrids with anchored gold nanoclusters (AuNCs), improving both conductivity and biocompatibility. Figure 8 shows that once AuNCs-GR@ZIF-8 was prepared and hemin/G-quadruplex DNAzyme decorated, an improved and efficient layered-branched hybridization chain reaction (LB-HCR) pattern was designed with the cascade-like assembly of four hairpins. HP1 and HP2 participated in the traditional HCR process to form long double-helix DNA, and AD1 and AD2 assembled alternately to facilitate the layer chain-branching growth of DNA

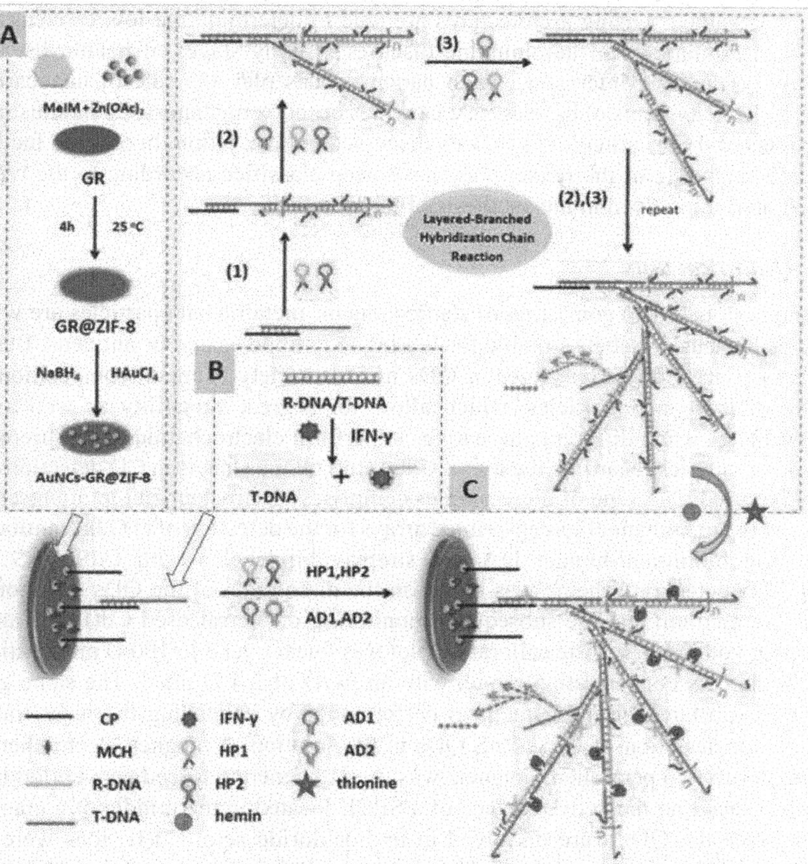

Figure 8. Schematic illustration of multiple-amplified electrochemical biosensor for the detection of IFN-γ: (A) synthesis of AuNCs-GR@ZIF-8, (B) target-induced release of T-DNA, and (C) cascade-like DNA assembly in LB-HCR process on the electrode. Reprinted from [36] with permission.

nanostructures. The target-triggered LB-HCR resulted in the generation of dendritic DNA nanostructures integrated with numerous *in-situ* formed hemin/G quadruplex DNAzyme as amplifying labels, which catalyzed the reduction of H_2O_2 with thionine. This multiple-amplified strategy exhibited a wide linear range from 1 fM to 50 pM IFN-γ.

3. Nanomaterials used as detection tags

Methodologies replacing conventional enzymatic labels with electrochemical tags using nanomaterials have advantages such as reducing the time required for the enzymatic reaction, or facilitating multiplexed detection. Moreover, these nanolabels are often adequate for increasing the sensitivity of detection, generating amplified electrochemical signals that endow the developed configuration with the analytical ability for application to complex samples which generally require sensitive methods.

The so designed sensing architectures are able to determine the low concentration levels of biomarkers of autoimmune diseases in early stages of health disorders. This is particularly interesting when nanomaterials play the role of nanocarriers, loading large amounts of bioreagents, enzymes or redox mediators. Some illustrative examples of these strategies have been discussed in this section. In order to facilitate the understanding of the reader, they have been classified according to the type of nanomaterial involved in the synthesis of the nanotag.

3.1 Quantum dots

Quantum dots (QDs) consisting of semiconductor metallic nanoparticles are widely used as biosensing labels suitable for linking to antibodies and/or antigens. Besides electroactivity, other advantages of QDs include variety of metal composition and sizing of these nanoparticles, which allow for a great versatility as well as the possibility of using different alternatives to perform electrochemical measurements. Various examples regarding the use of CdSe/ZnS QDs as labels in CD detection have been reported [37]. One of the examples comprises electrochemical immunosensing strategy using 8-channel screen-printed arrays for the detection of tTG IgA antibodies involving biotin-anti-human IgA and streptavidin labeled with CdSe/ZnS QDs [38]. This detection scheme was based on the dissolution of the QDs' components in acidic medium and the subsequent monitoring of the released Cd(II) by anodic stripping voltammetry. The calibration plot was linear over 3 to 100 U mL^{-1} anti-tTG IgA antibodies concentration range, with an LOD of 2.4 U mL^{-1}. The same group reported substantial improvement in performance by enhancing the mass transfer of ions for detection of CdSe/ZnS QDs in the solution. A magnetoelectrochemical support with two permanent magnets was used to generate forced convention in the droplet placed on the active surface of a SPCE located in the middle of a magnetic field. Once the QDs were dissolved in hydrochloride acid, Cd(II) ions were pre-concentrated on the electrode surface from a solution also containing Hg(II) and Fe(III), by applying a cathodic potential at which a mercury film is formed on the electrode and ferric ions are reduced simultaneously with cadmium deposition. In the presence of the magnetic field, the deposition efficiency was enhanced with Fe(III) acting as a pumping species generating a high cathodic current and a large Lorentz force that caused a convective effect in the solution. The application of this method for the determination of anti-tTG (IgA) levels provided a linear relationship with a slope value of 0.74 μA U^{-1} mL, and an LOD value of 1.7 U mL^{-1} [39].

Graphene quantum dots (GQDs) have also been used to improve the sensitivity of anti-tTG electrochemical biosensors. GQDs are composed of small graphene sheets with lateral dimensions smaller than 10 nm and less than 10 such graphene layers forming the nanoparticle [40]. They are characterized by a high surface area with a large length-to-diameter ratio (the ratio of length to thickness), low intrinsic toxicity, chemical inertness, mechanical stiffness, excellent solubility, high stability, easier grafting of their surface with receptors, and greater electrical conductivity than conventional semiconductor QDs [41]. Researchers have modified different substrates with GQDs for increasing the rate of electrochemical reactions. For instance, Gupta et al. combined GQDs and poly(amidoamine) (PAMAM) dendrimers to prepare a biosensor for the detection of anti-tTG IgA in human serum [42]. Figure 9 shows screen-printed electrodes modified with AuNPs and MWCNTs, used for the covalent attachment of GQDs and PAMAM. GQDs are characterized by high surface area to volume ratio and good conductivity, whereas PAMAM dendrimers possess a branched structure containing many amino groups for conjugation. Therefore, once tissue transglutaminase (tTG) antigen is immobilized, the binding with the specific anti-tTG antibody provides significant electrochemical changes measured by DPV, with a sensitivity of 1297.14 A cm^{-2} pg^{-1} and an LOD of 0.1 fg per 6 μL.

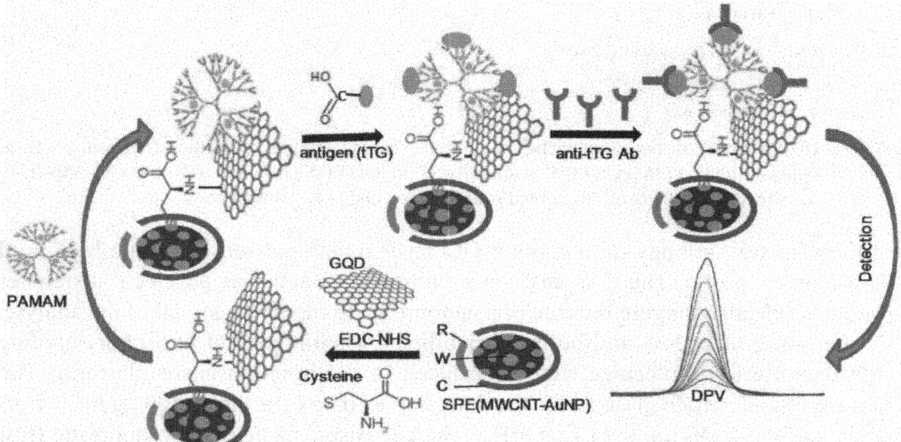

Figure 9. Scheme of preparation and fundamentals of the tTG-PAMAM-GQD-AuNPs/MWCNTs/SPCE biosensor for the detection of anti-tTG antibody. Reproduced from [42] with permission.

3.2 CNTs and graphene

CNTs and graphene or their derivatives have also been employed as nanocarriers for the detection of interleukins through immunoassay strategies. Wang et al. designed an electrochemical immunosensor for IL-6, using AuNPs, rGO and silica sol–gel (tetraethyl orthosilicate, TEOS) prepared *in situ* on an ITO electrode as the immobilization platform for anti-IL-6 capture antibody, and AuNPs-polydopamine (PDA)@CNTs as the carrier tag for immobilizing HRP-labeled detection antibodies (Figure 10A) [43]. The combination of the said three components—AuNPs, PDA and CNTs—in the nanolabel provided amperometric responses of magnitudes larger than any of the possible double combinations, probably due to the high electron transfer capability of CNTs and the higher surface-to-volume ratio of the resulting composite in the presence of PDA, which in turn yields a larger amount of immobilized HRP-antibodies. A dynamic working range of 1–40 pg mL^{-1} with a low LOD of 0.3 pg mL^{-1} IL-6 was achieved. Also, for the detection of IL-6, another immunosensor has been reported; it involves gold electrodes modified with a single layer of GO, followed by attachment of the capture antibody (Figure 10B) [44]. In this case, after a blocking step, a sandwich-type configuration was implemented by covalent immobilization

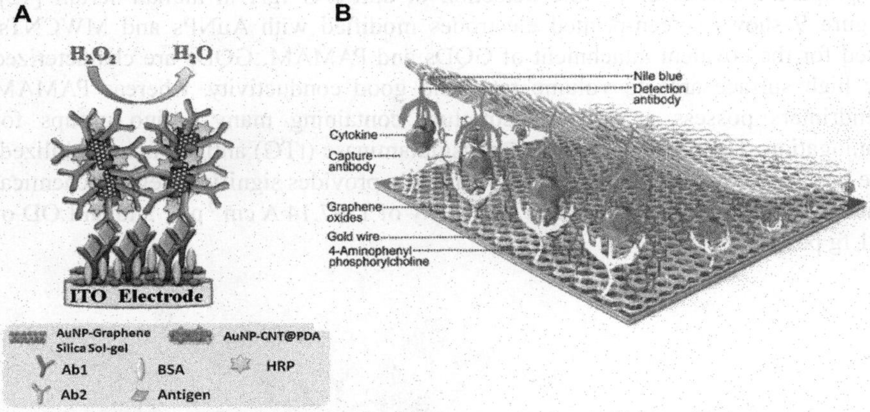

Figure 10. Schemes of two sandwich-type immunosensors prepared for the detection of IL-6: (A) HRP-dAb-(AuNPs-PDA@CNTs)-IL-6-cAb-AuNPs/rGO/TEOS/ITO, and (B) NB-GO-dAb-IL-6-cAb-GO/PPC/AuE. Reprinted from [43] (A) and [44] (B) with permission.

of the detector antibody onto another GO layer functionalized with Nile blue used as the redox probe. Thus, in this configuration, GO acted as an electron transfer bridge, while also playing the role of a nanoprobe, to report the signal of the analyte. Furthermore, to endow antifouling capabilities, 4-aminophenyl phosphorylcholine (PPC) zwitterion molecules were introduced in the immunosensor platform. The prepared bioelectrode allowed for the detection of IL-6 over the 1−300 pg mL^{-1} IL-6 range, with an LOD value of 1 pg mL^{-1}. The biosensor was used for monitoring IL-6 secretion in raw cells and in live mice [44].

3.3 Polymer/metal core-shell nanocomposites

Polymer/metal core-shell nanocomposites prepared with polystyrene (PS) and polydopamine (PDA) were proposed by Shi et al. for preparing metal labels for the simultaneous detection of IL-6 and MMP-9 (matrix metalloproteinase-9) [45]. MMP-9 is an enzyme involved in the pathological processes occurring in autoimmune diseases—such as degradation of collagen type 4, which is a major component of the basement membrane, thus enabling the invasion of T cells. Like IL-6, the detection of MMP-9 is of interest in the diagnosis and monitoring of autoimmune diseases such as RA, MS, systemic sclerosis (SSc), or systemic lupus erythematosus (SLE), among others [46]. In the said study, PS@PDA modified with AgNPs or Cd^{2+} ions was used as the carrier after immobilization of detected antibodies of IL-6 and MMP-9 respectively, whereas the corresponding anti-IL-6 and anti-MM-9 capture antibodies were immobilized onto a graphene nanoribbon (GNR)-modified heated screen-printed carbon electrode (HSPCE). The electrochemical detection was performed by square wave voltammetry (SWV) recording the stripping responses appearing at +0.1 V and –0.7 vs Ag/AgCl, which corresponded respectively with AgNPs oxidation and Cd^{2+} reduction. The developed method aided the detection of IL-6 and MMP-9 in concentration ranges from 10^{-3} to 10^{3} ng mL^{-1} and

10^{-5} to 10^{3} ng mL^{-1} respectively, with LODs of 0.1 pg mL^{-1} and 5 fg mL^{-1} respectively. This method was successfully applied to the analysis of serum samples from patients.

4 Conclusions and perspectives

Although in a few applications till date, electrochemical biosensors involving the use of nanomaterials have demonstrated their usefulness in the detection of biomarkers of autoimmune diseases, against more sophisticated and expensive techniques such as mass spectrometry, or less accurate and often time-consuming ELISA methodologies. The combination of electrochemical techniques with nanomaterials such as CNTs or AuNPs endows the biosensors with well-known special features, such as improved sensitivity and selectivity, simplicity and multiplexing capabilities, and point-of-care application. Moreover, other nanomaterials such as metal oxides, bimetallic nanoparticles, or frameworks provide benefits which are yet to be exploited to develop biosensing methods with better performance in terms of sensitivity and accuracy, while also offering greater stability and better selectivity.

Versatility is a particularly noteworthy feature of electrochemical biosensors for their applications in the domain of disease biomarkers detection. As noted in the works cited above, different techniques, electrolytic or otherwise, can be used for the said purpose. These techniques include amperometry, voltammetry and electrochemical impedance spectroscopy, all of which are very easy in terms of implementation; another technique mentioned involves stripping voltammetry, a highly sensitive and practically specific technique when using optimal experimental conditions. Likewise, electrochemical sensing platforms can be developed with bioreagents of different natures, whether antibodies or antigens, DNA chains or peptides, depending on the availability associated with the type of analyte.

Considering these aspects, it is expected that new nanomaterials would be discovered in the forthcoming years, making it possible to improve the analytical characteristics of detection aimed at ultra-sensitivity in order to detect biomarkers in even earlier stages of an illness. In addition, it is necessary to discover new biomolecules capable of interacting with biomarkers that do not yet have specific reagents, such as autoantibodies involved in several autoimmune diseases, and it should be extended to research in the design of biosensors based on new peptides and proteins, their metabolites, and nucleic acids, specifically micro-RNA, etc.

It is important to understand that the benefits achieved must be demonstrated by applying the implemented methods to clinical samples obtained from patients suffering from different types of diseases. The collaboration between technical and medical research will hence be crucial in the development of electrochemical biosensors for the detection of biomarkers in near future.

Acknowledgments

The financial support from PID2019-103899RB-I00 (Spanish Ministerio de Ciencia e Innovación) and RTI2018-096135-B-I00 (Ministerio de Ciencia, Innovación y

Universidades) Research Projects and the TRANSNANOAVANSENS-CM Program from Comunidad de Madrid (Grant S2018/NMT-4349) is gratefully acknowledged.

References

[1] Smolen, J.S., D. Aletaha, A. Barton, G.R. Burmester, P. Emery, G.S. Firestein, A. Kavanaugh, I.B. McInnes, D.H. Solomon, V. Strand and K. Yamamoto. 2018. Rheumatoid arthritis. Nat. Rev. Dis. Primers 4: 18001. Doi: 10.1038/nrdp.2018.1.

[2] Sweet, R.A., J.L. Cullen and M.J. Shlomchik. 2013. Rheumatoid factor B cell memory leads to rapid, switched antibody-forming cell responses. J. Immunol. 190: 1974–1981.

[3] Chinnadayyala, S.R., J. Park, M.A. Abbasi and S. Cho. 2019. Label-free electrochemical impedimetric immunosensor for sensitive detection of IgM rheumatoid factor in human serum. Biosens. Bioelectron. 143: 111642. Doi: 10.1016/j.bios.2019.111642.

[4] Pruijn, G.J., A. Wiik and W.J. van Venrooij. 2010. The use of citrullinated peptides and proteins for the diagnosis of rheumatoid arthritis. Arthritis Res. Ther. 12: 203. Doi: 10.1186/ar2903.

[5] Rossi, G., F. Real-Fernández, F. Panza, F. Barbetti, F. Pratesi, P. Rovero and P. Migliorini. 2014. Biosensor analysis of anti-citrullinated protein/peptide antibody affinity. Anal. Biochem. 465: 96–101.

[6] Villa, M.G., C. Jiménez-Jorquera, I. Haro, M.J. Gomara, R. Sanmartí, C. Fernández-Sánchez and E. Mendoza. 2011. Carbon nanotube composite peptide-based biosensors as putative diagnostic tools for rheumatoid arthritis. Biosens. Bioelectron. 27: 113–118.

[7] Lu, M.C., N.S. Lai, H.C. Yu, H.B. Huang, S.C. Hsieh and C.L. Yu. 2010. Anti–citrullinated protein antibodies bind surface-expressed citrullinated Grp78 on monocyte/macrophages and stimulate tumor necrosis factor α production. Arthritis Rheum. 62: 1213–1223.

[8] Yagati, A.K., M.-H. Lee and J. Min. 2018. Electrochemical immunosensor for highly sensitive and quantitative detection of tumor necrosis factor-α in human serum. Bioelectrochem. 122: 93–102.

[9] Mazloum-Ardakani, M., L. Hosseinzadeh and A. Khoshroo. 2015. Label-free electrochemical immunosensor for detection of tumor necrosis factor α based on fullerene-functionalized carbon nanotubes/ionic liquid. J. Electroanal. Chem. 757: 58–64.

[10] Chen, H., T.K. Choo, J. Huang, Y. Wang, Y. Liu, M. Platt, A. Palaniappan, B. Liedberg and A.I.Y. Tok. 2016. Label-free electronic detection of interleukin-6 using horizontally aligned carbon nanotubes. Mater. Des. 90: 852–857.

[11] Liu, N., Y. Huan, L. Yanyu, H. Zheng, X. Zheng, D. Lin and H. Dai. 2018. Combined electrochemiluminescent and electrochemical immunoassay for interleukin 6 based on the use of TiO_2 mesocrystal nanoarchitectures. Microchim. Acta 185: 277. Doi: 10.1007/s00604-018-2802-x.

[12] Lou, Y., T. He, F. Jiang, J.-J. Shi and J.-J. Zhu. 2014. A competitive electrochemical immunosensor for the detection of human interleukin-6 based on the electrically heated carbon electrode and silver nanoparticles functionalized labels. Talanta 122: 135–139.

[13] Kirkham, B.W., M.N. Lassere, J.P. Edmonds, K.M. Juhasz, P.A. Bird, C. Soon Lee, R. Shnier and I.J. Portek. 2006. Synovial membrane cytokine expression is predictive of joint damage progression in rheumatoid arthritis: A two-year prospective study (the DAMAGE study cohort). Arthritis Rheum. 54: 1122–1131.

[14] Graber, J.J., S.R. Allie, K.M. Mullen, M.V. Jones, T. Wang, C. Krishnan, A.I. Kaplin, A. Nath, D.A. Kerr and P.A. Calabresi. 2008. Interleukin-17 in transverse myelitis and multiple sclerosis. J. Neuroimmunol. 196: 124–132.

[15] Jo, H., S-K. Kim, H. Youn, H. Lee, K. Lee, J. Jeong, J. Mok, S.-H. Kim, H.-S. Park and C. Ban. 2016. A highly sensitive and selective impedimetric aptasensor for interleukin-17 receptor A. Biosens. Bioelectron. 81: 80–86.

[16] Bhavsar, K., A. Fairchild, E. Alonas, D.K. Bishop, J.T. La Belle, J. Sweeney, T.L. Alford and L. Joshi. 2009. A cytokine immunosensor for multiple sclerosis detection based upon label-free electrochemical impedance spectroscopy using electroplated printed circuit board electrodes. Biosens. Bioelectron. 25: 506–509.

[17] Derkus, B., E. Emregul, K.C. Emregul and C. Yucesan. 2014. Alginate and alginate-titanium dioxide nanocomposite as electrodematerials for anti-myelin basic protein immunosensing. Sens. Actuators B 192: 294−302.

[18] Derkus, B., P.A. Bozkurt, M. Tulu, K.C. Emregul, C. Yucesan and E. Emregul. 2017. Simultaneous quantification of myelin basic protein and Tau proteins in cerebrospinal fluid and serum of multiple sclerosis patients using nanoimmunosensor. Biosens. Bioelectron. 89: 781−788.

[19] Barthélemy, N.R., F. Fenaille, C. Hirtz, N. Sergeant, S. Schraen-Maschke, J. Vialaret, L. Buée, A. Gabelle, C. Junot, S. Lehmann and F. Becher. 2016. Tau protein quantification in human cerebrospinal fluid by targeted mass spectrometry at high sequence coverage provides insights into its primary structure heterogeneity. J. Proteome Res. 15: 667−676.

[20] Dutt, C.A. and E.Y. Park. 2019. Methylene blue-encapsulated liposomal biosensor for electrochemical detection of sphingomyelinase enzyme. Sens. Actuators B 301: 127153. Doi: 10.1016/j.snb.2019.127153.

[21] Kitley, J., P. Waters, M. Woodhall, M.I. Leite, A. Murchison, J. George, W. Küker, S. Chandratre, A. Vincent and J. Palace. 2014. Neuromyelitis optica spectrum disorders with aquaporin-4 and myelin-oligodendrocyte glycoprotein antibodies: A comparative study. JAMA Neurol. 71(3): 276–283.

[22] Son, M., D. Kim, K. Seok, S. Hong and T.H. Park. 2016. Detection of aquaporin-4 antibody using aquaporin-4 extracellular loop-based carbon nanotube biosensor for the diagnosis of neuromyelitis optica. Biosens. Bioelectron. 8: 87−91.

[23] Sakurai, A., J. Gavard, Y. Annas-Linhares, J.R. Basile, P. Amornphimoltham, T.R. Palmby, H. Yagi, F. Zhang, P.A. Randazzo, X. Li, R. Weigert and J.S. Gutkind. 2010. Semaphorin 3E initiates antiangiogenic signaling through plexin D1 by regulating Arf6 and R-Ras. Mol. Cell Biol. 30: 3086−3098.

[24] Mazzotta, C., E. Romano, C. Bruni, M. Manetti, G. Lepri, S. Bellando-Randone, J. Blagojevic, L. Ibba-Manneschi, M. Matucci-Cerinic and S. Guiducci. 2015. Plexin-D1/Semaphorin 3E pathway may contribute to dysregulation of vascular tone control and defective angiogenesis in systemic sclerosis. Arthritis Res Ther. 21: 221. Doi: 10.1186/s13075-015-0749-4.

[25] Yuan, Z., J. Chen, Y. Wen, C. Zhang, Y. Zhou, Z. Yang and C. Yu. 2019. A trimetallic CuAuPd nanowire as a multifunctional nanocomposites applied to ultrasensitive electrochemical detection of Sema3E. Biosens. Bioelectron. 145: 111677. Doi: 10.1016/j.bios.2019.111677.

[26] Lindfors, K., K. Kaukinen and M. Mäki. 2009. A role for anti-transglutaminase 2 autoantibodies in the pathogenesis of coeliac disease? Amino Acids 36: 685–691.

[27] Pasinszki, T. and M. Krebsz. 2018. Biosensors for non-invasive detection of celiac disease biomarkers in body fluids. Biosensors 8: 55. Doi: 10.3390/bios8020055.

[28] Gupta, S., A. Kaushal, A. Kumar and D. Kumar. 2016. Multiwalled carbon nanotubes based immunosensor for diagnosis of celiac disease. Mol. Cell Biol. 62: 3.

[29] Habtamu, H.B., T. Not, L. De Leo, S. Longo, L.M. Moretto and P. Ugo. 2019. Electrochemical immunosensor based on nanoelectrode ensembles for the serological analysis of IgG-type tissue transglutaminase. Sensors 19: 1233. Doi: 10.3390/s19051233.

[30] Giannetto, M., M. Mattarozzi, E. Umiltà, A. Manfredi, S. Quaglia and M. Careri. 2014. An amperometric immunosensor for diagnosis of celiac disease based on covalent immobilization of open conformation tissue transglutaminase for determination of anti-tTG antibodies in human serum. Biosens. Bioelectron. 62: 325–330.

[31] Leffler, D.A., K. Pallav, M. Bennett, S. Tariq, H. Xu and T. Kabbani. 2012. Open conformation tissue transglutaminase testing for celiac dietary assessment. Dig. Liver Dis. 44: 375–378.

[32] Ziegler, A.-G. and G.T. Nepom. 2010. Prediction and pathogenesis in type 1 diabetes. Immunity 32: 468–478.

[33] Pociot, F. and Å. Lernmark. 2016. Genetic risk factors for type 1 diabetes. Lancet 387: 2331−2339.

[34] Premaratne, S.G., J. Niroula, M.K. Patel, W. Zhong, S.L. Suib, A.K. Kalkan and S. Krishnan. 2018. Electrochemical and surface-plasmon correlation of a serum autoantibody immunoassay with binding insights: Graphenyl surface versus mercapto-monolayer. Anal. Chem. 90: 12456−12463.

[35] Chu, Z., H. Dai, Y. Liu and Y. Lin. 2017. Development of a semiconductor-based electrochemical sensor for interferon-γ detection. Int. J. Electrochem. Sci. 12: 9141−9149.

[36] Bao, T., M. Wen, W. Wen, X. Zhang and S. Wang. 2019. Ultrasensitive electrochemical biosensor of interferon-gamma based on gold nanoclusters-graphene@zeolitic imidazolate framework-8 and layered-branched hybridization chain reaction. Sens. Actuators B. 296: 1266062. 10.1016/j.snb.2019.05.083.

[37] Scherf, K.A., R. Ciccocioppo, M. Pohanka, K. Rimarova, R. Opatrilova, L. Rodrigo and P. Kruzliak. 2016. Biosensors for the diagnosis of celiac disease: Current status and future perspectives. Mol. Biotechnol. 58: 381–392.

[38] Martín-Yerga, D. and A. Costa-García. 2015. Towards a blocking-free electrochemical immunosensing strategy for anti-transglutaminase antibodies using screen-printed electrodes. Bioelectrochem. 105: 88–94.

[39] Martín-Yerga, D., P. Fanjul-Bolado, D. Hernández-Santos and A. Costa-García. 2017. Enhanced detection of quantum dots by the magnetohydrodynamic effect for electrochemical biosensing. Analyst 142: 1591–1600.

[40] Li, M., T. Chen, J.J. Gooding and J. Liu. 2019. Review of carbon and graphene quantum dots for sensing. ACS Sens. 4: 1732−1748.

[41] Campuzano, S., P. Yáñez-Sedeño and J.M. Pingarrón. 2019. Carbon dots and graphene quantum dots in electrochemical biosensing. Nanomaterials 9: 634. Doi: 10.3390/nano9040634.

[42] Gupta, S., A. Kaushala, A. Kumar and D. Kumar. 2017. Ultrasensitive transglutaminase based nanosensor for early detection of celiac disease in human. Int. J. Biol. Macromol. 105: 905−911.

[44] Qi, M., J. Huang, H. Wei, C. Cao, S. Feng, Q. Guo, E.M. Goldys, R. Li and G. Liu. 2017. Graphene oxide thin film with dual function integrated into a nanosandwich device for *in vivo* monitoring of interleukin-6. ACS Appl. Mater. Interfaces 9: 41659−41668.

[45] Shi, J.-J., T.-T. He, F. Jiang, E.S. Abdel-Halim and J.-J. Zhu. 2014. Ultrasensitive multi-analyte electrochemical immunoassay based on GNR-modified heated screen-printed carbon electrodes and PS@PDA-metal labels for rapid detection of MMP-9 and IL-6. Biosens. Bioelectron. 55: 51–56.

[46] Ram, M., Y. Sherer and Y. Shoenfeld. 2006. Matrix metalloproteinase-9 and autoimmune diseases. J. Clin. Immunol. 26: 299−307.

5

Immunosensors and Genosensors Based on Voltammetric Detection of Metal-Based Nanoprobes

*Anastasios Economou** and *Christos Kokkinos*

1. Introduction

Nowadays, there is a strong demand for on-site or point-of-care (POC) assays based on portable and disposable instrumentation for the detection or the determination of different analytes in samples of environmental, nutritional, pharmaceutical and clinical interest. The necessary requirements of such methodologies are operational and instrumental simplicity, rapidity, high specificity and sensitivity, low cost, small size, wide scope of applications and potential for multiplexed detection of different analytes.

Biosensors are fit-for-purpose (bio)analytical platforms that can fulfill these analytical and operational requirements and possess many attractive features [1]. They are based on the specific interaction between the target analyte and one or more biomolecular probe(s), the latter being biomolecules (such as enzymes, antibodies, oligonucleotides) that can selectively interact with the analyte; their interaction induces physical or chemical changes that can be monitored [2].

Affinity biosensors include immunosensors (which utilize an antibody specific to the target protein or cell as the biological recognition element) and genosensors (that use an oligonucleotide complementary to the target DNA or RNA fragment as the biomolecular recognition element) [3–5]. In affinity biosensors, either the target analyte or at least one of the probes is immobilized on a suitable transducer and the probe-analyte interaction is designed to generate a measurable signal.

Department of Chemistry, National and Kapodistrian University of Athens, Athens, 157 71 Greece.

Electrochemical biosensors use an electrochemical transducer to monitor the probe-analyte interactions. The advantages of electrochemical detection in biosensing include high specificity and sensitivity, scope for automation, miniaturization and multiplexing capabilities, combined with simple and portable instrumentation. So far, different types of electrochemical biosensors have been designed for clinical analysis [6, 7], food analysis and quality control [8, 9], environmental monitoring [10] and pathogen or toxin detection [11, 12]. In electrochemical biosensors, detection is performed by the transducer either directly (in a label-free format) [13, 14] or indirectly via a suitable label (in a labeled format) [15–21]. Although the use of labels increases the complexity of the analytical protocol, the assay time-scale and the overall cost, it provides enhanced detection sensitivity and selectivity and also enables multiplexed detection schemes.

Electrochemical labels are molecules, ions or atoms which possess redox properties and serve as "barcodes" for the indirect detection of the target biomolecule. The most widely used metal-based nanoparticles (NPs) in biosensing, serving as electrochemically active labels, are metal NPs (such as gold NPs (AuNPs) and silver NPs (AgNPs)) and quantum dots (QDs, nanocrystals composed of single or mixed metal salts) [16, 17, 21–27]. In voltammetric bioassays using metal-based NP labels, the target biomolecule is selectively bound with the recognition probe, which is conjugated with an appropriate NP label. After the affinity interaction, the metal-based nanoprobe is detected and quantified using a voltammetric technique.

AuNPs are the most widely used noble metal NPs, thanks to their stability, simple synthesis, biocompatibility and scope for effective conjugation with biomolecules; other metal NPs include AgNPs, copper NPs (CuNPs) and platinum NPs (PtNPs) [16, 26, 28, 29]. QDs are nanocrystals with sizes between 1 and 20 nm, composed of single or mixed metal salts (such as PbX, CdX, ZnX, where X=S, Se, Te) with interesting optical and electrochemical properties [30–32]. Several reviews on the synthesis of noble metal NPs and QDs have been published [32, 33]. The most common route for the synthesis of metal NPs is based on the reduction of the respective cations in the presence of a suitable reducing agent (such as sodium borohydride, D-glucose or sodium citrate) in a way that the size distribution of the resulting NPs is carefully controlled [24, 33]. The synthesis of QDs is based on different bottom-up or top-down methodologies, depending on the desired QDs type, properties and core material [24, 32].

A critical factor for effective utilization of noble metal NPs and QDs as redox labels is their surface modification with functional groups (such as amino and carboxyl groups) to enable further conjugation with specific biomolecules. Different NP functionalization schemes and conjugation strategies with biomolecules have been described in various relevant reviews [34, 35]. Immobilization of biomacromolecules on NPs can be achieved through physical adsorption using hydrophobic, electrostatic, hydrogen binding and van der Waals interactions. However, in this case, the adsorption process is poorly controlled. Moreover, covalent binding strategies are rather sophisticated and require several steps, including the derivatization of the NPs surface by introduction of different functional groups (e.g., –COOH and $-NH_2$). In the case of AuNPs, biomolecules can be attached to the gold surface through Au-SH bonds; the inherently strong interaction between thiol compounds (–SH, mercaptans or sulfydryl groups) and noble metal enables the formation of self-assembled monolayers (SAM) of a wide variety of thiol compounds on the surface of AuNPs. Primary amine groups are probably the most used functional group to link molecules covalently. NPs modified with carboxylic groups (–COOH) on their surface can react directly with primary amines existing on the surface of biomolecules after an activation step with 1-ethyl-3-(3-dimethylaminopropyl) carbodiimide (EDC) and N-hydroxysuccinimie (NHS). Attachment of molecules such as oligonucleotides or antibodies via amino groups is advantageous, for these biomolecules have many available reactive amino groups for conjugation, located at their surface. Furthermore, there are many examples involving linkage between biomolecules and NPs through an affinity protein-protein interaction, most commonly the biotin-streptavidin bond. The strategy of biomolecule binding via biotin-binding proteins takes advantage of the natural strong interaction between (strept)avidin and biotin (which is significantly resistant to a wide range of pH, elevated temperatures and extreme chemical conditions). In this case, NPs and biomolecule have to be modified with (strept)avidin and biotin respectively. Following different strategies (i e.g., EDC), (strep)tavidin attachment to NPs surfaces can be achieved by amino groups or carboxyl groups of the protein. In the case of biomolecule derivatization, biotin group can be introduced through amine groups of the biomolecule with biotin-NHS ester or biotin sulfo-NHS ester.

Metal-based NP labels can be detected voltammetrically using different methodologies. The two most widely used detection techniques are solution-phase voltammetry and anodic stripping voltammetry; the latter affords higher sensitivity thanks to the preconcentration deposition step that precedes the voltammetric scan [36]. AuNPs can be chemically oxidized in a Br_2/HBr medium to bromo-Au(III) complexes or electrochemically oxidized in HCl to chloro-Au(III) complexes. The released Au(III) is detected by cathodic voltammetry or, for higher sensitivity, by anodic stripping voltammetry after preconcentration (Figure 1(A)). A common strategy to increase the detection sensitivity is the catalytic or enzymatic growth of an Ag shell on AuNPs; after acidic dissolution of the Ag shell, Ag(I) is quantified by anodic stripping voltammetry (Figure 1(B)). Other metal NPs can be detected in the same way by direct voltammetry or stripping voltammetry. The detection of AuNPs and other metal NPs is commonly performed at carbon-based working electrodes.

On the other hand, QDs, which contain a Pb, Cd or Zn core, can be readily dissolved and chemically oxidized in an acidic medium (containing HCl or HNO_3) to the respective metal cations (Figure 1(C)); the released cations are detected by anodic stripping voltammetry, preferably at the Hg, Bi or Sn electrodes [36].

The synergistic effects of a high loading of NP moities at each recognition probe and the highly sensitive (stripping) voltammetric detection lead to ultrasensitive bioassays, sometimes reaching sub-picomolar limits of detection [22–24].

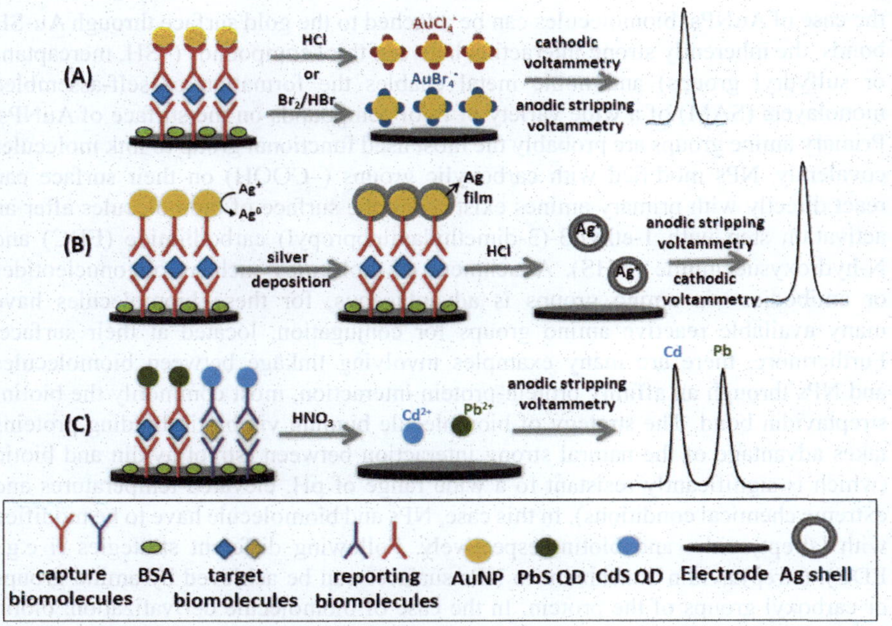

Figure 1. Biosensing based on voltammetric detection schemes for: (A) AuNPs, (B) AgNPs grown on AuNPs and (C) heavy metal QDs.

2. Immunosensors using metal-based nanoprobes

Typically, immunoassays follow either a sandwich or a competitive format [37]. In the case of the sandwich mode of operation, an immobilized capture antibody binds the target protein or cell in the sample, followed by an interaction of the target with a second reporting antibody labeled with a detectable tag. On the other hand, in the competitive mode, the unlabeled analyte competes either with the labeled analyte to bind an immobilized capture antibody or with the labeled reporting antibodies to bind the immobilized analyte.

So far, numerous interesting electrochemical immunosensing devices have been developed with metal NPs as labels. A competitive immunoassay has been developed for the determination of human serum albumin, using AuNPs as labels, with the detection of the AuNPs taking place voltammetrically at a 3-electrode screen-printed cell, following their electrochemical oxidation in HCl [38]. In recent years, carbon nanomaterials (e.g., graphene and carbon nanotubes) have been widely exploited for the fabrication of biosensors with enhanced performance. In one study, a sandwich immunoassay was described for the detection of human chorionic gonadotropin, using AuNPs as the label at a graphene-modified electrode [39]. After the immunoreaction, the AuNP labels were oxidized to $AuCl_4^-$ and quantified by voltammetry. In another study, a glassy carbon electrode was modified with graphene, single walled carbon nanotubes and AuNPs in a chitosan matrix [40]; this transducer formed the basis for a sandwich assay for the detection of prolactin, using AuNPs as signal tags immobilized on secondary antibodies and biotinylated DNA.

In addition to AuNPs, other metal NPs have also been reported as labels in voltammetric biosensing. An electrochemical immunosensor for the determination of antibodies to tick-borne encephalitis virus was developed by Khristunova et al. [41]. Covalent immobilization of the antigen was performed via thiolation and glutarization of the electrode's surface and detection was based on the silver reduction signal after dissolution of the Ag-antibodies bioconjugates. A sandwich immunosensor for the detection of *Staphylococcus aureus*, utilizing magnetic beads (MBs) and AgNPs, has also been reported in literature [42]. In this work, streptavidin-modified MBs were used for the immobilization of biotinylated primary anti-*S. aureus* aptamer, while a secondary anti-*S. aureus* aptamer was conjugated to the AgNPs used as labels. After magnetic separation and dissolution of AgNPs in HNO_3, the voltammetric detection of Ag(I) was carried out at a screen-printed carbon working electrode. AgNPs have also been used as labels for the detection of human IgG and IgE using 3-electrode screen-printing cells with bare carbon and AuNPs-modified working electrodes respectively [43, 44]. In both the cases, AgNPs were used as labels for the voltammetric determination of human IgGs, after their conjugation with the reporting antibody. Following the immunoassay, the AgNPs were dissolved in HNO_3 and the released Ag(I) was detected using a three-electrode screen-printed device with a carbon working electrode. In another study, silver dendrimer nanocomposites were synthesized and used as labels in immunoassays with stripping voltammetric detection of Ag(I) on a carbon electrode [45]. Other metal NPs can also serve as useful voltammetric labels in biosensing; relevant examples include CuNPs for the detection of glutathione [46], PtNPs for the detection of prostate specific antigen [47], Au-PdNPs for carbohydrate antigen 19-9 [48] and Ag/Au bimetallic NPs for the detection of *Escherichia coli* [49].

An important property of AuNPs is that they can serve as a catalyst to mediate the reduction of silver and copper ions. The reduction reaction leads to the *in situ* deposition of silver or copper on the AuNPs, the surface of which effectively serves as a nucleation site and results in the increase of the amount of the deposited copper or silver, with remarkable signal amplification. For example, an electrochemical sandwich-type immunosensor was reported for the detection of human cytomegalovirus glyco protein B, using Ag deposition catalyzed by AuNPs [50], while another immunosensor for the detection of hepatitis B surface antigen (HBsAg) employed MBs loaded with copper-encapsulated AuNPs [51]. In an attractive signal amplification process developed for the detection of carcinoembryonic antigen, involving silver deposition on polybead-loaded AuNPs, poly(styrene-co-acrylic acid) microbeads loaded with AuNPs were used as tags to label the signaling antibody, while AuNPs induced the deposition of silver [52]. In all these examples, the nucleated metal nanoclusters were detected voltametrically.

QDs (composed of heavy metal sulfide, selenide or telluride salts) have been widely used as voltammmetric labels in immunosensing. In respective studies, a competitive immunoassay was developed for the detection of 17b-estradiol, using CdSe QD labels detected at a bismuth-coated carbon electrode [53]; and a CdTe QD-based voltammetric biosensor was reported for the detection of neutravidin [54]. In another work, a PbS QDs-based sandwich-type immunoassay of C-reactive protein was used at a bismuth citrate-loaded graphite screen-printed working electrode [55]. Therein, bismuth citrate served as a precursor compound for the *in situ* generation of bismuth nanostractured film on the surface of the transducer during the electrolytic preconcentration step. In yet another study, an immunosensor with a ZrO_2-modified screen-printed carbon electrode was reported for the detection of organophosphorylated butyrylcholinesterase [56]. In this approach, the ZrO_2 NPs were used to capture the organophosphate moiety of organophosphorylated butyrylcholinesterase adducts, followed by the introduction of CdSe QD-tagged anti-butyrylcholinesterase conjugate to form a sandwich-type complex on the sensor's surface; Cd(II) released from the QDs were detected by anodic stripping voltammetry. Sharma et al. developed a novel electrochemical immunosensor for a sensitive and specific detection of anthrax protective antigen [57]. The immunosensor consisted of: (i) a glassy carbon electrode modified with a Nafion-multiwalled carbon nanotubes–bismuth nanocomposite film (BiNPs/Nafion-MWCNTs/GCE) as a sensing platform, and (ii) titanium phosphate NPs/cadmium(II)/mouse anti-protective antigen antibodies as signal amplification tags. Therein, the Cd(II) ions loaded on titanium phosphate NPs were detected *in situ* by cathodic voltammetry. Zhong et al. used CdS QDs-encapsulated metal-organic frameworks as signal-amplifying tags for ultrasensitive electrochemical detection of *Escherichia coli* O157:H7 in milk samples [58] (Figure 2). The tags were modified with anti-*E. coli* O157:H7 and a sandwich type electrochemical immunosensor for *E. coli* O157:H7 was fabricated, using voltammetric detection of the released Cd(II) ions.

Multiplexed detection is highly desirable in biosensing, for multiparametric information (e.g., monitoring of more than one biomarkers in cancer diagnosis)

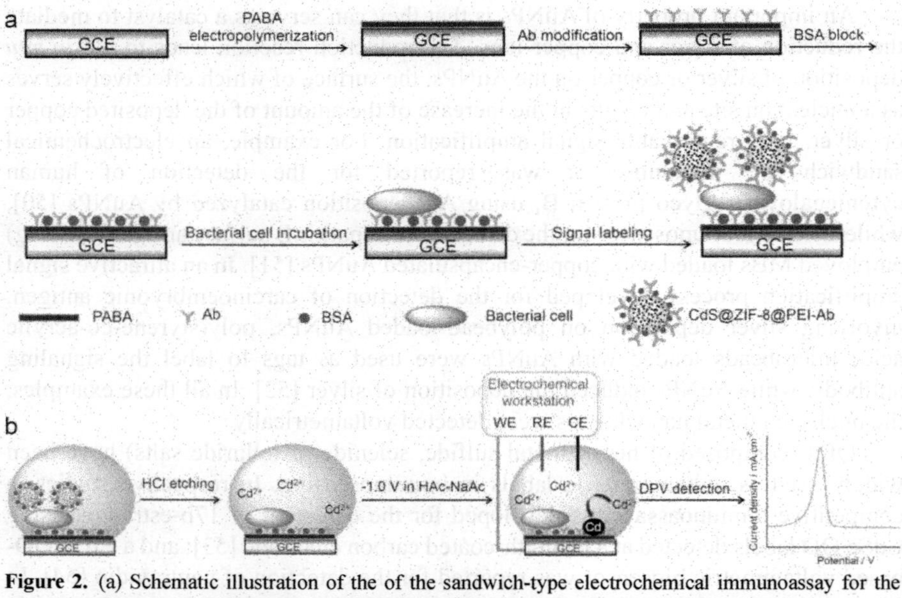

Figure 2. (a) Schematic illustration of the of the sandwich-type electrochemical immunoassay for the detection of *E. coli* O157:H7 using CdS@ZIF-8 as signal tags; (b) Schematic illustration of the steps for voltammetric detection. From [58], Figure 1.

is required in many cases. Two routes exist for multiplexed detection and will be discussed in the succeeding paragraphs.

The first approach is based on a single type of metallic nanoprobe using a spatially separated array of electrodes, each devoted to a single bioassay. For example, Leng et al. developed a screen-printed device with two separate graphite working electrodes (and silver reference and graphite counter electrodes) for the preparation of two sandwich-type assays using AuNP labels for the parallel detection of human imunoglobulin and goat imunoglobulin [59]; the AuNPs were electrooxidized in HCl to produce $AuCl_4^-$ which was detected voltammetrically. Another screen-printed chip with two working carbon electrodes was deployed for the quantification of carcinoembryonic antigen and a-fetoprotein, using antibody-labeled AuNPs and silver deposition, by Lai et al. [60].

The second, and comparatively more elegant, multiplexing approach relies on using different types of QDs (e.g., PbS, CdS and ZnS) as specific "bio-tags" to label two or more biomolecules. In this case, detection can be carried out simultaneously at a single sensor and differentiation at the detection stage is based on the different voltammetric peak potentials of the respective QD metals. Multiplexed voltammetric immunoassays with multiple QD labels have been developed for the detection of carcinoembryonic antigen and cetuximab using CdSe and ZnO QDs [61], for B-cell lymphoma 2 and Bcl-2-associated X protein using CdSeTe/CdS QDs and Ag nanoclusters [62], for immunoglobulin G and carcinoembryonic antigen using CdSe and PbS QDs [63], for tumor cells using CdTe and ZnSe QDs [64], for bovine casein and bovine IgG using CdSe/ZnS and PbS QDs [65], for tetracycline and chloramphenicol with CdS and PbS QDs [66], for *Escherichia coli* O157:H7, *campylobacter* and *salmonella* with CuS, PbS and CdS QDs [67], and for carcinoembryonic antigen and a-fetoprotein using CdS/DNA and PbS/DNA nanochains [68]. The design of a sandwich-type immunoassay protocol has also been reported for the simultaneous detection of multiple biomarkers (CA 125, CA 15-3 and CA 19-9) [69], wherein the capture probes were prepared by co-immobilizing capture antibodies on MBs. PAMAM dendrimer-metal sulfide QD nanolabels containing CdS, ZnS and PbS were synthesized and used to tag polyclonal rabbit probe antibodies. Cd(II). Zn(II) and Pb(II) released by the acidic dissolution of the corresponding QD nanolabels were quantified by stripping voltammetry at a mercury-film electrode (Figure 3).

Many types of microfluidic, flexible and lab-on-a-chip devices have been developed for the ultrasensitive detection of proteins and cells, using labels with metal-based nanoprobes. Martin-Yerga et al. reported an 8-channel screen-printed carbon electrochemical array controlled by a multipotentiostat for the detection of biotin and antitransglutaminase IgG antibodies, using CdSe/ZnS QDs as labels [70, 71]. After the immunoreactions, the QDs were dissolved in HCl and the released Cd(II) was detected by stripping voltammetry. Another study reported a PDMS flow injection microfluidic platform for the simultaneous detection of cardiac troponin I and C-reactive protein, using CdTe and ZnSe QDs as labels [72] (Figure 4). Oliveira et al. developed a simple, low-cost and disposable microfluidic device for *Salmonella typhimurium* (*S. typhi*) detection in milk, using AuNPs as a label [73]. *S. typhi* cells

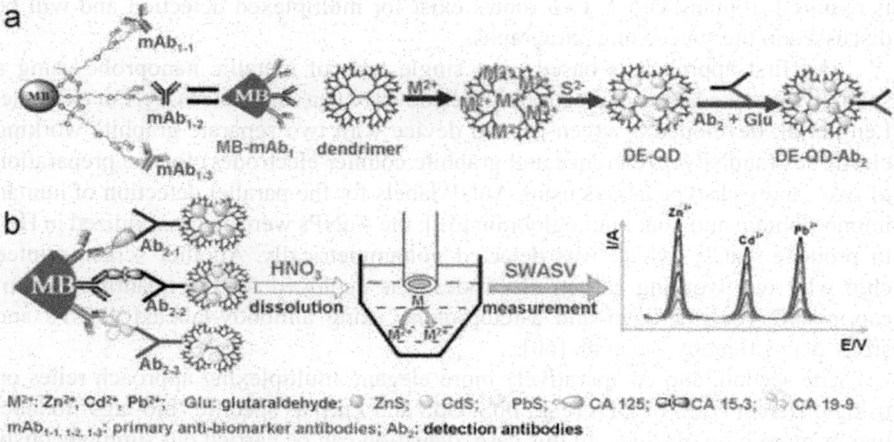

Figure 3. Schematic illustration of the multiplexed stripping voltammetric immunoassay protocol using dendrimer-metal sulfide QD nanolabels and trifunctionalized MBs: (a) preparation process and (b) measurement principle. From [69], Scheme 1.

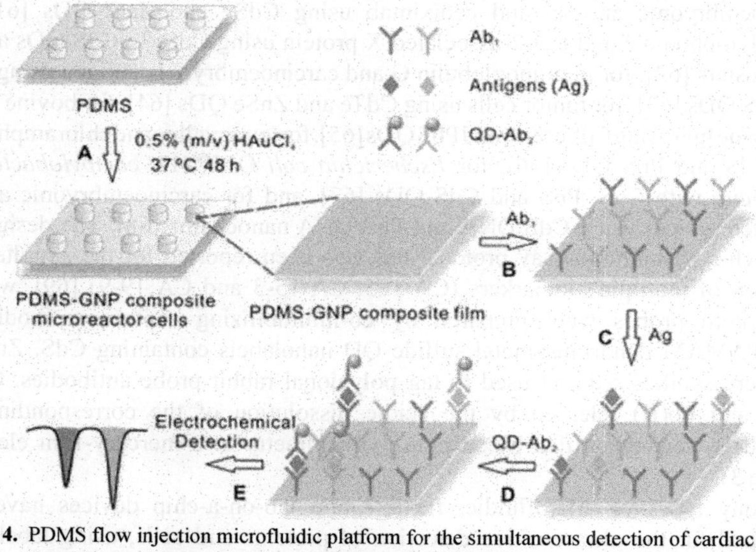

Figure 4. PDMS flow injection microfluidic platform for the simultaneous detection of cardiac troponin I and C-reactive protein using CdTe and ZnSe QDs as labels. From [72], Figure 2.

were captured using MBs modified with anti-*Salmonella* antibody, followed by the tagging with a probe anti-*Salmonella* antibody labeled with AuNPs. The MBs bearing the immunoconjugate were injected into the device and magnetically captured on the working electrode surface. Detection of the AuNPs was performed after oxidation and voltammetric detection. Medina-Sánchez et al. developed an on-chip magneto-immunoassay for apolipoprotein E, using CdSe/ZnS QDs as labeling carriers [74]. The immunoassay was performed in a flow mode, using MBs as a preconcentration platform in a flexible microfluidic chip with integrated screen-printed electrodes. The detection of Cd(II) was performed by stripping voltammetry.

3. Genosensors using metal-based nanoprobes

The operation of genosensors is based on the selective hybridization event of a single-stranded complementary probe oligonucleotide with the target DNA or RNA. Usually, the probe oligonucleotide or the target DNA or RNA is modified with metal-based nanoprobes which can be monitored electrochemically [75].

AuNPs have been widely used as labels for the development of electrochemical genosensors. For instance, a biotinylated oligonucleotide probe DNA was immobilized in a streptavidin-modified microwell [76]. The biotinylated target DNA hybridized with the capture probe DNA and streptavidin-modified AuNPs were used as labels. The AuNPs labels were dissolved in HBr/Br_2 solution and detected using stripping voltammetry at a glassy carbon working electrode. Daneshpour et al. developed a novel chip for the detection of RASSF1A tumor suppressor gene methylation, using Fe_3O_4/N-trimethylchitosan/AuNPs as tags to label a DNA probe [77] (Figure 5). Electrochemical detection involved the oxidation of AuNPs, along with cathodic voltammetry. In order to enhance the sensitivity of AuNPs-based DNA bioassays, different methodologies have been adopted, for instance conjugation of AuNPs with latex microspheres and with MBs [78]. AgNPs are another popular label in DNA biosensors. An relevant example is the development of a voltammetric biosensor for

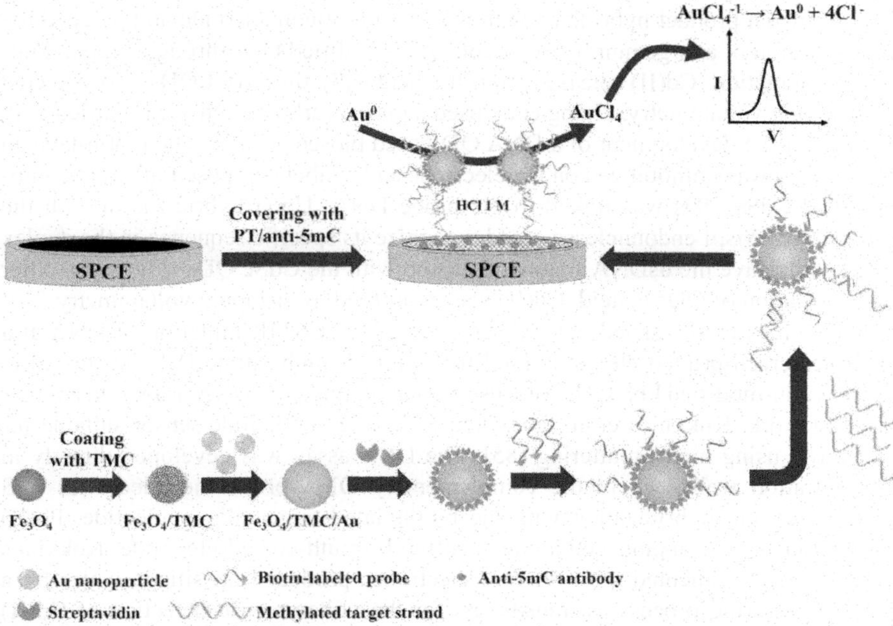

Figure 5. Schematic representation of the electrochemical DNA assay using Fe_3O_4/N-trimethylchitosan/AuNPs as labels. From [77], Figure 1.

the detection of sequence-specific DNA by Hu et al. [79]. Initially, thiolated peptide nucleic acid probes were immobilized onto gold surface and the target DNA was hybridized. Subsequently, hematin was added for catalyzing the reduction of Ag ions in the presence of catechol, leading to the formation of AgNPs, detected by anodic stripping voltammetry. In a study by Hakimian and Ghourchian, another genosensor was developed for the detection of miR-155 as a breast biomarker [80]. Therein, a thiolated probe was immobilized on the gold electrode surface and allowed to hybridize with the target; thereafter, positively charged polyethyleneimine-AgNPs were absorbed onto the negatively charged probe-target hybrid; finally, the anodic current produced due to the oxidation of AgNPs was recorded.

Different QD-based DNA assays have been reported in the literature. A DNA assay in microwells, developed by Sun et al., was based on a target-induced strand displacement reaction with blocker DNA (labeled with CdS QDs) from a biotinylated hairpin DNA [81]. For this assay, a hairpin-blocker DNA was immobilized on the surface of the microwell through biotin-streptavidin interaction. On addition of target DNA, the CdS-labeled blocker DNA was displaced by target DNA from the hairpin-blocker to form a new target-blocker DNA. Then, Cd(II) was released from the QDs using HNO_3 and detected by stripping voltammetry at a mercury-film sensor. Another interesting QDs-based electrochemical biosensor was developed by Li et al. for detection of telomerase activity at the single-cell level in humans [82], wherein a thiol-modified capture DNA was attached to an Au surface. The presence of telomerase enabled the addition of telomere repeats to the 3' end of the primer, accompanied by the incorporation of abundant biotins in the extension product. The extension product hybridized with the capture oligonucleotide and induced the concentration of a large number of streptavidin-modified QDs through streptavidin-biotin interaction. Cd(II) released from the acidically dissolved QDs was detected by stripping voltammetry. Zhang et al. used a glassy carbon electrode modified with AuNPs for the development of a DNA QD-based biosensor [83]. The biotin-labeled probe DNA was immobilized on the electrode and a label composed of streptavidin-modified CdSe QDs was linked to the probe DNA. Upon hybridization with the target DNA, MspI endonuclease could recognize its specific sequence in the duplex DNA and cleave the dsDNA fragments linked with the CdSe QDs (Figure 6). After the dissolution of QDs in acid, Cd(II) was monitored by stripping voltammetry.

Kokkinos et al. carried out the detection of the C634R mutation of DNA at a graphite screen-printed electrode modified with bismuth citrate [84]. In this case, streptavidin-modified PbS QDs were used to tag biotinylated DNA probes. In another piece of work, Kokkinos et al. fabricated a flexible 3-electrode sensor suitable for bioassays, using metal sputtering [85]. The DNA assay was developed directly in microtitration wells, where the complementary DNA probe modified with PbS QDs underwent hybridization with the biotinylated target oligonucleotide. Pb(II) was detected by stripping voltammetry at the bismuth-coated electrode. Kokkinos et al. also experimented with tin film electrodes produced on silicon wafer by a microengineering process as transducers for the voltammetric detection of Cd(II) liberated from QDs tags, enabling the detection of DNA at nanomolar levels [86].

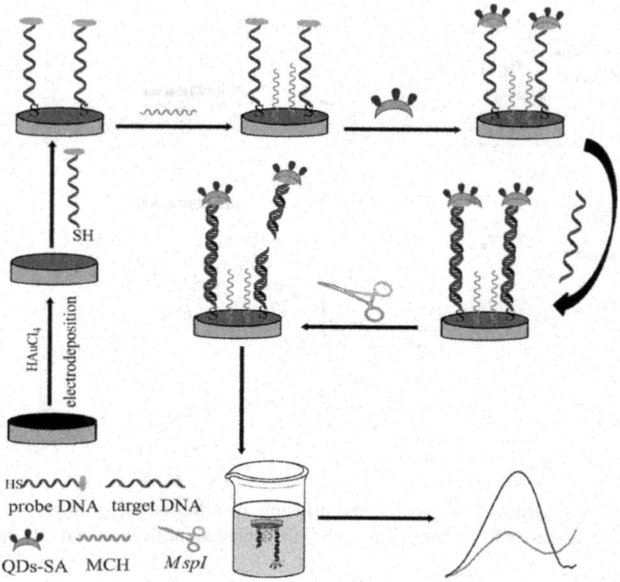

Figure 6. Schematic diagram of the operational principle of a QDs-based DNA biosensor using MspI endonuclease. From [83], Figure 1.

Several multiplexed voltammetric DNA and RNA assays using metal NPs and, more often, QD labels have also been reported in literature. For instance, Azzouzi et al. developed a neutravidin electrochemical biosensor for the simultaneous detection of miRNA-21 and miRNA-141 [87]. The assay was based on the fact that the target miRNA causes the opening of biotin-molecular beacons, followed by the capture of the biotinylated molecular beacons-metal NPs by the neutravidin biosensor and simultaneous detection of the captured AgNPs and AuNPs by stripping voltammetry. Vijian et al. developed an RNA genosensor for the detection of multiple pathogens using QDs tags [88]. Therein, RNA sequences were used as gene targets for *Vibrio cholerae*, *Salmonella* sp. and *Shigella* sp., while PbS, CdS and ZnS QDs were functionalized with DNA probes that were specific to each pathogen. The detection of released cations was performed by anodic stripping voltammetry. Krejcova et al. reported a multi-target barcode electrochemical assay for the detection of single point mutations in the H5N1 neuraminidase gene [89]. They used paramagnetic particles with covalently bound oligonucleotide probes for the isolation of complementary H5N1 chains labeled with QDs (CdS, ZnS and/or PbS). The metals in the QDs were detected by ASV at a mercury electrode. Rezai et al. developed an electrochemical assay for the simultaneous detection of two hemophilia A-related microRNAs (miR-1246 and miR-4521) [90] (Figure 7). The assay used heavy metal QDs encapsulated in metal-organic frameworks and catalytic hairpin assembly for signal amplification. In another piece of research, Zhang et al. reported a voltammetric bio-barcoded biosensor for the simultaneous detection of the protective antigen A gene of *Bacillus anthracis* and the insertion element gene of *Salmonella enteritidis* [91]. Their

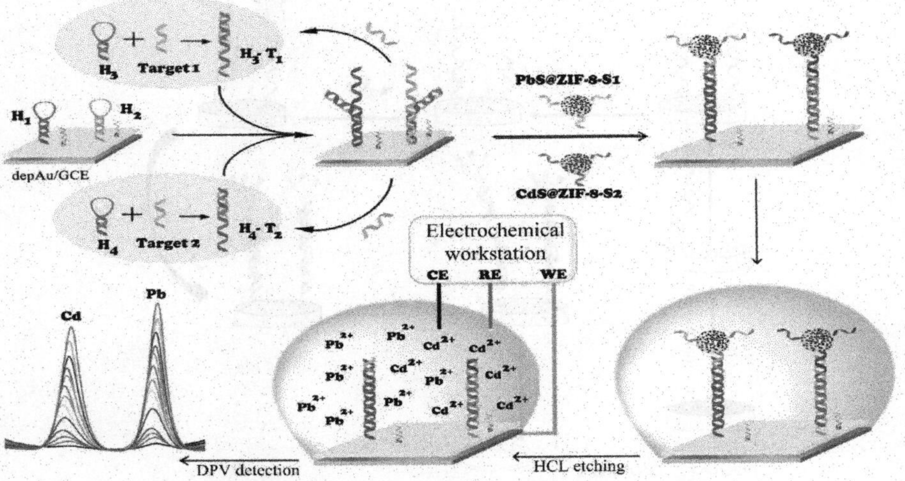

Figure 7. Schematic diagram of the operational principle of a biosensor for duplex detection of miR-1246 and miR-4521 using catalytic hairpin assembly for signal amplification and QD labels. From [90], Figure 1.

biosensor exploits three types of nanoparticles (AuNPs, MBs, and PbS and CdS QDs) for signal amplification, and a sandwich structure is formed: MBs-1st probeDNA/ target DNA/2nd probe DNA-AuNP-QDs barcoded DNA. A magnetic field is applied to separate the sandwich structure; the QDs are dissolved in nitric acid and the released cations are detected by anodic stripping voltammetry. Zhu et al. described an electrochemical assay for the multiplexed quantification of miRNAs, based on the combination of the high base-mismatch selectivity of ligase chain reaction and the significant electrochemical amplification of QDs barcodes [92]. Two reporting probes were prepared from oligonucleotides labeled with PbS and CdS QDs and two capture probes were immobilized on MBs. The miRNAs in the sample interacted with the reporting and capture probes, followed by the addition of T4 DNA ligase. After the release of the disjointed QDs barcodes from the MB-conjugates, the two target miRNAs of miR-155 and miR-27b could be simultaneously detected by stripping voltammetry.

Paper-based analytical devices (PADs) have lately gained immense popularity, thanks to the advantageous operational features of paper as an analytical platform. In one study, Kokkinos et al. introduced a fluidic electrochemical PAD (ePAD) for the voltammetric detection of DNA (associated with Multiple Endocrine Neoplasia Type 2), using CdSe QDs as the label [93]. The ePAD was patterned by wax-printing, and featured an assay zone, an inlet zone and a sink. On the reverse side of the paper, an electrochemical cell was formed by the deposition of sputtered metals. The DNA assay involved immobilization of the capture complementary oligonucleotide, hybridization with biotinylated target DNA and labeling with streptavidin-modified Cd-based QDs. The released Cd(II) was measured by stripping voltammetry at a tin-film sputtered electrode. In another research, a folding paper device for DNA sensing was introduced by Lu et al. [94]. The production procedure of the sensor consisted of wax-printing and screen-printing of electrodes. The device was modified with AuNPs and graphene in order to achieve an efficient DNA immobilization, and detection was based on a sandwich assay with AuNPs as labels.

4. Conclusion and challenges

Metal-based nanoprobes are very promising tools for the development of voltammetric biosensors in labelled format. Voltammetric detection is sensitive, selective and requires low-cost and portable instrumentation with minimal power consumption. In particular, stripping voltammetric detection allows for a further amplification step due to the initial accumulation of the target metal on the electrode surface. Furthermore, metal NPs and QDs can be easily prepared and functionalized with different chemical groups in order to facilitate conjugation with different types of biomolecules (proteins, cells and oligonucleotides). Their use as electrochemical labels enhances the detection sensitivity as a large number of potentially detectable label moieties can be attached to each target molecule. Also, signal amplification can be achieved using functionalized bulky carriers (e.g., metal-organic frameworks, carbon nanomaterials, MBs) with large surface areas, onto which an even larger amount of redox nanoprobes can be attached. Another alternative towards signal amplification is to initiate catalytic growth of metal nanoclusters.

The selection of a suitable working electrode is another critical factor. Typically, metal NPs are detected at carbon electrodes and the detection sensitivity can be enhanced by modification of the transducer with nanomaterials like metal NPs, metal oxides and carbon nanomaterials. In this context, "one-shot" low-cost 3-electrode sensors manufactured via screen-printing are ideal for the development of biosensors. Although heavy metals released from QDs can be detected at carbon electrodes, electrodes made of or modified with alloy-forming metals offer much higher sensitivity; while mercury is still being used, "green" bismuth- and tin-based electrodes are gaining popularity due to their low toxicity and satisfactory analytical performance. Microengineering techniques (e.g., photolithography and sputtering) and screen-printing are particularly useful for the fabrication of single metal-film electrodes, integrated 3-electrode sensors or array sensors for multiplexed detection.

In many cases, multiplexed analysis is required, in clinical analysis involving biomarker detection for instance. Although the use of electrode arrays for several spatially distinct and simultaneous bioassays is possible, this approach is limited by the requirement of a multi-potentiostat, while "cross-talk" between adjacent electrodes can occur. In this respect, the use of heavy metal QDs, which serve as "bio-barcodes", is highly advantageous. After labeling each biomolecule with a different type of QD, multiplexed detection is possible in a single voltammetric scan by exploiting the different redox potential of each metal-based QD.

Finally, new biosensing approaches based on the use of microfludic devices, paper-based devices and lab-on-a-chip devices have emerged; these platforms are useful for point-of-care and on-site applications due to their small size, low cost and disposability.

References

[1] Bahadır, E.B. and M.K. Sezginturk. 2015. Applications of commercial biosensors in clinical, food, environmental and biothreat/biowarfare analyses. Anal. Biochem. 478: 107–120.

[2] Justino, C.I.L., A.C. Freitas, R. Pereira, A.C. Duarte and T.A.P. Rosha-Santos. 2015. Recent developments in recognition elements for chemical sensors and biosensors. TrAC Trends Anal. Chem. 68: 2–17.

[3] Arugula, M.A. and A. Simonian. 2014. Novel trends in affinity biosensors: Current challenges and perspectives. Meas. Sci. Technol. 25: 032001.

[4] Holford, T.R.J., F. Davis and S.P.J. Higson. 2012. Recent trends in antibody based sensors. Biosens. Bioelectron. 34: 12–24.

[5] Tan, A., C. Lim, S. Zou, Q. Ma and Z. Gao. 2016. Electrochemical nucleic acid biosensors: From fabrication to application. Anal. Methods 8: 5169–5189.

[6] Hasanzadeh, M. and N. Shadjou. 2016. Electrochemical nanobiosensing in whole blood: Recent advances. TrAC Trends Anal. Chem. 80: 167–176.

[7] Wan, Y., Yan Su, X. Zhu, G. Liu and C. Fan. 2013. Development of electrochemical immunosensors towards point of care diagnostics. Biosens. Bioelectron. 47: 1–11.

[8] Martin-Fernandez, B., C.L. Manzanares-Palenzuela, M.S.P. Lopez, N. Santos- Alvarez and B. Lopez-Ruiz. 2017. Electrochemical genosensors in food safety assessment. Crit. Rev. Food Sci. Nutr. 13: 2758–2774.

[9] Duffy, G.F. and E.J. Moore. 2017. Electrochemical immunosensors for food analysis: A review of recent developments. Anal. Lett. 50: 1–32.

[10] Arduini, F., S. Cinti, V. Scognamiglio, D. Moscone and G. Palleschi. 2017. How cutting-edge technologies impact the design of electrochemical (bio)sensors for environmental analysis. A review. Anal. Chim. Acta 9: 15–42.

[11] Skladal, P., D. Kovar, V. Krajicek, P. Siskova, J. Pribyl and E. Svabenska. 2013. Electrochemical immunosensors for detection of microorganisms. Int. J. Electrochem. Sci. 8: 1635–1649.

[12] Orozco, J. and L.K. Medlin. 2013. Review: Advances in electrochemical genosensors based methods for monitoring blooms of toxic algae. Environ. Sci. Pollut. Res. 20: 6838–6850.

[13] Sang, S., Y. Wang, Q. Feng, Y. Wei, J. Ji and W. Zhang. 2016. Progress of new label-free techniques for biosensors: A review. Crit. Rev. Biotechnol. 36: 465–481.

[14] Luo, X. and J.J. Davis. 2013. Electrical biosensors and the label free detection of protein disease biomarkers. Chem. Soc. Rev. 42: 5944–5962.

[15] Pei, X., B. Zhang, J. Tang, B. Liu, W. Lai and D. Tang. 2013. Sandwich-type immunosensors and immunoassays exploiting nanostructure labels: A review. Anal. Chim. Acta 758: 1–18.

[16] Omidfar, K., F. Khorsand and M.D. Azizi. 2013. New analytical applications of gold nanoparticles as label in antibody based sensors. Biosens. Bioelectron. 43: 336–347.

[17] Ding, L., A.M. Bond, J. Zhai and J. Zhang. 2013. Utilization of nanoparticle labels for signal amplification in ultrasensitive electrochemical affinity biosensors: A review. Anal. Chim. Acta 797: 1–12.

[18] Huo, X., X. Liu, J. Liu, P. Sukumaran, S. Alwarappan and D.K.Y. Wong. 2016. Strategic applications of nanomaterials as sensing platforms and signal amplification markers at electrochemical immunosensors. Electroanalysis 28: 1730–1749.

[19] Lei, J. and H. Ju. 2016. Signal amplification using nanomaterials for biosensing. pp. 17–41. *In*: Tuantranont, A. (ed.). Springer Series on Chemical Sensors and Biosensors, Applications of Nanomaterials in Sensors and Diagnostics, Volume 14, Springer-Verlag, Heidelberg.

[20] Fenzl, C., T. Hirsch and A.J. Baeumner. 2016. Nanomaterials as versatile tools for signal amplification in (bio)analytical applications. TrAC Trends Anal. Chem. 79: 306–316.

[21] Koyappayil, A. and M.H. Lee. 2021. Ultrasensitive materials for electrochemical biosensor labels. Sensors 21: 89.

[22] Malekzad, H., P.S. Zangabad, H. Mirshekari, M. Karimi and M.R. Hamblin. 2017. Noble metal nanoparticles in biosensors: Recent studies and applications. Nanotechnol. Rev. 6: 301–329.

[23] Kokkinos, C. and A. Economou. 2017. Emerging trends in biosensing using stripping voltammetric detection of metal-containing nanolabels—A review. Anal. Chim. Acta 961: 12–32.

[24] Valera, E., A. Hernandez-Albors and M.P. Marco. 2016. Electrochemical coding strategies using metallic nanoprobes for biosensing applications. TrAC Trends Anal. Chem. 79: 9–22.
[25] Campuzano, S., P. Yáñez-Sedeño and J.M. Pingarrón. 2019. Nanoparticles for nucleic-acid-based biosensing: Opportunities, challenges, and prospects. Anal. Bioanal. Chem. 411: 1791–1806.
[26] Iglesias-Mayor, A., O. Amor-Gutiérrez, A. Costa-García and A. de la Escosura-Muñiz. 2019. Nanoparticles as emerging labels in electrochemical immunosensors. Sensors 19: 5137.
[27] Merkoci, A., M. Aldavert, S. Marın and S. Alegret. 2005. New materials for electrochemical sensing V: Nanoparticles for DNA labeling. TrAC Trends Anal. Chem. 24: 341–349.
[28] Zhou, W., X. Gao, D. Liu and X. Chen. 2015. Gold nanoparticles for *in vitro* diagnostics. Chem. Rev. 115: 10575–10636.
[29] Wang, J. 2012. Electrochemical biosensing based on noble metal nanoparticles. Microchim. Acta 177: 245–270.
[30] Pedrero, M., S. Campuzano and J.M. Pingarron. 2017. Electrochemical (bio)sensing of clinical markers using quantum dots. Electroanalysis 29: 24–37.
[31] Huang, H. and J.J. Zhu. 2013. The electrochemical applications of quantum dots. Analyst 138: 5855–5865.
[32] Farzin, M.A and H. Abdoos. 2021. A critical review on quantum dots: From synthesis toward applications in electrochemical biosensors for determination of disease-related biomolecules. Talanta 224: 121828.
[33] Jamkhande, P.G., N.W. Ghule, A.H. Bamer and M.G. Kalaskar. 2019. Metal nanoparticles synthesis: An overview on methods of preparation, advantages and disadvantages, and applications. J. Drug Deliv. Sci. Technol. 53: 101174.
[34] Karakoti, A.S., R. Shukla, R. Shanker and S. Singh. 2015. Surface functionalization of quantum dots for biological applications. Adv. Colloid Interface Sci. 215: 28–45.
[35] Yüce, M. and H. Kurt. 2017. How to make nanobiosensors: Surface modification and characterisation of nanomaterials for biosensing applications. RSC Adv. 7: 49386–49403.
[36] Economou, A. and C. Kokkinos 2016. Advances in stripping analysis of metals. pp. 1–18. *In*: Arrigan, D.W.M. (ed.). Electrochemical Strategies in Detection Science, RSC Detection Science Series No. 6, Royal Society of Chemistry, Cambridge, UK.
[37] Kokkinos, C., A. Economou and M.I. Prodromidis. 2016. Electrochemical immunosensors: Critical survey of different architectures and transduction strategies. TrAC Trends Anal. Chem. 79: 88–105.
[38] Omidfar, K., H. Zarei, F. Gholizadeh and B. Larijani. 2012. A high-sensitivity electrochemical immunosensor based on mobile crystalline material-41-polyvinyl alcohol nanocomposite and colloidal gold nanoparticles. Anal. Biochem. 421: 649–656.
[39] Lim, S.A., H. Yoshikawa, E. Tamiya, H.M. Yasin and M.U. Ahmed. 2014. Highly sensitive gold nanoparticle bioprobe based electrochemical immunosensor using screen printed graphene biochip. RSC Adv. 4: 58460–58466.
[40] Li, S., Y. Yan, L. Zhong, P. Liu, Y. Sang, W. Cheng and S. Ding. 2015. Electrochemical sandwich immunoassay for the peptide hormone prolactin using an electrode modified with graphene, single walled carbon nanotubes and antibody-coated gold nanoparticles. Microchim. Acta 182: 1917–1924.
[41] Khristunova, Y., E. Korotkova, B. Kratochvil, J. Barek, E. Dorozhko, V. Vyskocil, E. Plotnikov, O. Voronova and V. Sidelnikov. 2019. Preparation and investigation of silver nanoparticle–antibody bioconjugates for electrochemical immunoassay of tick-borne encephalitis. Sensors 19: 2103.
[42] Abbaspour, A., F. Norouz-Sarvestani, A. Noori and N. Soltani. 2015. Aptamer-conjugated silver nanoparticles for electrochemical dual-aptamer-based sandwich detection of *staphylococcus aureus*. Biosens. Bioelectron. 68: 149–155.
[43] Hao, N., H. Li, Y. Long, L. Zhang, X. Zhao, D. Xu and H.Y. Chen. 2011. An electrochemical immunosensing method based on silver nanoparticles. J. Electroanal. Chem. 656: 50–54.
[44] Song, W., H. Li, H. Liu, Z. Wu, W. Qiang and D. Xu. 2013. Fabrication of streptavidin functionalized silver nanoparticle decorated graphene and its application in disposable electrochemical sensor for immunoglobulin E. Electrochem. Commun. 31: 16–19.
[45] Stofik, M., Z. Stryhal and J. Maly. 2009. Dendrimer-encapsulated silver nanoparticles as a novel electrochemical label for sensitive immunosensors. Biosens. Bioelectron. 24: 1918–1923.

[46] Wang, Z., P. Han, X. Mao, Y. Yin and Y. Cao. 2017. Sensitive detection of glutathione by using DNA-templated copper nanoparticles as electrochemical reporters. Sens. Actuat. B Chem. 238: 325–330.
[47] Spain, E., S. Gilgunn, S. Sharma, K. Adamson, E. Carthy, R.O. Kennedy and R.J. Forster. 2016. Detection of prostate specific antigen based on electrocatalytic platinum nanoparticles conjugated to a recombinant scFv antibody. Biosens. Bioelectron. 77: 759–766.
[48] Yang, F., Z. Yang, Y. Zhuo, Y. Chai and R. Yuan. 2015. Ultrasensitive electrochemical immunosensor for carbohydrate antigen 19-9 using Au/porous graphene nanocomposites as platform and Au@Pd core/shell bimetallic functionalized graphene nanocomposites as signal enhancers. Biosens. Bioelectron. 66: 356–362.
[49] Eksi, H., R. Guzel, B. Guven, I.H. Boyaci and A.O. Solak. 2015. Fabrication of an electrochemical *E. coli* biosensor in biowells using bimetallic nanoparticle-labelled antibodies. Electroanalysis 27: 343–352.
[50] Pires, F., H. Silva, O. Dominguez-Renedo, M.A. Alonso-Lomillo, M.J. Arcos-Martinez and A.C. Dias-Cabral. 2015. Disposable immunosensor for human cytomegalovirus glycoprotein B detection. Talanta 136: 42–46.
[51] Shen, G. and Y. Zhang. 2010. Highly sensitive electrochemical stripping detection of hepatitis B surface antigen based on copper-enhanced gold nanoparticle tags and magnetic nanoparticles. Anal. Chim. Acta 674: 27–31.
[52] Lin, D., J. Wu, M. Wang, F. Yan and H. Ju. 2012. Triple signal amplification of graphene film, polybead carried gold nanoparticles as tracing tag and silver deposition for ultrasensitive electrochemical immunosensing. Anal. Chem. 84: 3662–3668.
[53] Chaisuwan, N., H. Xu, G. Wu and J. Liu. 2013. A highly sensitive differential pulse anodic stripping voltammetry for determination of 17b-estradiol (E2) using CdSe quantum dots based on indirect competitive immunoassay. Biosens. Bioelectron. 46: 150–154.
[54] Dua, D., J. Ding, Y. Tao, H. Li and X. Chen. 2008. CdTe nanocrystal-based electrochemical biosensor for the recognition of neutravidin by anodic stripping voltammetry at electrodeposited bismuth film. Biosens. Bioelectron. 24: 863–868.
[55] Kokkinos, C., M. Prodromidis, A. Economou, P. Petrou and S. Kakabakos. 2015. Disposable integrated bismuth citrate-modified screen-printed immunosensor for ultrasensitive quantum dot-based electrochemical assay of C-reactive protein in human serum. Anal. Chim. Acta 886: 29–36.
[56] Lu, D., J. Wang, L. Wang, D. Du, C. Timchalk, R. Barry and Y. Lin. 2011. A novel nanoparticle-based disposable electrochemical immunosensor for diagnosis of exposure to toxic organophosphorus agents. Adv. Funct. Mater. 21: 4371–4378.
[57] Sharma, M.K., J. Narayanan, S. Upadhyay and A.K. Goel. 2015. Electrochemical immunosensor based on bismuth nanocomposite film and cadmium ions functionalized titanium phosphates for the detection of anthrax protective antigen toxin. Biosens. Bioelectron. 74: 299–304.
[58] Zhong, M., L. Yang, H. Yang, C. Cheng, W. Deng, Y. Tan, Q. Xie and S. Yao. 2019. An electrochemical immunobiosensor for ultrasensitive detection of *Escherichia coli* O157:H7 using CdS quantum dots-encapsulated metal-organic frameworks as signal-amplifying tags. Biosens. Bioelectron. 126: 493–500.
[59] Leng, C., G. Lai, F. Yan and H. Ju. 2010. Gold nanoparticle as an electrochemical label for inherently crosstalk-free multiplexed immunoassay on a disposable chip. Anal. Chim. Acta 666: 97–101.
[60] Lai, G., L. Wang, J. Wu, H. Jua and F. Yan. 2012. Electrochemical stripping analysis of nanogold label-induced silver deposition for ultrasensitive multiplexed detection of tumor markers. Anal. Chim. Acta 721: 1–6.
[61] Yang, C., C. Xu, X. Wang and X. Hu. 2012. Quantum-dot-based biosensor for simultaneous detection of biomarker and therapeutic drug: First steps toward an assay for quantitative pharmacology. Analyst 137: 1205–1209.
[62] Zhou, S., Y. Wang and J.J. Zhu. 2016. Simultaneous detection of tumor cell apoptosis regulators Bcl-2 and Bax through a dual-signal-marked electrochemical immunosensor. ACS Appl. Mater. Interfaces 8: 7674–7682.
[63] Qian, J., H. Dai, X. Pan and S. Liu. 2011. Simultaneous detection of dual proteins using quantum dots coated silica nanoparticles as labels. Biosens. Bioelectron. 28: 314–319.

[64] Wu, Y., P. Xue, Y. Kang and K.M. Hui. 2013. Highly specific and ultrasensitive graphene-enhanced electrochemical detection of low-abundance tumor cells using silica nanoparticles coated with antibody-conjugated quantum dots. Anal. Chem. 85: 3166–3173.

[65] Kokkinos, C., M. Angelopoulou, A. Economou, M. Prodromidis, A. Florou, W. Haasnoot, P. Petrou and S. Kakabakos. 2016. Lab-on-a-membrane foldable devices for duplex drop-volume electrochemical biosensing using quantum dot tags. Anal. Chem. 88: 6897–6904.

[66] Liu, B., B. Zhang, G. Chen and D. Tang. 2014. Biotin-avidin-conjugated metal sulfide nanoclusters for simultaneous electrochemical immunoassay of tetracycline and chloramphenicol. Microchim. Acta 181: 257–262.

[67] Viswanathan, S., C. Rani and J.A. Ho. 2102. Electrochemical immunosensor for multiplexed detection of food-borne pathogens using nanocrystal bioconjugates and MWCNT screen-printed electrode. Talanta 94: 315– 319.

[68] Kong, F.Y., B.Y. Xu, J.J. Xu and H.Y. Chen. 2103. Simultaneous electrochemical immunoassay using CdS/DNA and PbS/DNA nanochains as labels. Biosens. Bioelectron. 39: 177–182.

[69] Tang, D., L. Hou, R. Niessner, M. Xu, Z. Gao and D. Knopp. 2013. Multiplexed electrochemical immunoassay of biomarkers using metal sulfide quantum dot nanolabels and trifunctionalized magnetic beads. Biosens. Bioelectron. 6: 37–43.

[70] Martin-Yerga, D., M.B. Gonzalez-Garcia and A. Costa-Garcia. 2013. Biosensor array based on the *in situ* detection of quantum dots as electrochemical label. Sens. Actuat. B Chem. 182: 184–189.

[71] Martin-Yerga, D., M.B. Gonzalez-Garcia and A. Costa-Garcia. 2014. Electrochemical immunosensor for anti-tissue transglutaminase antibodies based on the *in situ* detection of quantum dots. Talanta 130: 598–602.

[72] Zhou, F., M. Lu, W. Wang, Z.P. Bian, J.R. Zhang and J.J. Zhu. 2016. Electrochemical immunosensor for simultaneous detection of dual cardiac markers based on a poly(dimethylsiloxane)-gold nanoparticles composite microfluidic chip: A proof of principle. Clin. Chem. 56: 1701–1707.

[73] Oliveira, T.R., D.H. Martucci and R.C. Faria. 2018. Simple disposable microfluidic device for *Salmonella typhimurium* detection by magneto-immunoassay. Sens. Actuat. B Chem. 255: 684–69.

[74] Medina-Sánchez, M., S. Miserere, E. Morales-Narváez and A. Merkoçi. 2014. On-chip magneto-immunoassay for Alzheimer's biomarker electrochemical detection by using quantum dots as labels. Biosens. Bioelectron. 54: 279–284.

[75] Kokkinos, C. 2019. Electrochemical DNA biosensors based on labeling with nanoparticles. Nanomaterials 9: 1361.

[76] Liao, K.T., J.T. Cheng, C.L. Li, R.T. Liu and H.J. Huang. 2009. Ultra-sensitive detection of mutated papillary thyroid carcinoma DNA using square wave stripping voltammetry method and amplified gold nanoparticle biomarkers. Biosens. Bioelectron. 24: 1899–1904.

[77] Daneshpour, M., L. Syed Moradi, P. Izadi and K. Omidfar. 2016. Femtomolar level detection of RASSF1A tumor suppressor gene methylation by electrochemical nano-genosensor based on Fe_3O_4/TMC/Au nanocomposite and PT-modified electrode. Biosens. Bioelectron. 77: 1095–1103.

[78] Low, K.F., P. Rijiravanich, K.K.B. Singh, W. Surareungchai and C.Y. Yean. 2015. An electrochemical genosensing assay based on magnetic beads and gold nanoparticle-loaded latex microspheres for *Vibrio cholerae* detection. J. Biom. Nanotechn. 11: 702–710.

[79] Hu, Q., W. Hu, J. Kong and X. Zhang. 2015. Ultrasensitive electrochemical DNA biosensor by exploiting hematin as efficient biomimetic catalyst toward *in situ* metallization. Biosens. Bioelectron. 63: 269–275.

[80] Hakimian, F. and H. Ghourchian. 2020. Ultrasensitive electrochemical biosensor for detection of microRNA-155 as a breast cancer risk factor. Anal. Chim. Acta 1136: 1–8.

[81] Sun, A.L., Y.F. Zhang and X.N. Wang. 2015. Sensitive voltammetric determination of DNA via a target-induced strand-displacement reaction using quantum dot-labeled probe DNA. Microchim. Acta 182: 1403–1410.

[82] Li, C.C., J. Hu, M. Lu and C.Y. Zhang. 2018. Quantum dot-based electrochemical biosensor for stripping voltammetric detection of telomerase at the single-cell level. Biosens. Bioelectron. 122: 51–57.

[83] Zhang, C., J. Lou, W. Tu, J. Bao and Z. Dai. 2015. Ultrasensitive electrochemical biosensing for DNA using quantum dots combined with restriction endonuclease. Analyst 140: 506–511.

[84] Kokkinos, C., M. Prodromidis, A. Economou, P. Petrou and S. Kakabakos. 2015. Quantum dot-based electrochemical DNA biosensor using a screen-printed graphite surface with embedded bismuth precursor. Electrochem. Comm. 60: 47–5.

[85] Kokkinos, C., A. Economou, T. Speliotis, P. Petrou and S. Kakabakos. 2015c. Flexible microfabricated film sensors for the in situ quantum dot-based voltammetric detection of DNA hybridization in microwells. Anal. Chem. 87: 853−857.

[86] Kokkinos, C., A. Economou, P. Petrou and S. Kakabakos. 2013. Microfabricated Tin−film electrodes for protein and DNA sensing based on stripping voltammetric detection of Cd(II) released from quantum dots labels. Anal. Chem. 85: 10686−10691.

[87] Azzouzi, S., Z. Fredj, A.P.F. Turner, M. Ben Ali and W.C. Mak. 2019. Generic neutravidin biosensor for simultaneous multiplex detection of microRNAs via electrochemically encoded responsive nanolabels. ACS Sens. 4: 326–334.

[88] Vijian, D., S.V. Chinni, L.S. Yin, B. Lertanantawong and W. Surareungchai. 2016. Non-protein coding RNA-based genosensor with quantum dots as electrochemical labels for attomolar detection of multiple pathogens. Biosens. Bioelectron. 77: 805–811.

[89] Krejcova, L., D. Hynek, P. Kopel, M.A.M. Rodrigo, V. Adam, J. Hubalek, P. Babula, L. Trnkova and R. Kizek. 2013. Development of a magnetic electrochemical bar code array for point mutation detection in the H5N1 neuraminidase gene. Viruses 5: 1719–1739.

[90] Rezaei, H., M. Motovali-Bashi and S. Radfar. 2019. An enzyme-free electrochemical biosensor for simultaneous detection of two hemophilia A biomarkers: Combining target recycling with quantum dots-encapsulated metal-organic frameworks for signal amplification. Anal. Chim. Acta 1092: 66–74.

[91] Zhang, D., M.C. Huarng and E.C.A. Alocilja. 2010. Multiplex nanoparticle-based bio-barcoded DNA sensor for the simultaneous detection of multiple pathogens. Biosens. Bioelectron. 26: 1736–1742.

[92] Zhu, W., X. Su, X. Gao, Z. Dai and X. Zou. 2014. A label-free and PCR-free electrochemical assay for multiplexed microRNA profiles by ligase chain reaction coupling with quantum dots barcodes. Biosens. Bioelectron. 53: 414–419.

[93] Kokkinos, C.T., D.L. Giokas, A.S. Economou, P.S. Petrou and S.E. Kakabakos. 2018. Paper-based microfluidic device with integrated sputtered electrodes for stripping voltammetric determination of DNA via quantum dot labeling. Anal. Chem. 90: 1092−1097.

[94] Lu, J., S. Ge, L. Ge, M. Yan and J. Yu. 2012. Electrochemical DNA sensor based on three-dimensional folding paper device for specific and sensitive point-of-care testing. Electrochim. Acta 80: 334–341.

6

Applications of Sensitive Electrode Surfaces; Determination of Vitamins

*Stella Girousi** and *Panayiotis Zararis*

1. Introduction

The term vitamin was first used in 1912, and the associated capital letters as we know today began to be used in 1916. Today, vitamins are categorized as either water-soluble (vitamins C and B) or fat-soluble (A, D, E, K). Vitamins are a group of heterogeneous chemical compounds known for their usefulness in the normal functioning of organisms. Some vitamins, in addition to their use as dietary supplements, are used as food additives due to their antioxidant properties.

Vitamins are organic substances (without calories) that regulate the body's functions. More specifically, vitamins regulate:

1. the absorption of nutrients,
2. their metabolism,
3. the digestion of nutrients, and
4. the creation of body tissues.

They got their name from the Latin word 'vita', which means life, and from the word 'amines' because they were originally believed to be related to amines. International IU units are used to measure the amount of vitamins in food or in the body. The amount in grams of vitamin corresponding to 1 IU is different for each vitamin. For instance, 1 IU is equal to 0.3 μg of vitamin A, while 1 IU is equal to 0.025 μg of vitamin D. Their sources and their chemical structure have been summarized in the literature [1–6]. A thorough introduction about the physicochemical and electrochemical principles regarding the characterization of vitamins is also available in literature [7–12, 14] In Table 1 are shown the chemical structures of the vitamins.

Analytical Chemistry laboratory, School of Chemistry, Faculty of Sciences, 54124 Thessaloniki, Greece.
* Corresponding author: girousi@chem.auth.gr

Table 1. Chemical structure of vitamins.

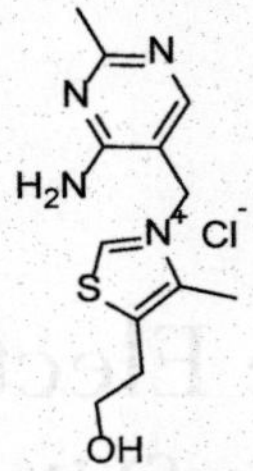 B1, thiamine	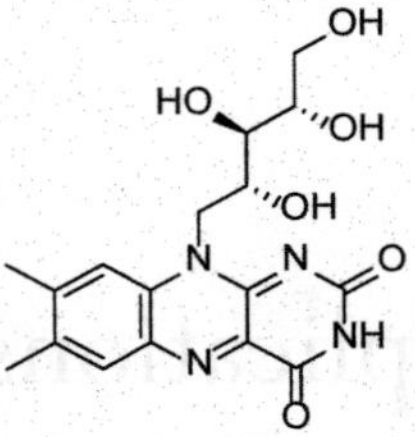 B2, rivoflavin
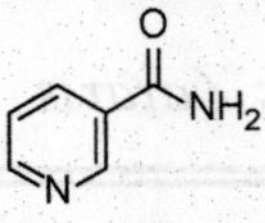 B_3, nicotinamide	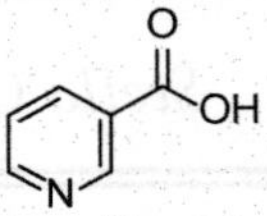 B_4, niacin
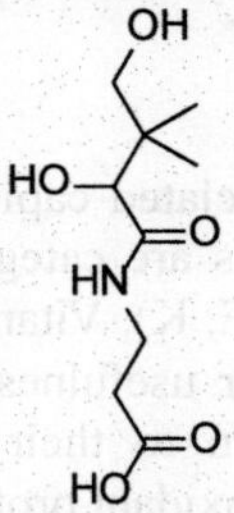 B_5, pantothenic acid	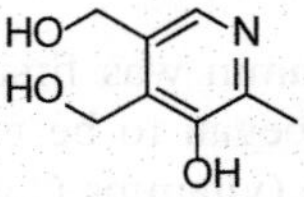 B_6, pyridoxine
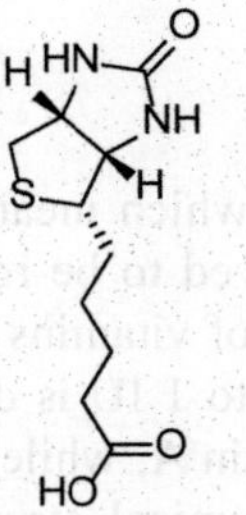 B_7, biotin	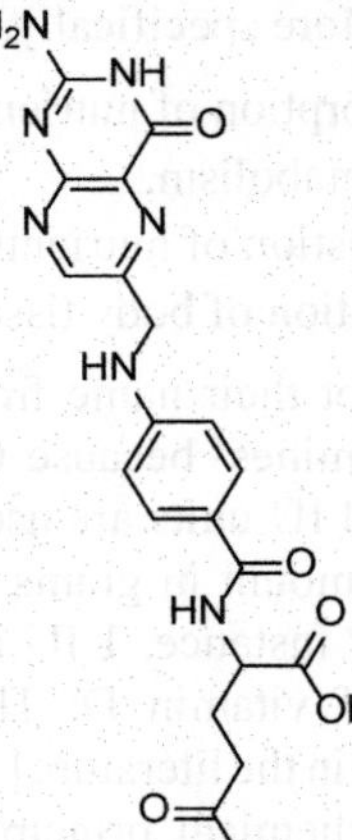 B9, folic acid

Table 1 contd. ...

...Table 1 contd.

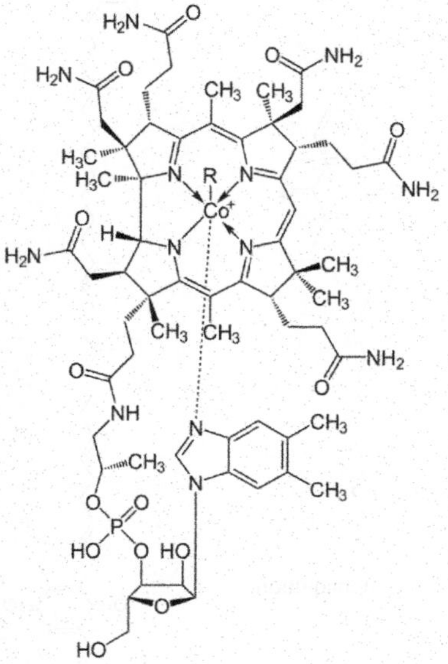 B_{12} cyanocobalamin	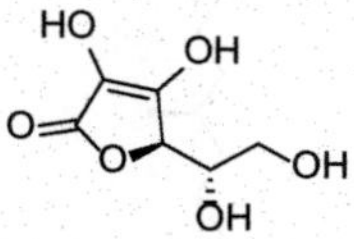 C, ascorbic acid
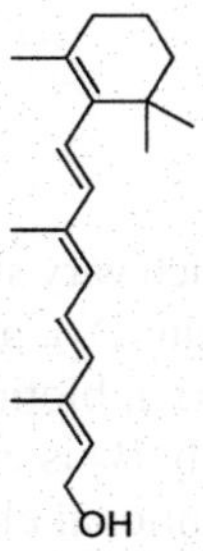 A_1, retinol	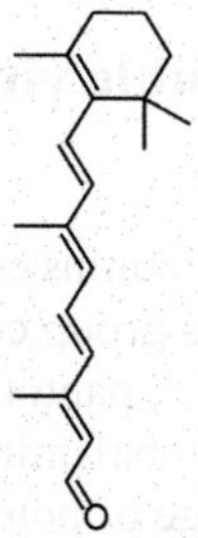 Retinal
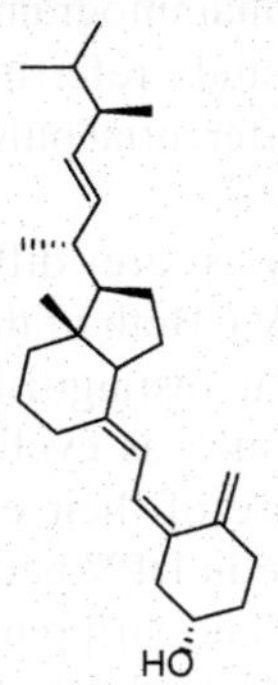 D_2, ergocalciferol	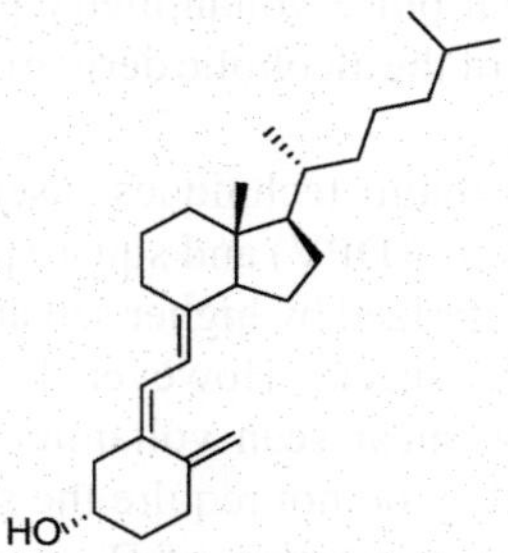 D_3, cholocalciferol

Table 1 contd. ...

...*Table 1 contd.*

 E, tocopherol	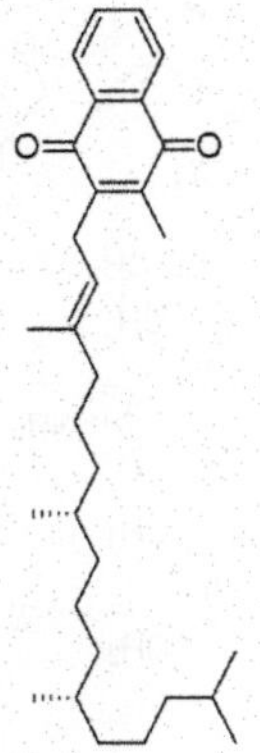 K_1, menaquinone 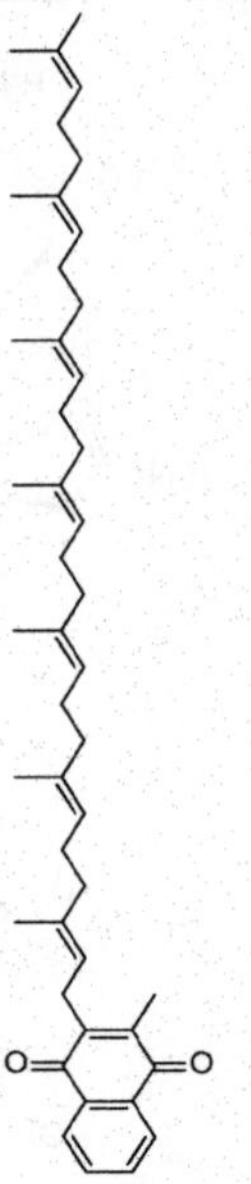 K_2, menaquinone

2. Water-soluble vitamins

2.1 Vitamin B

The group of B vitamins consists of various compounds which vary significantly from one another. This group consists of thiamine (B_1), riboflavin (B_2), nicotinic acid and nicotinamide (B_3), pantothenic acid and its calcium salt (B_5), biotin (B_7), folic acid (B_9), and cyanocobalamin (B_{12}). Of all these vitamins, only B_5 is electrochemically inactive to a range of potentials typically applied to conventional electrodes.

It has been observed that Vit B_5 cannot be detected in some electrolytic systems since no peaks appear in the corresponding cyclic voltammograms recorded by differential pulse voltammetry [15]. There is only one study referring to the redox behavior of the alcoholic derivative D-panthenol using differential pulse voltammetry [16].

The main techniques used in vitamin B analysis are differential pulse voltammetry (DPV) and square pulse voltammetry (SWV). Both of these techniques are characterized by higher sensitivity due to their ability to distinguish faradaic from capacitive current. However, less sensitive techniques such as cyclic voltammetry (CV) and linear scan voltammetry are still used because of their ease of use and because they do not require the same parameters as those in DPV and SWV.

For the detection of B vitamins, the use of cyclic and stripping voltammetry on glassy carbon or mercury electrodes, as well as on electrodes formed with gold, mercury and organic materials has been reported in literature [13].

When it comes to electrode selection, most methods rely on chemically modified electrodes. This is because a suitable modifier can reduce the potential and/or minimize contamination of the electrode surface, and increase the sensitivity and the separation of peaks.

This method is particularly useful in complex systems such as interfering biological samples. Glassy carbon, a material characterized by low porosity, good electrical conductivity and high stability even in corrosive media, is the most widely used substrate.

Among the most innovative modifiers, it is worth mentioning nanoparticles like carbon nanotubes (CNTs), gold nanoparticles, and graphene. Carbon paste electrodes are also still widely used. As a matter of fact, they are an easy way to make modified electrodes [17]. The modifier can be dissolved in a binder or mixed with the paste during homogenization. Carbon pastes are reusable materials that have excellent conductivity and electrocatalytic properties.

Despite the trend towards more "green" chemistry and concerns about the toxicity and waste of mercury, mercury-based electrodes are still widely used in the detection of vitamins. Bismuth electrodes, which appeared a decade ago to replace mercury, have also found some applications in the detection of vitamins. This trend probably reflects the fact that the first electrochemical methods proposed for vitamins were methods of redistribution.

Also noteworthy is the introduction of screen-printed electrodes (SPEs), which have attracted interest in recent years due to their low cost, commercial viability, high repeatability and the absence of time-consuming cleaning procedures (as required for common electrodes).

Regarding the performance of the proposed methods, the limits of detection and their ranges are variable. This is to be expected as it reflects the wide variety of substrates to be studied. As is well known, the concentration of vitamins varies in many orders of magnitude, from very high in pharmaceuticals and dietary supplements, to very low in food and biological fluids.

Furthermore, each substrate needs a specific treatment with the most appropriate method; too much dilution not only minimizes the effect of the substrate but also affects the final accuracy, as the uncertainty should be changed taking dilution into account.

Stripping voltammetric methods were developed for the determination of vitamin B1 by various electrodes such as glassy carbon modified with lead film, or mercury, with ring-shaped electrode with liquid silver amalgam modified (AgLAF-AgSAE) as well as paste electrode DNA/MWCNT [17–20].

Lead has been found to be valuable as a substitute for mercury, given its low toxicity, high potential range, ability to operate over a wide pH range and the fact that its surface can be easily renewed. Two of these methods allow for the simultaneous determination of vitamin B1 with other vitamins such as B_2 and C. The use of the AgLAF-AgSAE electrode gives well-separated peaks for B_1 and C, with a detection limit comparable to mercury electrodes [18–21]. The actual samples studied were biological, pharmaceutical and juices. The detection limits reported in the literature were similar.

The electrochemistry of vitamin B_2 (riboflavin) has been extensively studied in the last 10 years. Stripping voltammetric methods have been developed on Bi-Cu electrodes, amalgams of silver, mercury, hot graphite and DNA-modified electrodes. Carbonated glass electrodes modified with various materials have also been tested for stripping voltammetry. It is worth mentioning that the most novel nanomaterials include poly (3,4 ethylenedioxythiophene)/zirconia nanocomposites (PEDOT/ZrO2NPs), oleamine nanocomposites (OLA), NiO/polyunsaturates and Cr/SnO [21–24]. A single glassy carbon electrode has also been used to detect B_2 in breast milk [25]. The use of Co-Y zeolites on carbon paste electrodes in differential pulse voltammetry measurements has been found to increase electrode selectivity and sensitivity due to the interactions of riboflavin with zeolites [26].

Some of these methods rely on the selective adsorption of riboflavin on the modifier, thus allowing a lower LOD and avoiding interference from other substances. Nanocomposites show excellent catalytic activity for riboflavin detection. Co-determination of riboflavin with ascorbic acid and folic acid has also been studied by researchers [26, 27].

Modifiers help overcome the problems and limitations of simple electrodes. It is a characteristic of classical glassy carbon electrodes to not show stable voltammetric signals for many vitamins in a mixture due to the contamination caused by their oxidation products [27].

Nicotinic acid and nicotinamide (vitamin B_3) can be detected in pharmaceutical samples using cyclic voltammetry and gold electrodes [28]. Gold has also been used as a substrate for the deposition of thioglycolic acid monolayers and the resulting probe used to detect nicotinic acid in foods through stripping voltammetry [29].

Nicotinic acid can be detected in urine samples by amperometric detection at a carbon/TiO_2 nanotube electrode inoculated with La [30]. This electrode prevents interference from other vitamins due to the specialized interactions of the nanocomposite with nicotinic acid. Nicotinamide in pharmaceutical samples has also been detected by polarography using a surfactant system as well as with a carbon paste electrode modified with macrocyclic compounds [31, 32].

As far as vitamin B6 is concerned, many studies have focused on the detection of mainly pyridoxine. In pharmaceutical and urine samples, pyrodoxine has been detected using unmodified glassy carbon electrodes, through cyclic voltammetry and differential pulse voltammetry [33, 34].

Glassy carbon has also been studied as a substrate in pyridoxine sensors, with various modifiers such as DNA, carbon nanotubes, methylene green polymer with modified carbon nanotubes, and carbon nanotubes with Au-CuO, for differential pulse voltammetry; Carbon paste electrodes modified with nanoparticles and ionic liquids, ZnO or ZnO_2, crown ethers and polyurethane have also been found to be able to detect vitamin B6 through square pulse voltammetry [37–40].

It has been successfully detected in various types of samples, including multivitamin tablets, cereals, energy drinks, blood and cerebrospinal fluid, using graphite electrodes modified with molecularly imprinted polymer using bimetallic magnetic nanoparticles as the substrate.

The detection of biotin (Vitamin B_7) has been mentioned very few times in the literature, and only in the last decade. It is associated with the analysis of food, using biosensors. One of the investigations that took place used a composite liquid ion-chitin film with electrochemically deposited Pd-Fe-Ni nanoparticles as the sensor [42], while another reported a composite electrode sensor where the electrochemical signal was based on the enzymatic activity of the enzyme superoxide. Horseradish when adding H_2O_2 as a substrate and o-phenylenediamine as a co-substrate [43]. Square wave voltammetry was used in both the studies.

Folic acid (B_9) has been extensively studied due to its biological importance. Many examples with carbon paste electrodes have been reported for the detection of vitamin B_9. Among the materials used to make the electrodes are TiO_2 nanoparticles composite materials (with differential pulse voltammetry) [44] and carbon nanotubes (with circular and differential pulse voltammetry) [45].

Ru(II)-ZnO complex composites with carbon nanotubes for circular voltammetry and square pulse voltammetry [46], gold nanoparticles for circular voltammetry and square pulse voltammetry, Co for square pulse voltammetry [49], as well as various polymers for circular voltammetry and differential pulse voltammetry have also been studied [50, 51]. The resulting sensors were used to analyze real samples, such as pharmaceutical and biological samples, and food. Glassy carbon electrodes have also been modified with nanostructured materials for differential and quadratic pulse voltammetry [52, 53]; polymeric films for differential pulsed voltammetry [54] and metals for resorption adsorption voltammetry [55] have also been investigated for the detection of vitamins B9 and C. Furthermore, gold electrodes with nanostructured materials in differential pulse voltammetry have also been reported [56].

Stripping voltammetry after the accumulation of different materials such as silver amalgams [57], mercury [58], and bismuth film [59] has also proven to be a technique capable of detecting folic acid, mainly in pharmaceuticals.

Vitamin B12 can be detected using disposable graphite electrodes modified with carbon nanotubes and chitosan, through square pulse voltammetry and cyclic voltammetry [60]; peptide nanotubes also do a good job with square pulse voltammetry [61]. Both these methods have proven to be simple, fast, low-cost and capable of detecting the vitamin in pharmaceutical samples.

The same electrode proposed for use in refolding voltammetry for the detection of folic acid [57] has been shown to be capable of B12 detection as well [62]. Other associated studies with stripping voltammetry have used bismuth films [63] and printed disposable carbon mesh electrodes [64].

2.2 Vitamin C

Ascorbic acid is undoubtedly the most researched vitamin. This is because not only is it the most common antioxidant, but is also found in the brains of mammals, along with various neurotransmitters. Hence, its detection is particularly important for medical science. It is the most common electrochemically active biological compound and is easily oxidized; this forms the basis of its electrochemical detection.

Ascorbic acid forms an irreversible redox pair with dehydroascorbic acid [3]. Oxidation of ascorbic acid involves the release of two electrons and two protons, and the production of dehydroascorbic acid, followed by irreversible hydration when pH is less than 4 [65].

Irreversible hydration yields an electrochemically inert product, 2,3-dicetogulonic acid, which is formed after the dehydroascorbic acid lactone ring opens; the former is easily absorbed on the electrode surface, resulting in contamination.

Bibliographic data on the oxidation of ascorbic acid at pH < 8 reports two consecutive oxidation reactions of an electron, accompanied by immediate dehydration, making the oxidation irreversible. During the oxidation with gold electrodes, two signals corresponding to two stages have been found to be generated. The first stage is a two-electron process in which two protons are exchanged in the pH range 2–4.5, one proton in pH range 4.5–8 and two protons at pH > 8.

In particular, at pH values less than the pKa1 of L-ascorbic acid, which is about 4.5, two protons are exchanged. At higher pH levels, an electron is exchanged with the ascorbate anion as an electroactive species. These observations are consistent with the variations in peak dynamics as a function of pH for pH levels up to 8.

Identification of ascorbic acid oxidation products is difficult as the study of the second oxidation at pH >8 is not possible since ascorbic acid and dehydroascorbic acid are unstable in basic solutions. The intermediate, the ascorbate anion, has been found to be oxidized to diketolactone, and then dehydrated to dehydroascorbic acid which is rearranged into another diol further oxidized at higher potentials [66]. For this electrochemical pair, the height of the anodic peak relative to the concentration of the substance to be analyzed corresponds to the oxidation of the reduced form.

The stoichiometric formation of hydrogen peroxide from ascorbic acid has been studied with square pulse voltammetry in a standard wine system, with a hanging mercury drop electrode. Ascorbic acid and peroxide were detected in the same sample initially by an ascending scan for ascorbic acid, and a downstream for hydrogen peroxide. The potential was scanned from 0 to 400 mV for ascorbic acid. The resulting voltammograms were repeatable and the repeatability of the ascorbic acid peak current was considerably good, with a relative deviation of less than 1% [67].

Although L-ascorbic acid is one of the most electroactive biomolecules, it is generally difficult to determine its concentration with unmodified carbon electrodes or pure metal electrodes due to their susceptibility to contamination. However, several studies have been performed with the aim of sustainable detection of ascorbic acid using such electrodes, which include necessary stages of pre-treatment or repeated stages of electrochemical purification.

More specifically, ascorbic acid can be detected using a gold electrode with linear scan voltammetry, a gold monocrystalline electrode with cyclic voltammetry and differential pulse voltammetry, a platinum electrode with cyclic voltammetry (L = 0.25 mM), a carbon paste electrode with cyclic voltammetry (LOD = 0.21 mM LOQ = 0.068 mM), and a glassy carbon electrode with differential pulse voltammetry [67–69].

The need to reduce the potential and minimize electrode contamination has led to attempts at modifying electrodes to increase their sensitivity and enhance the separation of the peaks, especially for complex media such as biological samples, where ascorbic acid coexists with other electroactive compounds.

Linear scan voltammograms with gold electrodes modified with dimercaptothiadiazole layers have shown that a thin layer of dimercaptothiadiazole formed in the process separates the uric acid and ascorbic acid voltammetric signals [70]. In one study, ascorbic acid was detected simultaneously with dopamine, using a gold electrode modified with gold nanoparticles [71].

One of the most convenient and easy-to-use materials for preparing modified electrodes is undoubtedly carbon paste. The modifier can be dissolved in a binder or mixed with the paste during homogenization. Pastes are dynamic and renewable materials with excellent conductivity and electrocatalytic properties; hence, they have a wide range of applications in electrochemical or bioelectrochemical sensors [17, 72].

The response of ascorbic acid to cyclic voltammetry measurements has been studied with an unmodified carbon paste electrode and a carbon paste electrode modified with tetrabromo-p-benzoquinone. The superpotential was observed to be reduced by about 430 mV and the current peak was amplified with the modified electrode, which was effective not only in detecting ascorbic acid, dopamine and uric acid, but also in simultaneous voltammetric quantification of their concentration in the mixture.

The electrolytic behavior of a modified carbon paste electrode with methionine for ascorbic acid electro-oxidation has been studied through cyclic voltammetry. The results show a good amplitude of current at the anode peak for the modified electrode (compared to the unmodified one). This method has been adapted to detect ascorbic acid in real samples and the results have been considered satisfactory [74]. Carbon paste electrode modified with 2,2'-(1,8-octanedhyldisenitriloethylidine)-bis-hydroquinone show a high electrocatalytic activity towards ascorbic acid. The current is significantly amplified, compared to an unmodified electrode [75].

In a study, different porphyrins were used in the design of seven carbon paste and diamond paste electrodes for their use in the detection of ascorbic acid by differential pulse voltammetry. The detection limits were quite low – between 1.1×10^{-14} M and 5.1×10^{-7} M. Detection of ascorbic acid in pharmaceutical samples and beverages showed recovery rates greater than 92% and 91.5% respectively. The surface of the electrodes was easily renewable by polishing [76].

A graphene-inoculated carbon paste electrode was prepared by adding graphene to the carbon paste mixture in another study. Compared to the single electrode, an improved electrochemical response was found with the graphene electrode; this was attributed to the excellent electrical conductivity of graphene. This electrode was further used to detect ascorbic acid, with low hyperpotential, enhanced current response and good sensitivity [77].

The voltammetric response of manganese dioxide composite electrode with graphite, for ascorbic acid detection has been studied in a phosphate buffer of pH 7.2. Therein, the oxidation current was found to increase with increasing amount of ascorbic acid and the voltammetric peaks appeared for concentrations down to 250 μM [78].

Other carbon-based materials (e.g., glassy carbon) have been used as modifiable electrodes due to their high electrical conductivity, chemical resistance, wide current suitability, and compatibility with acidic media. The glassy carbon electrode has been modified by electrochemical oxidation in mildly acidic media (0.1 M H_2SO_4) for the single and simultaneous detection of ascorbic acid, dopamine and uric acid [79]. In this case, single and modified electrode cyclic voltammetry responses were recorded in a phosphate buffer. The modified electrode gave a redox pair, which was explained by the formation of functional groups during the electrochemical pre-treatment. In particular, the presence of these functional groups led to a reduction in the oxidation superpowers relative to the single glassy carbon electrode, allowing for a simple and sensitive simultaneous detection of the three substances [79].

Polymeric films have been found to enhance the electrocatalytic activity of ascorbic acid by increasing the rate of electron transfer between the glassy carbon electrode and the polymeric film. In one study, a fixed modified glassy carbon electrode based on poly (3-(5-chloro-2-hydroxyphenylase)-4,5-dihydroxynaphthalen-2,7-disulfonic acid) film was prepared by electrochemical polymerization to study its electrochemical behavior with cyclic voltammetry. The modified electrode showed electrolytic activity towards ascorbic acid, dopamine and uric acid, at pH 4, with a decrease in the superpowder by about 0.12, 0.35 and 0.50 V respectively [80].

An electrode formed by a poly (caffeic acid) film deposited on a glassy carbon electrode surface can be used to detect ascorbic acid, dopamine and their mixture through cyclic voltammetry [81]. This modified electrode has good sensitivity, selectivity and stability and can be applied to the analysis of drug samples, yielding satisfactory results.

Another voltammetric method for the detection of ascorbic acid is the use of poly (modified bromocresol violet) glass electrode. Cyclic voltammetry studies show that compared to the single electrode, the modified electrode showed better electrolytic activity towards ascorbic acid by reducing the oxidation potential by about 240 mV and increasing the current response. Ascorbic acid recoveries were also good – between 99.78 and 102.54% [82].

The use of conductive polymers offers unparalleled opportunities for translating analyst-recipient interactions into measurable responses. The advantages of conductive polymer sensors over chemical sensors stem from the ability of conductive polymers to exhibit properties that respond to even minor system disturbances [83].

A team of researchers prepared thin polyaniline films doped with polyvinylsulfonate and polystyrene sulphide anions on platinum microelectrodes and studied the oxidation of ascorbic acid in various media, including wine and port juice samples. The electrodes maintained their electroactivity in neutral pH solutions. The differences found in the amplitude of the currents demonstrate the effect of inoculation ions on the redox properties of polyaniline, showing possible means of modifying its conductivity and morphology [84].

In the presence of the cationic surfactant CTAB adsorbed on the surface of a glassy carbon electrode, a significant catalytic effect on the oxidation of ascorbic acid has been observed [85]. Another glassy carbon electrode with the surfactant was modified with a chitosan film with cetylpyridinium bromide and used for the detection of uric acid and ascorbic acid through differential pulse voltammetry. This electrode showed effective electrocatalytic activity and satisfactory selective resolution for both the acids [86].

A new glassy carbon electrode modified with copper diphenate complex has been reported for use in cyclic voltammetry [87]. It showed very effective electrocatalytic action against the anodic oxidation of dopamine and ascorbic acid, with a significant reduction in superconductivity. The electrode was found to be stable and reproducible, and yielded satisfactory results in the analysis of ascorbic acid and dopamine in pharmaceutical samples and food [87].

Standard analysis curves for ascorbic acid, dopamine and uric acid have been found to be linear across a range of concentrations using a glassy carbon electrode modified with LaFeO3 nanoparticles. Good correlation coefficient values were also observed therein [88].

In one study, a downstream pre-treated boron-grafted diamond electrode was used for the simultaneous ascending detection of ascorbic acid and caffeine by differential pulse voltammetry. This method was successfully deployed in pharmaceutical samples and the results were confirmed by high pressure liquid chromatography [89].

Carbon nanotubes have excellent electrical, chemical, mechanical and structural properties, which make them particularly attractive for applications in chemical sensors in general, and for electrochemical sensors in particular. In addition to their enhanced electrochemical activity, electrodes modified with carbon nanotubes have been found to be useful in the accumulation of biomolecules and in the reduction of surface pollution. Remarkable sensitivity and conductivity allow the use of carbon nanotubes as sensitive nanoscale sensors [90].

Following the above, a new electrochemical method was studied for *in vivo* measurements of ascorbic acid in mouse brains, with carbon fiber electrodes modified with multi-walled carbon nanotubes. The technique was based on the electrochemical property of nanotubes to promote the oxidation of ascorbic acid [91]. Simultaneous voltammetric detection of ascorbic acid and rutin on a carbon paste electrode modified with a carbon-chitosan nanotube composite film has also been achieved [92].

Simultaneous selective detection of ascorbic acid, paracetamol and tryptophan by square pulse voltammetry, with a carbon paste electrode modified with multi-walled carbon nanotubes, has been reported in literature [93]. The use of a carbon paste electrode modified with graphite-multi-walled carbon nanotubes and a cationic surfactant has also been proposed. The electrochemical response of this modified electrode to ascorbic acid was studied through cyclic voltammetry and differential pulse voltammetry at pH 2. Compared to activated carbon or graphite, the modified electrode not only shifted the oxidation potential of ascorbic acid to less positive potentials, but also amplified the peak current. Moreover, the oxidation of ascorbic acid was very stable, allowing for its quantification in pharmaceutical samples and food [94].

Differential pulse voltammetry has been applied to the simultaneous detection of ascorbic acid, dopamine and uric acid, with a sensor based on helical carbon nanotubes. A water-soluble cationic polymer, poly (diallyldimethylammonium chloride), was used to modify the surface of the nanotubes, with the aim of improving their dispersion and adhesion. The current responses with this modified electrode were much enhanced in differential pulse voltammetry than in cyclic voltammetry. The properties of the electrode facilitated the simultaneous determination of ascorbic acid, dopamine and uric acid in fetal bovine serum samples [95].

In another study, a new composite carbon electrode containing the ionic liquid n-octylpyridinium hexafluorophosphate and single-walled carbon nanotubes was prepared and studied. Compared to other graphite and paraffin oil composite electrodes, this electrode showed a significant increase in the electron transfer rate for the electroactive compound and a significant reduction in the superconductivity of the ascorbic acid reaction. Furthermore, the new electrode had a strong electrocatalytic effect and was used for the quantification of ascorbic acid in real food samples [96].

A new type of modified electrode sensor for ascorbic acid was prepared by depositing multi-walled and poly (Nile blue A) carbon nanotubes on a glassy carbon electrode surface in yet another study. Nile blue was electropolymerized either above the nanotube layer. The electrodes were studied through cyclic voltammetry and the best results were found with the modified polymer electrodes under the thinnest layer of nanotubes. Ascorbic acid was quantified using cyclic voltammetry; the modified electrodes exhibited good sensitivity, large linear range, a detection limit of 1.6 μM, and good stability [97].

Poly (xanthuric acid) and multi-walled carbon nanotube composites can be prepared to modify glass glass electrode for the detection of ascorbic acid [98]. A well-defined redox pair is identified as a result of the redox reaction of poly (xanthuric acid). The composite is found to be stable at different scan rates and pH values, and the electrode reduces the superpotential, while increasing the current responses. With a potential of +300 mV, its sensitivity to ascorbic acid detection is 160.2 μA/mmol cm^2 [98].

Recently, a graphene/platinum-modified glassy carbon electrode was prepared for the simultaneous quantification of ascorbic acid, dopamine and uric acid levels through cyclic voltammetry and differential pulse voltammetry. In the said case, platinum nanoparticles with an average diameter of 1.7 nm self-assembled on the surface of graphene [99].

A high response sensor for ascorbic acid, using oxidized polypyrrole and palladium nanoparticles, has also been reported [100]. In the presence of palladium nanoparticles, polypyrrole was plated on a gold electrode through cyclic voltammetry, and oxidized in NaOH solution. The results showed that the modified electrode had the ability to catalyze the oxidation of ascorbic acid by reducing its oxidation potential to 0 V, with a detection limit of 1 μM [100].

A simple and effective strategy proposed for the detection of ascorbic acid is the synthesis of nickel nanoparticles dispersed in poly (1,5-diaminonaphthalene). The electrochemical characterization of the electrode shows a stable redox behavior of the Ni(III)/Ni(II) pair in 0.1 μM NaOH solution. Here, the electro-oxidation of glucose, ascorbic acid and dopamine in alkaline solutions can be studied by quadratic pulse voltammetry; the limit of detection of ascorbic acid has been found to be 0.01 μM [101].

Carbon nanotubes are sophisticated, ideal materials for supporting metal nanoparticles in electrocatalysis. Thus, composite carbon nanoparticle-composite films combine the advantages of enhanced electrocatalytic action and large surface area. Following this, a new bimetallic film with platinum and gold nanoparticles in Nafion polymer and carbon nanotubes, developed potentiostatically, has been reported. The composite film was found to be able to detect ascorbate anion, epinephrine and urate anion through cyclic voltammetry and differential pulse voltammetry [102].

A carbon-ceramic material, $SiO_2/C/Nb_2O_5$, has also been used to develop electrodes, and studied using cyclic voltammetry and differential pulse voltammetry. It showed the ability to improve electron transfer between the electrode surface and ascorbic acid. The presence of Nb_2O_5 on the surface of SiO_2/C reduced the oxidation potential of ascorbic acid by about 180 mV. The peak current was significantly amplified compared to the ascorbic acid response on the surface of the SiO_2/C electrode. This behavior points towards the electrocatalytic effect of the $SiO_2/C/Nb_2O_5$ electrode on the oxidation of ascorbic acid. The voltammetric response of ascorbic acid to this electrode may be due to its interactions with Nb_2O_5, which involves the formation of covalent bonds, thus leading to very rapid oxidation [103].

Modified carbon-ceramic electrodes with single-walled carbon nanotubes have also been used for the simultaneous detection of paracetamol and ascorbic acid by differential pulse voltammetry. The electrodes showed excellent electrochemical catalytic activity compared to simple glassy carbon electrodes [104].

3. Fat-soluble vitamins

3.1 Vitamin A

The conversion between the different forms of vitamin A compounds involves oxidation and reduction processes, which is suitable for their study by electrochemical methods. There are some studies describing the voltammetric properties of retinol, retinal and retinoic acid, but most electrochemical studies have been performed as a means of quantifying their levels in biological samples.

Through cyclic voltammetry, retinol has been found to be oxidized at a potential of +0.8 V with an Ag/AgCl electrode in tetrahydrofuran with tetrabutylammonium perchloride, in methanol with acetate buffer, and in an acetonitrile/water mixture with $LiClO_4$ [105]. Its oxidation is irreversible since only one oxidation peak is observed and the electrochemical oxidation of retinol to tetrahydrofuran takes place with the exchange of one electron per molecule. However, the oxidation of retinal to tetrahydrofuran takes place at +1 V with Ag/AgCl electrode and is carried out by the transfer of up to 4 electrons per molecule, and is also irreversible.

Electrochemical studies suggest that retinoids can be reduced by platinum electrodes or electrodes coated with mercury and lead [106]. Retinol is reduced at about –2 V, whereas retinal is reduced at –1.42 V with respect to the reference black electrode. These reductions take place through the transfer of one electron at cyclic voltammetry times (seconds) but with a larger number of electrons at electrolysis times (minutes).

Furthermore, while retinol is reduced by a chemically irreversible process at a scan rate of 0.1 V/s, retinal reduction is found to be chemically reversible since the ratio of upward ipox to downward ipred is 1 for a scan rate of 0.1 V/s [107].

Carotenoids found in nature have been extensively studied by voltammetric methods (cyclic and quadratic pulse). They are subjected to oxidation by two processes of one electron transfer in CH_2Cl_2 between +0.5 and +1 V with respect to the reference calomel electrode [108, 109]. The voltammetric behavior of carotenoids has been studied mainly in CH_2Cl_2 due to their good solubility in it.

Many carotenoids can also be reduced to various stages in CH_2Cl_2 at potentials more negative than –1 V with respect to the reference black electrode. More attention has been paid to the oxidation mechanism because the important biological properties of these compounds (antioxidant properties and photoprotection) are considered to be related to cations. For most carotenoids, the first oxidation (E_1) occurs at +0.5 to +0.72 V and the second (E_2) between +0.52 and +0.95 V vs Ag/AgCl.

Some carotenoids show oxidation by 2 electrons in cyclic voltammetry experiments, while others (such as β-carotene) show only one oxidation, which corresponds to the transfer of two electrons in two processes of one electron transfer occurring almost simultaneously [108].

In the case of β-carotene, it is believed that the second oxidation of one electron during cyclic voltammetry takes place at a very similar potential to the first and, therefore, only one oxidation peak is observed, corresponding to the chemically reversible oxidation of two electrons of β-carotene in the oxidized form with oxidation number +2, through an intermediate with oxidation number +1 [109].

The oxidized forms of carotenoids are not very stable at room temperature, and only survive for a few seconds in CH_2Cl_2. Nevertheless, this is sufficient to observe the reverse reduction process in cyclic voltammetry.

3.2 Vitamin D

The electrochemical properties of vitamin D have not been extensively reported in the literature [111]. A study of vitamins D2 and D3 through cyclic voltammetry in methanol with acetate buffer concluded that both of these compounds can be irreversibly oxidized, although the number of electrons involved in the oxidation process was not reported [112]. The oxidation was considered irreversible because only one oxidation peak was detected during the scan, and no reduction peak was observed when the scan direction was reversed. The potential of the Epox oxidation peak for both the vitamins was found to be approximately +1.1 V with respect to the black reference electrode, and no reduction in negative potential values was detected.

Cyclic voltammetry experiments have also been performed on vitamin D2 in ethanol, with LiCO4 as the electrolyte, and a chemically irreversible oxidation peak at about +1.2 V relative to the Ag/AgCl reference electrode was detected [111]. Both the studies, in methanol and ethanol, were performed on a glassy carbon electrode, and these vitamins were found to be adsorbed during the oxidative scan. Adsorption involves the passivation of the electrode surface by oxidized products, leading to continuously reduced oxidative currents during continuous scans.

More recently, vitamins D_2 and D_3 have been studied through cyclic voltammetry using glassy carbon and platinum electrodes, and both the compounds showed similar behavior by transferring one electron during their irreversible oxidation [113]. Vitamins D_2 and D_3 have also been studied through cyclic voltammetry and differential pulse voltammetry, using a glassy carbon electrode; here, they exhibited different electrochemical behaviors due to differences in their solubility [114].

In one study, vitamin D_3 was detected in real samples by cyclic voltammetry and differential pulse voltammetry, using a modified glassy carbon electrode with $Ni(OH)_2$ particles in an inorganic SiO_2 matrix/graphene oxide [115].

3.3 *Vitamin E*

The voltammetric behavior of vitamin E, α-tocopherol in particular, has been extensively studied over the last 40 years. It has been found to undergo some interesting redox reactions where the intermediates have remarkably long life times compared to the oxidative intermediates of other phenols.

The general electrochemical mechanism of α-tocopherol (which also applies to β, γ and δ tocopherols) in organic solvents CH_3CN or CH_2Cl_2 [116]. The path followed depends on whether the oxidation takes place in the presence or absence of fat-soluble acids and bases. The R chain does not affect the electrochemical properties; so, when it is replaced by a methylene group in a standard compound, the resulting voltammograms are identical.

In the absence of an acid or base, α-tocopherol is oxidized by an electron at a potential of about +0.5 V relative to a standard F_c/F_c^+ electrode and forms α-TOH. The deprotonation rate constant calculated by simulating variable scan rate through cyclic voltammetry in CH_3CN was estimated to be $3 \pm 2 \times 104\ s^{-1}$ [117, 118].

Because the oxidation potential of the α-TO radical is lower than that of α-tocopherol, the α-TO radical oxidizes directly at the electrode surface and forms the α-TO^+ transmagnetic cation. In general, the oxidation reaction can be considered to take place through ECE mechanism (electron transfer, chemical reaction, electron transfer).

However, there is another possibility where the second electron transfer step follows a homogeneous redistribution mechanism, which also produces α-TO^+. Regardless of the exact course of the reaction, the oxidation to CH_3CN and CH_2Cl_2 is completely chemically reversible in the fast time frames of cyclic voltammetry and as well as during electrolysis; so, by applying the reduction potential, the starting material can be significantly recovered [119, 120]. The position and shape of the reduction peak are affected by the moisture traces present in the solvent [117].

Both the aforementioned processes involve the transfer of two electrons and one proton. The reason why there is a considerable difference between the reduction potential of the two phases is that the elementary phases of electron transfer take place at different potentials and the transfer of the proton takes place between them.

One difference observed in the electrochemical behavior of α-tocopherol compared to other phenols is that it shows a reduction peak after oxidation [117, 118].

Although all phenols follow the oxidation mechanism mentioned above, a majority of them form very unstable transmagnetic cations when oxidized and therefore show only the peak of oxidation. Moreover, electrochemical oxidation of phenols in aprotic solvents usually results in a large number of products, including dimers formed by the irreversible dimeric bonding of radicals [121–124].

When α-tocopherol is deposited directly on the surface of an electrode as drops and subsequent voltammetric oxidation takes place in aqueous solutions, paraquinone is formed immediately in large yields over a wide pH range (1–13), probably due to the reaction of α-TO^+ with water or hydroxide ions [125, 126].

3.4 Vitamin K

Electrochemical studies with aprotic organic solvents with low water content have shown that vitamin K behaves in a manner as expected for a quinone-structured compound and is, therefore, reduced in 2 steps involving one electron transfer each [127–130].

The mechanism of electrochemical reduction of vitamin K1 and other quinones in aprotic organic solvents has been described in the literature [125]. It has been noted that although it is known that water is always present as an impurity in organic solvents, electrochemists rarely measure or report the percentage of moisture in the solvents they use. Therefore, recent studies have suggested the use of the voltammetric behavior of vitamin K_1 as a method for estimating the amount of moisture in solvents.

The difference between the dynamics of the first and second electron transfer is measured and compared with the exact water content in the solvent as determined by Karl Fischer titration equations. Standard curves are generated for various solvents that allow voltammetric scanning in the presence of vitamin K_1 for determining the level of moisture traces in them [127, 130].

Voltammetric experiments have also been performed on vitamin K_1 in aqueous media over a wide range of pH values, either by depositing it as an oil on the electrode surface [131] or by incorporating it into the phospholipids or alkanethiols bound to the electrode surface [132–134].

When vitamin K_1 is reduced in an aqueous medium at a low pH, it is reduced by 2 electrons and 2 protons to form hydroquinone. The reduction is chemically reversible and occurs in a voltammetric step, although there might be elementary electron and proton transfers that occur simultaneously. In aqueous media, the reverse oxidation of hydroquinone occurs with 2 electrons and 2 protons and is evident from one voltammetric peak [135, 136].

An interesting question is whether the hemiquinone is important in the biological reactions of vitamin K, since in the presence of water or protons, its reduction always occurs with 2 electrons. However, in some cases, the anionic radical of hemiquinone has been observed in aqueous solutions, possibly due to a shift in the equilibrium of the reaction involving the dione [137].

In Table 2 are summarized the main features of voltammetric techniques for the detection of water-soluble vitamins. In Table 3 are summarized the main features of voltammetric techniques for the detection of fat-soluble vitamins.

Table 2. Main features of voltammetric techniques for the detection of water-soluble vitamins.

Vitamin	Technique	Electrode	Substrate	Reference
B_1	Stripping voltammetry	Glassy carbon/lead	Pharmaceuticals/ food	[18]
	Differential pulse voltammetry	Carbon nanotubes modified with DNA	Human blood	[19]
	Stripping voltammetry	Solid silver amalgam modified with liquid silver amalgam film	Pharmaceuticals, juice	[20]
B_2	Stripping voltammetry	Solid silver amalgam modified with liquid silver amalgam film	Pharmaceuticals, juice	[20]
	Stripping voltammetry	Copper plated with bismuth film	Pharmaceuticals	[22]
	Differential pulse voltammetry	Glassy carbon modified with Poly (3,4 ethylenedioxythiophene)/ Zirconia	Mixture B2, B6, C	[24]
	Differential pulse voltammetry	Glassy carbon	Breast milk	[25]
	Cyclic voltammetry	Carbon paste modified with zeolite Co^{2+} -Y	Polyvitamins formulation	[26]
	Differential pulse voltammetry	Glassy carbon modified with 3-amino-5-mercapto-1,2,4-triazole	Blood plasma	[27]
B_3	Cyclic voltammetry	Gold electrode	Pharmaceuticals	[28]
	Cyclic voltammetry	Gold coated with mercaptoacetic acid	Food	[29]
	Cyclic voltammetry	Carbon nanotube paste/TiO_2, Lanthanum doped	Urine	[30]
	Cyclic voltammetry, Differential pulse voltammetry	HMDE	Polyvitamins formulation	[31]
B_6	Cyclic voltammetry, Differential pulse voltammetry	Glassy carbon modified with Poly (3,4 ethylenedioxythiophene)/ Zirconia	Mixture B2, B6, C	[24]
	Cyclic voltammetry, Differential pulse voltammetry	Glassy carbon	B6 formulation	[33]
	Cyclic voltammetry	Glassy carbon	Ανάλυση βιταμερών B6	[34]
	Differential pulse voltammetry	Graphite modified with methylene green and carbon nanotubes	Pharmaceuticals	[35]
	Cyclic voltammetry, Differential pulse voltammetry	Modified glassy carbon with Au-CuO nanoparticles and carbon nanotubes	Polyvitamins	[36]
	Square wave voltammetry	Carbon paste modified with ZnO nanoparticles	Food	[37]

Table 2 contd. ...

...Table 2 contd.

Vitamin	Technique	Electrode	Substrate	Reference
	Square wave voltammetry	Polyurethane modified graphite	Pharmaceuticals	[38]
	Square wave voltammetry	Modified carbon paste with crown ethers	Polyvitamins	[39]
	Square wave voltammetry	Carbon paste modified with ZrO_2 and ionic liquids	Food	[40]
	Square wave voltammetry	Graphite modified with MIPs and nanoparticles Fe/Cu	Pharmaceuticals, blood	[41]
B_7	Square wave voltammetry	Glassy carbon modified with Pd-Fe-Ni nanoparticles	Infant milk, liver, eggs	[42]
B_9	Differential pulse voltammetry	Modified glassy carbon	Blood plasma	[27]
	Differential pulse voltammetry	Carbon paste modified with TiO_2 nanoparticles	Injections of deoxyholic acid	[44]
	Cyclic voltammetry, Differential pulse voltammetry	Glassy carbon modified with nanoparticles and ionic liquid	Food	[45]
	Cyclic voltammetry, Differential pulse voltammetry	Carbon paste modified with Ru(III) complex-ZnO/carbon nanotubes	Urine, food, pharmaceuticals	[46]
	Square wave voltammetry	Carbon paste modified with gold nanoparticles	Blood plasma	[47]
	Cyclic voltammetry, Square wave voltammetry	Carbon paste modified with Fe_3O_4 nanoparticles	Medicines, blood plasma, lemon juice	[48]
	Square wave voltammetry	Carbon paste modified with nanomalgam Pt: Co	Food	[49]
	Cyclic voltammetry, Differential pulse voltammetry	Graphite modified with cobalt (II) tetraaminophthalocyanine	Pharmaceuticals, urine	[50]
	Cyclic voltammetry	Modified carbon paste with nickel ions dispersed in poly (o-anisidine) film	Pharmaceuticals	[51]
	Cyclic voltammetry, Differential pulse voltammetry	Glassy carbon modified with platinum nanoparticles inoculated with carbon nanotubes	Pharmaceuticals, food	[52]
	Cyclic voltammetry, Differential pulse voltammetry	Glassy carbon modified with platinum nanoparticles inoculated with carbon nanotubes	Pharmaceuticals, food	[52]
	Cyclic voltammetry, Differential pulse voltammetry	Gold modified with polypyrrole/ polyoxetallic anions/gold nanoparticles	Polyvitamins, blood serum	[53]

Table 2 contd. ...

...Table 2 contd.

Vitamin	Technique	Electrode	Substrate	Reference
	Cyclic voltammetry, Differential pulse voltammetry	Glassy carbon modified with 5-amino-2-mercapto-1,3,4-thiodiazole	Blood serum	[54]
	Stripping voltammetry	Lead-coated glassy carbon	Pharmaceuticals	[55]
	Differential pulse voltammetry	Gold modified with gold nanoparticles	Folic acid tablets, food	[56]
	Differential pulse voltammetry	Solid silver amalgam	Juice	[57]
	Stripping voltammetry	Hanging mercury drop	Multivitamins	[58]
	Stripping voltammetry	Bismuth modified glassy carbon	Pharmaceuticals	[59]
B_{12}	Cyclic voltammetry, Square wave voltammetry	Graphite modified with chitosan/ carbon nanotubes	Pharmaceuticals	[60]
	Square wave voltammetry	Graphite modified with peptide nanotubes	Pharmaceuticals	[61]
	Cyclic voltammetry, differential pulse voltammetry, linear scan voltammetry	Solid silver amalgam	Multivitamins	[62]
	Cyclic voltammetry, adsorptive voltammetry	Glassy carbon plated with bismuth	Pharmaceuticals	[63]
	Stripping voltammetry	Screen printed electrode	Pharmaceuticals	[64]
C	Stripping voltammetry	Solid silver amalgam modified with liquid silver amalgam film	Pharmaceuticals, juices	[20]
	Differential pulse voltammetry	Poly (3,4 ethylenedioxythiophene) glass-modified glass/Zirconia	Mixture of vitamins B2, B6, C	[24]
	Differential pulse voltammetry	Glass-modified carbon with 3-amino-5-mercapto-1,2,4-triazole	Blood plasma	[27]
	Cyclic voltammetry, Square wave voltammetry	Carbon paste modified with Ru (III)-ZnO complex/carbon nanotubes	Urine, pharmaceuticals, food	[46]
	Cyclic voltammetry, Square wave voltammetry	Carbon paste modified with Fe_3O_4 nanoparticles	Pharmaceuticals, blood plasma, lemon juice	[48]
	Cyclic voltammetry, Differential pulse voltammetry	Glass-modified carbon with 5-amino-2-mercapto-1,3,4-thiodiazole	Blood serum	[54]

Table 2 contd. ...

...Table 2 contd.

Vitamin	Technique	Electrode	Substrate	Reference
	Cyclic voltammetry, linear scan voltammetry, Differential pulse voltammetry, Square wave voltammetry	Glassy carbon	Pharmaceuticals	[65]
	Square wave voltammetry	Hanging mercury drop	Wine	[67]
	Cyclic voltammetry, Differential pulse voltammetry	Glassy carbon	Pharmaceuticals, juices	[68]
	Linear sweep voltammetry	Gold modified with dimercaptothiodiazole	Simultaneous determination of uric and ascorbic acid	[70]
	Cyclic voltammetry, Differential pulse voltammetry	Gold modified with gold nanoparticles	Simultaneous determination of ascorbic acid and dopamine	[71]
	Cyclic voltammetry, Differential pulse voltammetry	Carbon paste modified with tetrabromo-p-benzoquinone	Simultaneous determination of uric, ascorbic acid and dopamine	[73]
	Cyclic voltammetry, Differential pulse voltammetry	Carbon paste modified with methionine	Juices	[74]
	Cyclic voltammetry, Differential pulse voltammetry	Carbon paste modified with hydroquinone	Pharmaceuticals	[75]
	Cyclic voltammetry, Differential pulse voltammetry	Carbon and diamond paste modified with porphyrins	Pharmaceuticals, drinks	[76]
	Cyclic voltammetry, Differential pulse voltammetry	Carbon paste modified with graphene	–	[77]
	Linear sweep voltammetry	Carbon powder modified with magnesium dioxide	–	[78]
	Cyclic voltammetry	Oxidized glassy carbon	Simultaneous determination of uric, ascorbic acid and dopamine	[79]

Table 2 contd. ...

...Table 2 contd.

Vitamin	Technique	Electrode	Substrate	Reference
	Differential pulse voltammetry	Glassy carbon modified with a disulfonic acid derivative	Simultaneous determination of uric, ascorbic acid and dopamine	[80]
	Cyclic voltammetry	Glassy carbon-modified polycarbonate (caffeic acid)	Simultaneous determination of uric, ascorbic acid and dopamine	[81]
	Cyclic voltammetry, Differential pulse voltammetry	Glassy carbon electrode modified with Poly bromocresol violet)	Pharmaceuticals	[82]
	Square wave voltammetry	Glassy carbon electrode modified with Poly (3,4- ethylene dioxythiophene)	Simultaneous determination of uric, ascorbic acid and dopamine	[83]
	Cyclic voltammetry	Platinum modified with polyaniline	Drinks	[84]
	Differential pulse voltammetry	Glassy carbon electrode modified with chitosan and surfactant	Simultaneous determination of uric and ascorbic acid	[85]
	Cyclic voltammetry	Glassy carbon electrode	Simultaneous determination of ascorbic acid, hydroquinone and nitrobenzene	[86]
	Cyclic voltammetry, Differential pulse voltammetry	Glassy carbon electrode modified with copper dinuclear complex	Simultaneous determination of ascorbic acid and dopamine in food and medicine	[87]
	Cyclic voltammetry	Glassy carbon electrode modified with $LaFeO_3$ nanoparticles	Simultaneous determination of ascorbic acid and dopamine in food and medicine	[88]
	Differential pulse voltammetry	Boron doped dimamond electrode	Simultaneous determination of ascorbic acid and caffeine in medicine	[89]
	Cyclic voltammetry	Carbon fiber modified with carbon nanotubes	Mice brains	[91]

Table 2 contd. ...

...Table 2 contd.

Vitamin	Technique	Electrode	Substrate	Reference
	Cyclic voltammetry, Differential pulse voltammetry	Carbon paste modified with chitosan/carbon nanotubes	Simultaneous determination of ascorbic acid and rutin in drugs	[92]
	Cyclic voltammetry, Square wave voltammetry	Carbon paste modified with carbon nanotubes	Simultaneous determination of ascorbic acid, paracetamol and tryptophan	[93]
	Cyclic voltammetry, Differential pulse voltammetry	Graphite paste modified with carbon nanotubes and surfactant	Pharmaceuticals, food	[94]
	Cyclic voltammetry, Differential pulse voltammetry	Glassy carbon modified with carbon nanotubes and cationic polymer	Simultaneous determination of uric, ascorbic acid and dopamine in fetal bovine serum	[95]
	Cyclic voltammetry, Differential pulse voltammetry	Carbon/ionic liquid/carbon nanotube composite electrode	Food	[96]
	Cyclic voltammetry, Differential pulse voltammetry	Glass-modified carbon with carbon nanotubes and Nile blue	Pharmaceuticals	[97]
	Linear sweep voltammetry	Glassy carbon modified with poly(xanthuric acid) and carbon nanotubes	–	[98]
	Cyclic voltammetry, Differential pulse voltammetry	Glassy carbon modified with platinum/graphene	Simultaneous determination of uric, ascorbic acid and dopamine	[99]
	Cyclic voltammetry	Gold modified with polypyrrole and Pd nanoparticles	–	[100]
	Cyclic voltammetry, Square wave voltammetry	Glassy carbon modified with nickel nanoparticles	Simultaneous determination of ascorbic acid, glucose and dopamine	[101]
	Cyclic voltammetry, Differential pulse voltammetry	Glassy carbon modified with carbon nanotubes/Nafion/nano Pt/nano Au	Simultaneous determination of uric, ascorbic acid and epinephrine	[102]

Table 2 contd. ...

...Table 2 contd.

Vitamin	Technique	Electrode	Substrate	Reference
	Cyclic voltammetry, Differential pulse voltammetry	SiO_2/C modified with Niobium oxide	–	[103]
	Cyclic voltammetry, Differential pulse voltammetry	Carbon-ceramic modified with carbon nanotubes	Simultaneous determination of ascorbic acid, paracetamol in drugs	[104]

Table 3. Main features of voltammetric techniques for the detection of fat-soluble vitamins.

Vitamin	Technique	Electrode	Subtrate	Reference
A	Cyclic voltammetry	Glassy carbon	Creams, food	[105]
	Cyclic voltammetry	Mercury coated with phospholipids	–	[106]
	Cyclic voltammetry	Hanging mercury drop electrode	–	[107]
	Cyclic voltammetry, Square wave voltammetry	Platinum	–	[108]
	Cyclic voltammetry	Platinum	–	[109]
	Cyclic voltammetry	Platinum	–	[110]
D	Cyclic voltammetry	Glassy carbon	Pharmaceuticals	[112]
	Cyclic voltammetry, Square wave voltammetry	Glassy carbon/ platinum	–	[113]
	Cyclic voltammetry, Differential pulse voltammetry	Glassy carbon	Detection in a mixture of vitamins	[114]
	Cyclic voltammetry, Differential pulse voltammetry	Glassy carbon modified with SiO_2/ graphene oxide/ $(NiOH)_2$	–	[115]
E	Cyclic voltammetry	Platinum	–	[110]
	Cyclic voltammetry	Platinum	–	[117]
	Cyclic voltammetry	Platinum	–	[118]
	Cyclic voltammetry	Platinum	–	[119]
	Cyclic voltammetry, Square wave voltammetry	Glassy carbon/ platinum	–	[120]
	Cyclic voltammetry	Platinum	–	[124]
	Cyclic voltammetry	Glassy carbon modified with lecithin and Nafion	Detection of low molecular weight biological molecules	[125]
	Cyclic voltammetry	Pyrolytic graphite	–	[126]

Table 3 contd. ...

...Table 3 contd.

Vitamin	Technique	Electrode	Subtrate	Reference
K	Cyclic voltammetry, Square wave voltammetry	Glassy carbon/ platinum	–	[127]
	Steady state voltammetry	Gold	–	[128]
	Cyclic voltammetry	Platinum	–	[129]
	Cyclic voltammetry, Square wave voltammetry	Glassy carbon	Quantification of traces of moisture in organic solvents	[130]
	Cyclic voltammetry	Gold	–	[131]
	Cyclic voltammetry	Gold	–	[132]
	Cyclic voltammetry	Gold	Menadione permeability study	[133]
	Cyclic voltammetry	Glassy carbon modified with lipids	–	[134]
	Cyclic voltammetry	Pyrolytic graphite	Study of electrochemical behavior of phylloquinone	[136]
	Cyclic voltammetry	Glassy carbon	Study of electrochemical behavior of quinones	[137]

4. Conclusion

The use of electrochemical analysis methods, given their much higher speeds and lower costs, is an attractive alternative to conventional analytical techniques. The simultaneous identification of many vitamins is a challenge but the results of recent studies are promising. Many new electrode materials, especially nanostructured materials, have been introduced and have improved the selectivity and sensitivity of detection, while also aiding the quantification of vitamins in real samples, in concentrations much smaller than that of the other components.

The range of applications of electrochemical methods is particularly wide—from pharmaceutical and nutritional supplements to more complex samples such as food and biological fluids. However, most of the proposed methods still require pre-treatment and/or dilution of the actual sample for analysis.

The challenge for future studies will be to develop sensors capable of identifying vitamins directly in actual samples; this will further increase the speed of analysis. The development of mobile devices in combination with the use of consumable sensors will open a new avenue of prospects. Electrochemical techniques have proven to be excellent candidates for such applications, and small portable potentiostats are already available in the market.

References

[1] Combs, G. The Vitamins. Fundamental Aspects in Nutrition and Health, 3rd ed., Academic Press, n.d.

[2] Azzi, A. 2007. Molecular mechanism of α-tocopherol action. Free Radic. Biol. Med. 43: 16–21.

[3] Pisoschi, A.M., A. Pop, A.I. Serban and C. Fafaneata. 2014. Electrochemical methods for ascorbic acid determination. Electrochim. Acta 121: 443–460. Doi: 10.1016/j.electacta.2013.12.127.

[4] Ingrid T. Loessing (ed.). 2007. Vitamin A: New Research, Nova.

[5] Frankenburg, F.R. 2009. Vitamin discoveries and disasters: History, science, and controversies. ABC-CLIO.

[6] Litwack, G. 2007. Vitamin E. Academic Press.

[7] Atkins, P. and J. De Paula. 2011. Physical Chemistry for the Life Sciences. Oxford University Press, USA.

[8] Li, G. and P. Miao. 2012. Electrochemical Analysis of Proteins and Cells. Springer Science & Business Media.

[9] Dahmen, E.A.M.F. (ed.). 1986. Chapter 1 Introduction. pp. 3–9. *In*: Electroanal. Theory Appl. Aqueous Non-Aqueous Media Autom. Chem. Control, Elsevier. Doi: https://doi.org/10.1016/S0167-9244(08)70252-X.

[10] Heinze, J. 1984. Cyclic voltammetry—"electrochemical spectroscopy". New analytical methods (25). Angew. Chemie Int. Ed. English 23: 831–847. Doi: 10.1002/anie.198408313.

[11] Girousi, S. 2015. Biosensors. pp. 95–122. *In*: Bioanalytical Chemistry, 1st ed., Hellenic Academic Libraries Association Athens. https://repository.kallipos.gr/handle/11419/3667.

[12] Herzog, G. and V. Beni. 2013. Stripping voltammetry at micro-interface arrays: A review. Anal. Chim. Acta. 769: 10–21. Doi: 10.1016/j.aca.2012.12.031.

[13] Brunetti, B. 2016. Recent advances in electroanalysis of vitamins. Electroanalysis 28: 1930–1942. Doi: 10.1002/elan.201600097.

[14] Blake, C.J. 2007. Analytical procedures for water-soluble vitamins in foods and dietary supplements: A review. Anal. Bioanal. Chem. 389: 63–76.

[15] Brunetti, B. and E. Desimoni. 2014. Voltammetric determination of vitamin B 6 in food samples and dietary supplements. J. Food Compos. Anal. 33: 155–160.

[16] Wang, L.-H. and S.-W. Tseng. 2001. Direct determination of d-panthenol and salt of pantothenic acid in cosmetic and pharmaceutical preparations by differential pulse voltammetry. Anal. Chim. Acta. 432: 39–48.

[17] Vytřas, K., I. Švancara and R. Metelka. 2009. Carbon paste electrodes in electroanalytical chemistry. J. Serbian Chem. Soc. 74: 1021–1033.

[18] Tyszczuk-Rotko, K. 2012. New voltammetric procedure for determination of thiamine in commercially available juices and pharmaceutical formulation using a lead film electrode. Food Chem. 134: 1239–1243.

[19] Brahman, P.K., R.A. Dar and K.S. Pitre. 2013. DNA-functionalized electrochemical biosensor for detection of vitamin B1 using electrochemically treated multiwalled carbon nanotube paste electrode by voltammetric methods. Sensors Actuators B Chem. 177: 807–812.

[20] Baś, B., M. Jakubowska and Ł. Górski. 2011. Application of renewable silver amalgam annular band electrode to voltammetric determination of vitamins C, B 1 and B 2. Talanta 84: 1032–1037.

[21] Yosypchuk, B., M. Heyrovský, E. Palecek and L. Novotný. 2002. Use of solid amalgam electrodes in nucleic acid analysis. Electroanalysis 14: 1488–1493.

[22] Sá, É.S., P.S. da Silva, C.L. Jost and A. Spinelli. 2015. Electrochemical sensor based on bismuth-film electrode for voltammetric studies on vitamin B 2 (riboflavin). Sensors Actuators B Chem. 209: 423–430.

[23] Kumar, D.R., D. Manoj and J. Santhanalakshmi. 2014. Au–ZnO bullet-like heterodimer nanoparticles: Synthesis and use for enhanced nonenzymatic electrochemical determination of glucose. RSC Adv. 4: 8943–8952.

[24] Nie, T., K. Zhang, J. Xu, L. Lu and L. Bai. 2014. A facile one-pot strategy for the electrochemical synthesis of poly (3, 4-ethylenedioxythiophene)/Zirconia nanocomposite as an effective sensing platform for vitamins B 2, B 6 and C. J. Electroanal. Chem. 717: 1–9.

[25] Mikheeva, E.V., O.A. Martynyuk, G.B. Slepchenko and L.S. Anisimova. 2009. Study of the voltammetric behavior of vitamin B 2 and the development of a procedure for its determination in breast milk. J. Anal. Chem. 64: 731–734.

[26] Nezamzadeh-Ejhieh, A. and P. Pouladsaz. 2014. Voltammetric determination of riboflavin based on electrocatalytic oxidation at zeolite-modified carbon paste electrodes. J. Ind. Eng. Chem. 20: 2146–2152.

[27] Revin, S.B. and S.A. John. 2012. Simultaneous determination of vitamins B 2, B 9 and C using a heterocyclic conducting polymer modified electrode. Electrochim. Acta. 75: 35–41.

[28] Wang, X., N. Yang and Q. Wan. 2006. Cyclic voltammetric response of nicotinic acid and nicotinamide on a polycrystalline gold electrode. Electrochim. Acta 52: 361–368.

[29] Yang, N. and X. Wang. 2008. Thin self-assembled monolayer for voltammetrically monitoring nicotinic acid in food. Colloids Surfaces B Biointerfaces 61: 277–281.

[30] Wu, J., H. Liu and Z. Lin. 2008. Electrochemical performance of a carbon nanotube/La-doped TiO2 nanocomposite and its use for preparation of an electrochemical nicotinic acid sensor. Sensors 8: 7085–7096.

[31] Kotkar, R.M. and A.K. Srivastava. 2008. Polarographic behavior of nicotinamide in surfactant media and its determination in cetyltrimethylammonium bromide surfactant system. Anal. Sci. 24: 1093–1098.

[32] Kotkar, R.M. and A.K. Srivastava. 2008. Electrochemical behavior of nicotinamide using carbon paste electrode modified with macrocyclic compounds. J. Incl. Phenom. Macrocycl. Chem. 60: 271–279.

[33] Wu, Y.-H. and F.-J. Song. 2008. Voltammetric investigation of vitamin B_6 at a glassy carbon electrode and its application in determination. Bull. Korean Chem. Soc. 29: 38–42.

[34] Gonzalez-Rodriguez, J., J.M. Sevilla, T. Pineda and M. Blazquez. 2012. Electrochemical analysis on compounds of the vitamin B6 family using glassy carbon electrodes. Int. J. Electrochem. Sci. 7: 2221–2229.

[35] Barsan, M.M., C.T. Toledo and C.M.A. Brett. 2015. New electrode architectures based on poly (methylene green) and functionalized carbon nanotubes: Characterization and application to detection of acetaminophen and pyridoxine. J. Electroanal. Chem. 736: 8–15.

[36] Kumar, D.R., D. Manoj, J. Santhanalakshmi and J.-J. Shim. 2015. Au-CuO core-shell nanoparticles design and development for the selective determination of Vitamin B 6. Electrochim. Acta 176: 514–522.

[37] Raoof, J.B., N. Teymoori and M.A. Khalilzadeh. 2015. ZnO nanoparticle ionic liquids carbon paste electrode as a voltammetric sensor for determination of Sudan I in the presence of vitamin B6 in food samples. Food Anal. Methods 8: 885–892.

[38] Fonseca, C.A., G.C.S. Vaz, J.P.A. Azevedo and F.S. Semaan. 2011. Exploiting ion-pair formation for the enhancement of electroanalytical determination of pyridoxine (B 6) onto polyurethane-graphite electrodes. Microchem. J. 99: 186–192.

[39] Desai, P.B., R.M. Kotkar and A.K. Srivastava. 2008. Electrochemical behaviour of pyridoxine hydrochloride (vitamin B6) at carbon paste electrode modified with crown ethers. J. Solid State Electrochem. 12: 1067–1075.

[40] Baghizadeh, A., H. Karimi-Maleh, Z. Khoshnama, A. Hassankhani and M. Abbasghorbani. 2015. A voltammetric sensor for simultaneous determination of vitamin C and vitamin B6 in food samples using ZrO2 nanoparticle/ionic liquids carbon paste electrode. Food Anal. Methods. 8: 549–557.

[41] Patra, S., E. Roy, R. Das, P. Karfa, S. Kumar, R. Madhuri and P.K. Sharma. 2015. Bimetallic magnetic nanoparticle as a new platform for fabrication of pyridoxine and pyridoxal-5′-phosphate imprinted polymer modified high throughput electrochemical sensor. Biosens. Bioelectron. 73: 234–244.

[42] Gholivand, M.-B., A.R. Jalalvand, H.C. Goicoechea, G. Paimard and T. Skov. 2015. Surface exploration of a room-temperature ionic liquid-chitin composite film decorated with electrochemically deposited PdFeNi trimetallic alloy nanoparticles by pattern recognition: An elegant approach to developing a novel biotin biosensor. Talanta 131: 249–258.
[43] Kergaravat, S.V., M.I. Pividori and S.R. Hernandez. 2012. Evaluation of seven cosubstrates in the quantification of horseradish peroxidase enzyme by square wave voltammetry. Talanta 88: 468–476.
[44] Ardakani, M.M., M.A.L.I. Sheikhmohseni, H. Beitollahi, A. Benvidi and H. Naeimi. 2011. Simultaneous determination of dopamine, uric acid, and folic acid by a modified TiO_2 nanoparticles carbon paste electrode. Turkish J. Chem. 35: 573–585.
[45] Xiao, F., C. Ruan, L. Liu, R. Yan, F. Zhao and B. Zeng. 2008. Single-walled carbon nanotube-ionic liquid paste electrode for the sensitive voltammetric determination of folic acid. Sensors Actuators B Chem. 134: 895–901.
[46] Karimi-Maleh, H., F. Tahernejad-Javazmi, M. Daryanavard, H. Hadadzadeh, A.A. Ensafi and M. Abbasghorbani. 2014. Electrocatalytic and simultaneous determination of ascorbic acid, nicotinamide adenine dinucleotide and folic acid at Ruthenium (II) Complex-ZnO/CNTs nanocomposite modified carbon paste electrode. Electroanalysis 26: 962–970.
[47] Arvand, M. and M. Dehsaraei. 2013. A simple and efficient electrochemical sensor for folic acid determination in human blood plasma based on gold nanoparticles–modified carbon paste electrode. Mater. Sci. Eng. C. 33: 3474–3480.
[48] Kingsley, M.P., P.B. Desai and A.K. Srivastava. 2015. Simultaneous electro-catalytic oxidative determination of ascorbic acid and folic acid using Fe3O4 nanoparticles modified carbon paste electrode. J. Electroanal. Chem. 741: 71–79.
[49] Jamali, T., H. Karimi-Maleh and M.A. Khalilzadeh. 2014. A novel nanosensor based on Pt: Co nanoalloy ionic liquid carbon paste electrode for voltammetric determination of vitamin B 9 in food samples. LWT-Food Sci. Technol. 57: 679–685.
[50] Georgescu, R., J.F. van Staden, R.-I. Stefan-van Staden and C. Boscornea. 2015. Evaluation of amperometric dot microsensors for the analysis of folic acid in pharmaceutical tablets and urine samples. J. Porphyr. Phthalocyanines. 19: 679–687.
[51] Ojani, R., J. Raoof and S. Zamani. 2009. Electrocatalytic oxidation of folic acid on carbon paste electrode modified by nickel ions dispersed into poly (o-anisidine) film. Electroanalysis 21: 2634–2639.
[52] Kun, Z., Z. Ling, H. Yi, C. Ying, T. Dongmei, Z. Shuliang and Z. Yuyang. 2012. Electrochemical behavior of folic acid in neutral solution on the modified glassy carbon electrode: Platinum nanoparticles doped multi-walled carbon nanotubes with Nafion as adhesive. J. Electroanal. Chem. 677: 105–112.
[53] Babakhanian, A., S. Kaki, M. Ahmadi, H. Ehzari and A. Pashabadi. 2014. Development of α-polyoxometalate–polypyrrole–Au nanoparticles modified sensor applied for detection of folic acid. Biosens. Bioelectron. 60: 185–190.
[54] Kalimuthu, P. and S.A. John. 2009. Selective electrochemical sensor for folic acid at physiological pH using ultrathin electropolymerized film of functionalized thiadiazole modified glassy carbon electrode. Biosens. Bioelectron. 24: 3575–3580.
[55] Korolczuk, M. and K. Tyszczuk. 2007. Determination of folic acid by adsorptive stripping voltammetry at a lead film electrode. Electroanalysis 19: 1959–1962.
[56] Mirmoghtadaie, L., A.A. Ensafi, M. Kadivar, M. Shahedi and M.R. Ganjali. 2013. Highly selective, sensitive and fast determination of folic acid in food samples using new electrodeposited gold nanoparticles by differential pulse voltammetry. Int. J. Electrochem. Sci. 8: 3755–3767.
[57] Bandžuchová, L. and R. Šelešovská. 2011. Voltammetric determination of folic acid using liquid mercury free silver amalgam electrode. Acta Chim. Slov. 58.
[58] Farias, P.A.M., M.C. Rezende and J.C. Moreira. 2012. Folic acid determination in neutral pH electrolyte by adsorptive stripping voltammetry at the mercury film electrode. IOSR J. Pharm. 2: 302–311.

[59] Rutyna, I. 2015. Determination of folic acid at a bismuth film electrode by adsorptive stripping voltammetry. Anal. Lett. 48: 1593–1603.
[60] Kuralay, F., T. Vural, C. Bayram, E.B. Denkbas and S. Abaci. 2011. Carbon nanotube–chitosan modified disposable pencil graphite electrode for Vitamin B 12 analysis. Colloids Surfaces B Biointerfaces 87: 18–22.
[61] Pala, B.B., T. Vural, F. Kuralay, T. Çırak, G. Bolat, S. Abacı and E.B. Denkbaş. 2014. Disposable pencil graphite electrode modified with peptide nanotubes for Vitamin B 12 analysis. Appl. Surf. Sci. 303: 37–45.
[62] Bandžuchová, L., R. Šelešovská, T. Navrátil and J. Chýlková. 2013. Silver solid amalgam electrode as a tool for monitoring the electrochemical reduction of hydroxocobalamin. Electroanalysis 25: 213–222.
[63] Kreft, G.L., O.C. de Braga and A. Spinelli. 2012. Analytical electrochemistry of vitamin B 12 on a bismuth-film electrode surface. Electrochim. Acta 83: 125–132.
[64] Michopoulos, A., A.B. Florou and M.I. Prodromidis. 2015. Ultrasensitive determination of vitamin B12 Using disposable graphite screen-printed electrodes and anodic adsorptive voltammetry. Electroanalysis 27: 1876–1882.
[65] Erdurak-Kiliç, C.S., B. Uslu, B. Dogan, U. Ozgen, S.A. Ozkan and M. Coskun. 2006. Anodic voltammetric behavior of ascorbic acid and its selective determination in pharmaceutical dosage forms and some Rosa species of Turkey. J. Anal. Chem. 61: 1113–1120.
[66] Rueda, M., A. Aldaz and F. Sanchez-Burgos. 1978. Oxidation of L-ascorbic acid on a gold electrode. Electrochim. Acta 23: 419–424.
[67] Bradshaw, M.P., P.D. Prenzler and G.R. Scollary. 2002. Square-wave voltammetric determination of hydrogen peroxide generated from the oxidation of ascorbic acid in a model wine base. Electroanalysis 14: 546–550.
[68] Yilmaz, S., M. Sadikoglu, G. Saglikoglu, S. Yagmur and G. Askin. 2008. Determination of ascorbic acid in tablet dosage forms and some fruit juices by DPV. Int. J. Electrochem. Sci. 3: 1534–1542.
[69] Wang, J. and B.A. Freiha. 1983. Evaluation of differential pulse voltammetry at carbon electrodes. Talanta 30: 317–322.
[70] Kalimuthu, P. and S.A. John. 2005. Solvent dependent dimercaptothiadiazole monolayers on gold electrode for the simultaneous determination of uric acid and ascorbic acid. Electrochem. Commun. 7: 1271–1276.
[71] Raoof, J.B., A. Kiani, R. Ojani, R. Valiollahi and S. Rashid-Nadimi. 2010. Simultaneous voltammetric determination of ascorbic acid and dopamine at the surface of electrodes modified with self-assembled gold nanoparticle films. J. Solid State Electrochem. 14: 1171–1176.
[72] Švancara, I., A. Walcarius, K. Kalcher and K. Vytřas. 2009. Carbon paste electrodes in the new millennium. Open Chem. 7: 598–656.
[73] Zare, H.R., N. Nasirizadeh and M.M. Ardakani. 2005. Electrochemical properties of a tetrabromo-p-benzoquinone modified carbon paste electrode. Application to the simultaneous determination of ascorbic acid, dopamine and uric acid. J. Electroanal. Chem. 577: 25–33.
[74] Chethana, B.K. and Y.A. Naik. 2012. Electrochemical oxidation and determination of ascorbic acid present in natural fruit juices using a methionine modified carbon paste electrode. Anal. Methods 4: 3754–3759.
[75] Mazloum-Ardakani, M., F. Habibollahi, H.R. Zare, H. Naeimi and M. Nejati. 2009. Electrocatalytic oxidation of ascorbic acid at a 2, 2′-(1, 8-octanediylbisnitriloethylidine)-bis-hydroquinone modified carbon paste electrode. J. Appl. Electrochem. 39: 1117–1124.
[76] Stefan-van Staden, R.-I., S.C. Balasoiu, J.F. van Staden and G.-L. Radu. 2012. Microelectrodes based on porphyrins for the determination of ascorbic acid in pharmaceutical samples and beverages. J. Porphyr. Phthalocyanines 16: 809–816.
[77] Li, F., J. Li, Y. Feng, L. Yang and Z. Du. 2011. Electrochemical behavior of graphene doped carbon paste electrode and its application for sensitive determination of ascorbic acid. Sensors Actuators B Chem. 157: 110–114.

[78] Langley, C.E., B. ŠLJUKIC, C.E. Banks and R.G. Compton. 2007. Manganese dioxide graphite composite electrodes: Application to the electroanalysis of hydrogen peroxide, ascorbic acid and nitrite. Anal. Sci. 23: 165–170.

[79] Thiagarajan, S., T.-H. Tsai and S.-M. Chen. 2009. Easy modification of glassy carbon electrode for simultaneous determination of ascorbic acid, dopamine and uric acid. Biosens. Bioelectron. 24: 2712–2715.

[80] Ensafi, A.A., M. Taei and T. Khayamian. 2009. A differential pulse voltammetric method for simultaneous determination of ascorbic acid, dopamine, and uric acid using poly (3-(5-chloro-2-hydroxyphenylazo)-4, 5-dihydroxynaphthalene-2, 7-disulfonic acid) film modified glassy carbon electrode. J. Electroanal. Chem. 633: 212–220.

[81] Li, N.B., W. Ren and H.Q. Luo. 2008. Simultaneous voltammetric measurement of ascorbic acid and dopamine on poly (caffeic acid)-modified glassy carbon electrode. J. Solid State Electrochem. 12: 693–699.

[82] Zhang, R., S. Liu, L. Wang and G. Yang. 2013. Electroanalysis of ascorbic acid using poly (bromocresol purple) film modified glassy carbon electrode. Measurement 46: 1089–1093.

[83] Kumar, S.S., J. Mathiyarasu, K.L.N. Phani and V. Yegnaraman. 2006. Simultaneous determination of dopamine and ascorbic acid on poly (3, 4-ethylenedioxythiophene) modified glassy carbon electrode. J. Solid State Electrochem. 10: 905–913.

[84] Kilmartin, P.A., A. Martinez and P.N. Bartlett. 2008. Polyaniline-based microelectrodes for sensing ascorbic acid in beverages. Curr. Appl. Phys. 8: 320–323.

[85] Cao, X., Y. Xu, L. Luo, Y. Ding and Y. Zhang. 2010. Simultaneous determination of uric acid and ascorbic acid at the film of chitosan incorporating cetylpyridine bromide modified glassy carbon electrode. J. Solid State Electrochem. 14: 829–834.

[86] Sripriya, R., M. Chandrasekaran and M. Noel. 2006. Voltammetric analysis of hydroquinone, ascorbic acid, nitrobenzene and benzyl chloride in aqueous, non-aqueous, micellar and microemulsion media. Colloid Polym. Sci. 285: 39–48.

[87] Wang, M., X. Xu and J. Gao. 2007. Voltammetric studies of a novel bicopper complex modified glassy carbon electrode for the simultaneous determination of dopamine and ascorbic acid. J. Appl. Electrochem. 37: 705–710.

[88] Wang, G., J. Sun, W. Zhang, S. Jiao and B. Fang. 2009. Simultaneous determination of dopamine, uric acid and ascorbic acid with $LaFeO_3$ nanoparticles modified electrode. Microchim. Acta. 164: 357–362.

[89] Lourencao, B.C., R.A. Medeiros, R.C. Rocha-Filho and O. Fatibello-Filho. 2010. Simultaneous differential pulse voltammetric determination of ascorbic acid and caffeine in pharmaceutical formulations using a boron-doped diamond electrode. Electroanalysis 22: 1717–1723.

[90] Wang, J. 2005. Carbon-nanotube based electrochemical biosensors: A review. Electroanalysis 17: 7–14.

[91] Zhang, M., K. Liu, L. Xiang, Y. Lin, L. Su and L. Mao. 2007. Carbon nanotube-modified carbon fiber microelectrodes for *in vivo* voltammetric measurement of ascorbic acid in rat brain. Anal. Chem. 79: 6559–6565.

[92] Deng, P., Z. Xu and J. Li. 2013. Simultaneous determination of ascorbic acid and rutin in pharmaceutical preparations with electrochemical method based on multi-walled carbon nanotubes–chitosan composite film modified electrode. J. Pharm. Biomed. Anal. 76: 234–242.

[93] Keyvanfard, M., R. Shakeri, H. Karimi-Maleh and K. Alizad. 2013. Highly selective and sensitive voltammetric sensor based on modified multiwall carbon nanotube paste electrode for simultaneous determination of ascorbic acid, acetaminophen and tryptophan. Mater. Sci. Eng. C 33: 811–816.

[94] Noroozifar, M., M.K. Motlagh and H. Tavakoli. 2012. Determination of ascorbic acid by a modified multiwall carbon nanotube paste electrode using cetrimonium iodide/iodine. Turkish J. Chem. 36: 645–658.

[95] Zhang, B., D. Huang, X. Xu, G. Alemu, Y. Zhang, F. Zhan, Y. Shen and M. Wang. 2013. Simultaneous electrochemical determination of ascorbic acid, dopamine and uric acid with helical carbon nanotubes. Electrochim. Acta. 91: 261–266.

[96] Ping, J., Y. Wang, J. Wu, Y. Ying and F. Ji. 2012. Determination of ascorbic acid levels in food samples by using an ionic liquid–carbon nanotube composite electrode. Food Chem. 135: 362–367.

[97] Kul, D., M.E. Ghica, R. Pauliukaite and C.M.A. Brett. 2013. A novel amperometric sensor for ascorbic acid based on poly (Nile blue A) and functionalised multi-walled carbon nanotube modified electrodes. Talanta 111: 76–84.

[98] Lin, K.-C., P.-C. Yeh and S.-M. Chen. 2012. Electrochemical determination of ascorbic acid using poly (xanthurenic acid) and multi-walled carbon nanotubes. Int. J. Electrochem. Sci. 7: 12752–12763.

[99] Sun, C.-L., H.-H. Lee, J.-M. Yang and C.-C. Wu. 2011. The simultaneous electrochemical detection of ascorbic acid, dopamine, and uric acid using graphene/size-selected Pt nanocomposites. Biosens. Bioelectron. 26: 3450–3455.

[100] Shi, W., C. Liu, Y. Song, N. Lin, S. Zhou and X. Cai. 2012. An ascorbic acid amperometric sensor using over-oxidized polypyrrole and palladium nanoparticles composites. Biosens. Bioelectron. 38: 100–106.

[101] Hathoot, A.A., U.S. Yousef, A.S. Shatla and M. Abdel-Azzem. 2012. Voltammetric simultaneous determination of glucose, ascorbic acid and dopamine on glassy carbon electrode modified byNiNPs@ poly 1, 5-diaminonaphthalene. Electrochim. Acta 85: 531–537.

[102] Yogeswaran, U., S. Thiagarajan and S.-M. Chen. 2007. Nanocomposite of functionalized multiwall carbon nanotubes with nafion, nano platinum, and nano gold biosensing film for simultaneous determination of ascorbic acid, epinephrine, and uric acid. Anal. Biochem. 365: 122–131.

[103] Arenas, L.T., P.C.M. Villis, J. Arguello, R. Landers, E.V. Benvenutti and Y. Gushikem. 2010. Niobium oxide dispersed on a carbon–ceramic matrix, SiO 2/C/Nb 2 O 5, used as an electrochemical ascorbic acid sensor. Talanta 83: 241–248.

[104] Habibi, B., M. Jahanbakhshi and M.H. Pournaghi-Azar. 2011. Differential pulse voltammetric simultaneous determination of acetaminophen and ascorbic acid using single-walled carbon nanotube-modified carbon–ceramic electrode. Anal. Biochem. 411: 167–175.

[105] Ziyatdinova, G., E. Giniyatova and H. Budnikov. 2010. Cyclic voltammetry of retinol in surfactant media and its application for the analysis of real samples. Electroanalysis 22: 2708–2713.

[106] Nelson, A. 1992. Voltammetry of retinal in phospholipid monolayers adsorbed on mercury. J. Electroanal. Chem. 335: 327–343.

[107] Powell, L.A. and R.M. Wightman. 1981. Mechanism of electrodimerization of retinal and cinnamaldehyde. J. Electroanal. Chem. Interfacial Electrochem. 117: 321–333.

[108] Liu, D., Y. Gao and L.D. Kispert. 2000. Electrochemical properties of natural carotenoids. J. Electroanal. Chem. 488: 140–150.

[109] Hapiot, P., L.D. Kispert, V. V Konovalov and J.-M. Savéant. 2001. Single two-electron transfers vs successive one-electron transfers in polyconjugated systems illustrated by the electrochemical oxidation and reduction of carotenoids. J. Am. Chem. Soc. 123: 6669–6677.

[110] Tan, Y.S. and R.D. Webster. 2011. Electron-transfer reactions between the diamagnetic cation of α-tocopherol (vitamin E) and β-carotene. J. Phys. Chem. B 115: 4244–4250.

[111] Liu, X., K. Liu, G. Cheng and S. Dong. 2001. Electrochemical and spectroscopic study of vitamin D 2 by *in situ* thin layer CD spectroelectrochemistry. J. Electroanal. Chem. 510: 103–107.

[112] Hart, J.P., M.D. Norman and C.J. Lacey. 1992. Voltammetric behaviour of vitamins D2 and D3 at a glassy carbon electrode and their determination in pharmaceutical products by using liquid chromatography with amperometric detection. Analyst 117: 1441–1445.

[113] Chan, Y.Y., Y. Yue and R.D. Webster. 2014. Voltammetric studies on vitamins D2 and D3 in organic solvents. Electrochim. Acta 138: 400–409.

[114] Cincotto, F.H., T.C. Canevari and S.A.S. Machado. 2014. Highly sensitive electrochemical sensor for determination of vitamin D in mixtures of water-ethanol. Electroanalysis 26: 2783–2788.

[115] Canevari, T.C., F.H. Cincotto, R. Landers and S.A.S. Machado. 2014. Synthesis and characterization of α-nickel (II) hydroxide particles on organic-inorganic matrix and its application in a sensitive electrochemical sensor for vitamin D determination. Electrochim. Acta 147: 688–695.

[116] Webster, R.D. 2012. Voltammetry of the liposoluble vitamins (A, D, e and K) in organic solvents. Chem. Rec. 12: 188–200. Doi: 10.1002/tcr.201100005.

[117] Tan, Y.S., S. Chen, W.M. Hong, J.M. Kan, E.S.H. Kwek, S.Y. Lim, Z.H. Lim, M.E. Tessensohn, Y. Zhang and R.D. Webster. 2011. The role of low levels of water in the electrochemical oxidation of α-tocopherol (vitamin E) and other phenols in acetonitrile. Phys. Chem. Chem. Phys. 13: 12745–12754.

[118] Yao, W.W., H.M. Peng, R.D. Webster and P.M.W. Gill. 2008. Variable scan rate cyclic voltammetry and theoretical studies on tocopherol (vitamin E) model compounds. J. Phys. Chem. B. 112: 6847–6855.

[119] Williams, L.L. and R.D. Webster. 2004. Electrochemically controlled chemically reversible transformation of α-tocopherol (vitamin E) into its phenoxonium cation. J. Am. Chem. Soc. 126: 12441–12450.

[120] Lee, S.B., C.Y. Lin, P.M.W. Gill and R.D. Webster. 2005. Transformation of α-tocopherol (vitamin E) and related chromanol model compounds into their phenoxonium ions by chemical oxidation with the nitrosonium cation. J. Org. Chem. 70: 10466–10473.

[121] Speiser, B. and A. Rieker. 1980. Electrochemical oxidations: Part V. Electrochemical investigations into the behaviour of 2, 6-di-tert-butyl-4-(4-dimethylamino-phenyl)-phenol Part 2: Anodic oxidation in the presence of 2, 6-dimethylpyridine and the mechanism of the formation of the pheno. J. Electroanal. Chem. Interfacial Electrochem. 110: 231–246.

[122] Vigalok, A., B. Rybtchinski, Y. Gozin, T.S. Koblenz, Y. Ben-David, H. Rozenberg and D. Milstein. 2003. Metal-stabilized phenoxonium cation. J. Am. Chem. Soc. 125: 15692–15693.

[123] Webster, R.D. 2007. New insights into the oxidative electrochemistry of vitamin E. Acc. Chem. Res. 40: 251–257.

[124] Wilson, G.J., C.Y. Lin and R.D. Webster. 2006. Significant differences in the electrochemical behavior of the α-, β-, γ-, and δ-tocopherols (Vitamin E). J. Phys. Chem. B. 110: 11540–11548.

[125] Yao, W.W., C. Lau, Y. Hui, H.L. Poh and R.D. Webster. 2011. Electrode-supported biomembrane for examining electron-transfer and ion-transfer reactions of encapsulated low molecular weight biological molecules. J. Phys. Chem. C 115: 2100–2113.

[126] Wain, A.J., J.D. Wadhawan, R.R. France and R.G. Compton. 2004. Biphasic redox chemistry of α-tocopherol: Evidence for electrochemically induced hydrolysis and dimerization on the surface of and within femtolitre droplets immobilized onto graphite electrodes. Phys. Chem. Chem. Phys. 6: 836–842.

[127] Hui, Y., E.L.K. Chng, C.Y.L. Chng, H.L. Poh and R.D. Webster. 2009. Hydrogen-bonding interactions between water and the one-and two-electron-reduced forms of vitamin K1: Applying quinone electrochemistry to determine the moisture content of non-aqueous solvents. J. Am. Chem. Soc. 131: 1523–1534.

[128] Rees, N.V., A.D. Clegg, O.V. Klymenko, B.A. Coles and R.G. Compton. 2004. Marcus theory for outer-sphere heterogeneous electron transfer: Predicting electron-transfer rates for quinones. J. Phys. Chem. B 108: 13047–13051.

[129] Chih, T., S.Y. Yeh and C.M. Wang. 2003. Evidence and applications for electron transfer between vitamin K 3 and oxygen. J. Electroanal. Chem. 543: 135–142.

[130] Hui, Y., E.L.K. Chng, L.P.-L. Chua, W.Z. Liu and R.D. Webster. 2010. Voltammetric method for determining the trace moisture content of organic solvents based on hydrogen-bonding interactions with quinones. Anal. Chem. 82: 1928–1934.

[131] Bulovas, A., N. Dirvianskytė, Z. Talaikytė, G. Niaura, S. Valentukonytė, E. Butkus and V. Razumas. 2006. Electrochemical and structural properties of self-assembled monolayers of 2-methyl-3-(ω-mercaptoalkyl)-1, 4-naphthoquinones on gold. J. Electroanal. Chem. 591: 175–188.

[132] Jeuken, L.J.C., R.J. Bushby and S.D. Evans. 2007. Proton transport into a tethered bilayer lipid membrane. Electrochem. Commun. 9: 610–614.

[133] Cannes, C., F. Kanoufi and A.J. Bard. 2002. Cyclic voltammetric and scanning electrochemical microscopic study of menadione permeability through a self-assembled monolayer on a gold electrode. Langmuir 18: 8134–8141.

[134] Yang, J.-E., J.-H. Yoon, M.-S. Won and Y.-B. Shim. 2010. Electrochemical and spectroelectrochemical behaviors of vitamin K 1/lipid modified electrodes and the formation of radical anion in aqueous media. Bull. Korean Chem. Soc. 31: 3133–3138.

[135] Wadhawan, J.D., A.J. Wain and R.G. Compton. 2003. Electrochemical probing of photochemical reactions inside femtolitre droplets confined to electrodes, ChemPhysChem. 4: 1211–1215.

[136] Wain, A.J., J.D. Wadhawan and R.G. Compton. 2003. Electrochemical studies of vitamin K1 microdroplets: electrocatalytic hydrogen evolution. ChemPhysChem. 4: 974–982.

[137] Gupta, N. and H. Linschitz. 1997. Hydrogen-bonding and protonation effects in electrochemistry of quinones in aprotic solvents. J. Am. Chem. Soc. 119: 6384–6391.

7

Fiber-optic Sensors with Microsphere

Paulina Listewnik,[1] *Valery V. Tuchin*[2,3,4]
and *Małgorzata Szczerska*[1,*]

1. Introduction

Fiber-optic sensors are devices used for the measurement of various physical (such as temperature, pressure, vibration [3–6]) and chemical quantities ([7, 8]). Furthermore, their structure allows them to be used in places in which common electric sensors cannot be inset, like remote and hard to access locations, hazardous places, potentially explosive areas (ATEX directives), places endangered by fire, and regions with severe climatic and corrosive conditions [9–11]. These sensors are manufactured as small and compact devices resistant to external influences like electromagnetic interference or ionizing radiation [1, 2]. Because of their numerous advantages and wide application range, the development of fiber-optic sensors has been distinctly visible throughout the last decade.

State-of-the-art devices used in fiber-optics technology enable the manufacture of microstructures of various geometrical parameters which can be integrated into a fiber-optic sensor. There are a multitude of fiber-optic structures, including tapers, microrings, microcavities, microdiscs [12–15], as well as various techniques and types of applications of these structures [16]. However, they are mainly used as microresonators [17, 18]. And then, there are microspheres – the most common and

[1] Department of Metrology and Optoelectronics, Faculty of Electronics, Telecommunications and Informatics, Gdańsk University of Technology, Narutowicza Street 11/12, 80-233 Gdańsk, Poland.

[2] Research-Educational Institute of Optics and Biophotonics, Saratov State University (National Research University of Russia), Astrakhanskaya 83, 410012 Saratov, Russian Federation.

[3] Laboratory of Laser Diagnostics of Technical and Living Systems, Institute of Precision Mechanics and Control RAS, Rabochaya 24, 410028 Saratov, Russian Federation.

[4] Interdisciplinary Laboratory of Biophotonics, National Research Tomsk, State University, Lenin's av. 36, 634050 Tomsk, Russian Federation.

* Corresponding author: malszcze@pg.edu.pl

readily used of the structures. To date, the majority of research focusing on sensors with optical microspheres deals with highly coherent light sources (i.e., lasers) for inciting a resonance through the phenomenon of Whispering Mode Gallery [19–21].

Furthermore, utilizing any of the mentioned microstructures often requires them to be coupled with other structures [22, 23], as a means to introduce a signal, and uses specialized types of optical fibers [24, 25]. The sensors discussed in this chapter are designed to act as standalone devices, made from a standard telecommunication fiber, therefore reducing the dimensions and complexity of the system, thus in turn minimizing the possible impacts on the accuracy of the performed measurements and ultimately ensuring the availability of components and materials.

In this chapter, the problems occurring during the manufacture of an optical microsphere have been elucidated, along with the theoretical possibility of modification of their properties by altering their structure or coating. Furthermore, representative sensors with microspheres and the way to tune their parameters have been presented.

2. Microsphere as a standalone device

In most of the fiber-optic microsphere sensors described in the literature, light is coupled directly with the microspherical resonator through an external setup and transmitted to the detector via attached fibers [26]. The working principle of a standalone microspherical sensor is based on the idea to use the fiber (on which the microsphere is fabricated) as a way to feed the light to the structure as well as for the transmission of the output signal.

The theoretical rationale for such a device was presented by Hirsch [27]. The microsphere is created on the tip of a commonly used single mode fiber (like SMF-28) by short-time localized heating (by an electrical arc discharge employed in fiber fusion splicers, for instance). If fabricated carefully, a microsphere may preserve the original structure of the fiber, with a distinct boundary between the material of the fiber core and the cladding (Figure 1).

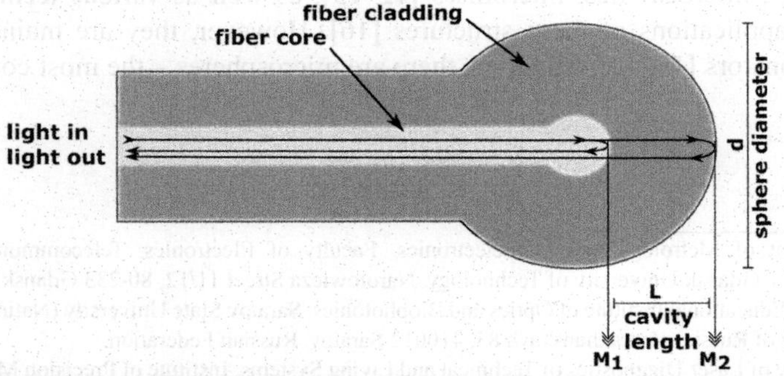

Figure 1. Standalone fiber-optic microsphere device: schematic diagram of the structure (M1, M2 – boundaries between different materials upon reflection of light; L – cavity length; d – sphere diameter).

When the light source illuminates the device through the optical fiber, the light is reflected off the boundary between media of different refractive indices – first between the core and the cladding of the optical fiber, and then between the cladding and the examined environment in which the sensor is enveloped, therefore inciting interference. The wave reflected off the first boundary is always constant, whereas the wave reflected on the second boundary is dependent on changes to the surrounding medium. Such a structure has been described as a low-finesse Fabry-Pérot interferometer, where the interferometer cavity is made of fiber cladding material and its length is related to the size of the microsphere [27].

Sensors employing a microsphere structure can be used to detect changes in the refractive index of an external medium by measuring the reflected signal intensity. Moreover, owing to the intrinsic Fabry-Pérot interferometer, it is also possible to monitor the state of the sensor structure by measuring the interference component visible in the reflected spectrum. The position of the interference fringes is directly related to the length of the interferometer cavity and, thus, for instance, can be used to detect deformation of the sensor head or for distinguishing between sensors of different sizes in a distributed sensing system.

Figure 2 presents the typical measurement setup that is used for the sensor.

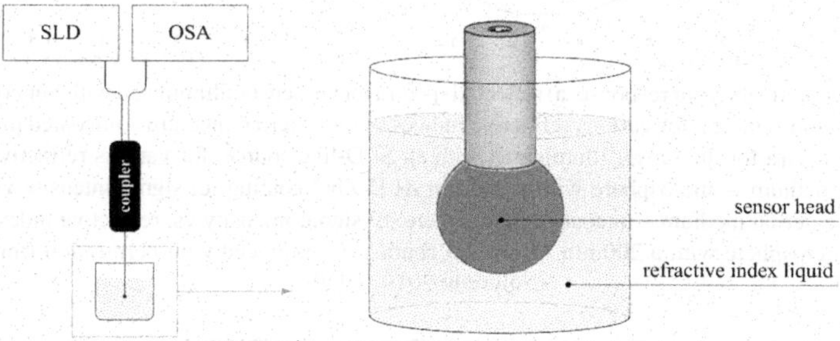

Figure 2. Typical measurement setup.

The results of the simulation of sensor response for a microsphere of 240 μm diameter have been displayed in Figure 3 [28].

3. Microsphere manufacturing—what can go wrong?

When opting to use an optical microsphere in the fiber-optic sensor, the goal is to create a clean and perfectly round structure for highly precise results. Paying close attention to all the details, like predetermined fusion current supplied to the electrodes, fiber prefuse time and the type of fiber during the fabrication of a microsphere makes the sensors reproducible, thus rendering it easy to modify them for tuning to the selected measurements. Figure 4 shows an example of the ideal microsphere; the image has been obtained by a Scanning Electron Microscope (Quanta FEG 250, OR, USA).

However, in the quest for a perfect microsphere, numerous obstacles are met along the way. It is important to pay attention to every element in order to ensure the production of the best microstructure possible.

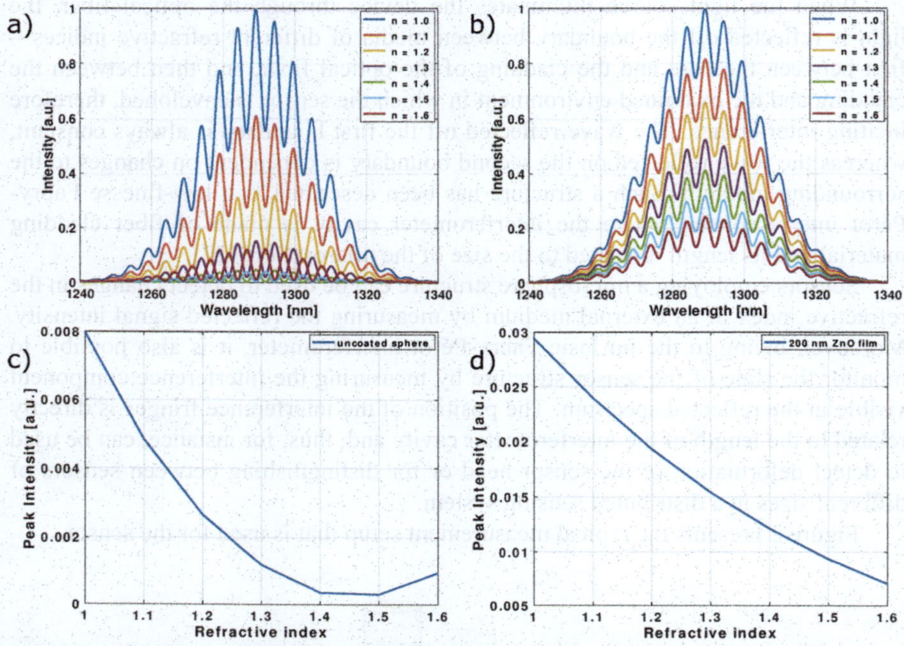

Figure 3. Theoretical sensor response: a) reflected spectra for the sensor illuminated with Superluminescent Diode (SLD)-like sources, for various refractive indices of the external medium – uncoated microspheres; b) reflected spectra for the sensor illuminated with an SLD-like source, for various refractive indices of the external medium – microsphere with a 200 nm ALD ZnO coating; c) signal intensity vs. refractive index of the external medium – uncoated microsphere; d) signal intensity vs. refractive index of external medium – microsphere with a 200 nm ALD ZnO coating. Reproduced with permission from Ref. [28], Copyright 2019, MDPI.

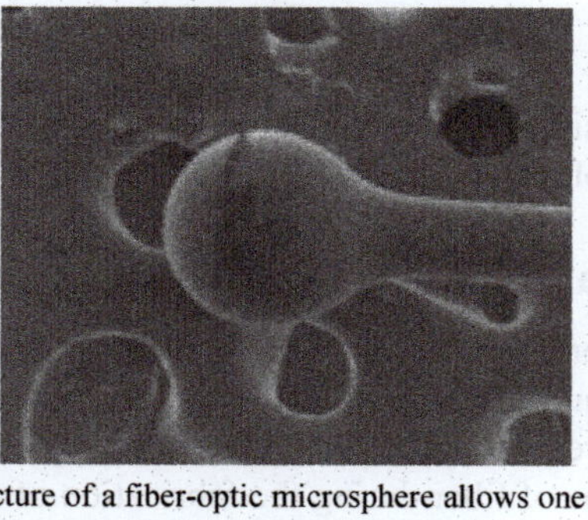

Figure 4. The exemplary structure of a fiber-optic microsphere allows one to obtain highly accurate and reproducible measurement results.

During the process of makingan optical microsphere on the end-face of an optical fiber, a slew of issues are likely to occur; these include:

- deformity in the microsphere,
- offset between the center of a microsphere and the core of a fiber,
- inconsistent diameter of the microsphere,

- damage to the inner structure integrity,
- appearance of defects in the material structure, and
- improper application of the coating.

Each of the above mentioned issues introduces uncertainty on its own. However, any combination of these invalidates the measurements completely.

The most criticial aspect during the fabrication of the microsphere, visible at the very first glance, is obtaining the structure which is round and centered vis-á-vis the fiber propagating the signal. Each deformity – whether an elongation of the microsphere or a gravitational drop (like the ones shown in Figure 5) – causes a considerable alteration of the optical spectrum. Figure 5 presents an image obtained by an optical microscope (Olympus CX31, Japan).

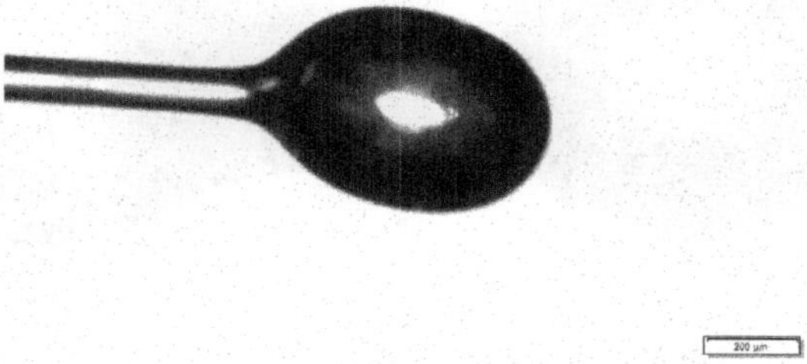

Figure 5. Elongation and/or offset of the microsphere during the fabrication process are the easiest mistakes to spot.

Secondly, obtaining the selected diameter of the microsphere is of utmost importance. One of the features of microsphere-based fiber-optic sensors is the presence of an intrinsic fixed cavity. Changing the diameter of the microstructure influences the length of the cavity; therefore, using a microsphere of a diameter different from the one specifically selected results in a vastly different spectrum of measured signals, compared to that expected, which in turn causes erroneous data analysis. Figure 6 shows examples of a variety of microsphere dimensions.

Another issue shown in Figure 6 is the preservation of the inner structure of the microsphere. It can be observed that the core of the fiber was disturbed during its production. The microsphere made at the core of an optical fiber was detached, resulting in the rest of the core to form a taper. Figures 7 and 8 were obtained by a Stimulated Emmision Depletion microscopy (Leica TCS SP8 STED, Germany).

Magnification of the microsphere surface, as presented in Figure 8, shows certain damage to the fiber structure. This indicates that the parameters of the splicer during the production of the microspheres were improperly set, therefore causing the emergence of air bubbles within the material of the optical fiber. Such a damage distorts the signal during measurement.

Additional issues emerge during the deposition of the coating on the surface of the microsphere. Therefore, the ALD method is chosen to obtain a uniform coating and to avoid run-off of the deposited material which occurs in methods like dip coating (presented in Figure 9). The images in Figure 9 were obtained by a digital microscope (Keyence VHX-7000 Series, Japan).

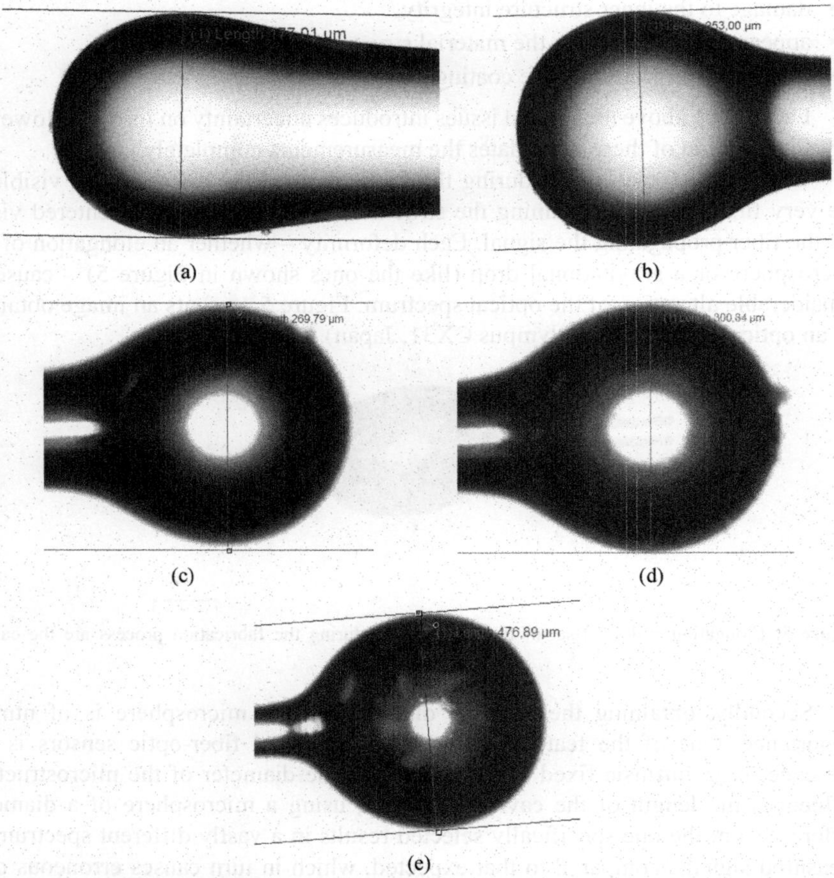

Figure 6. Altering the diameter of the microsphere results in changing of the cavity dimensions. Examples of microspheres with varying diameters: (a) 177 µm, (b) 253 µm, (c) 270 µm, (d) 301 µm and (e) 477 µm.

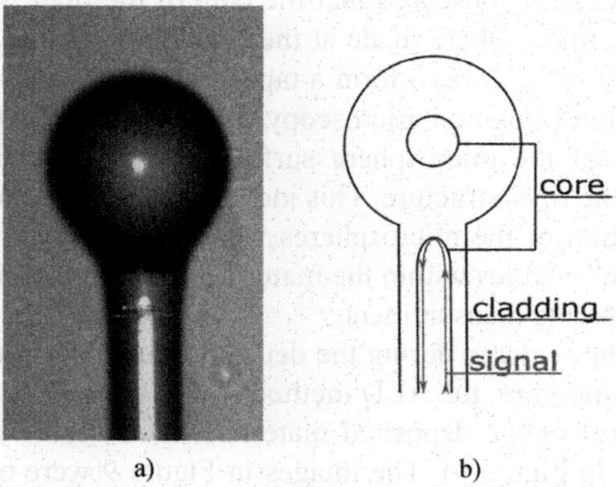

Figure 7. Impaired inner structure integrity of the microsphere: a) STED microscopy image, b) schematic explanation of the issue.

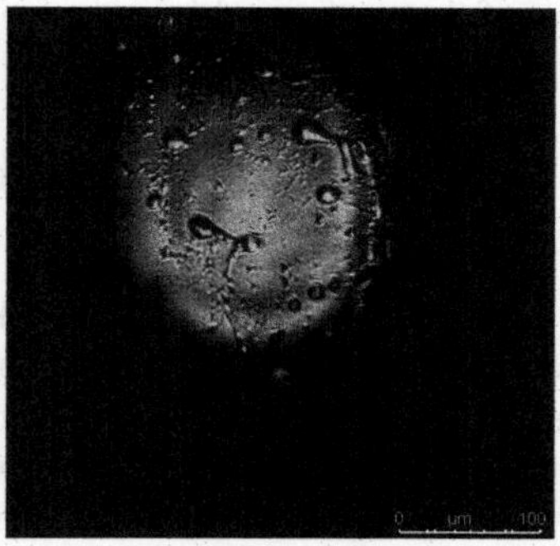

Figure 8. Magnification of the microsphere surface with affected material structure.

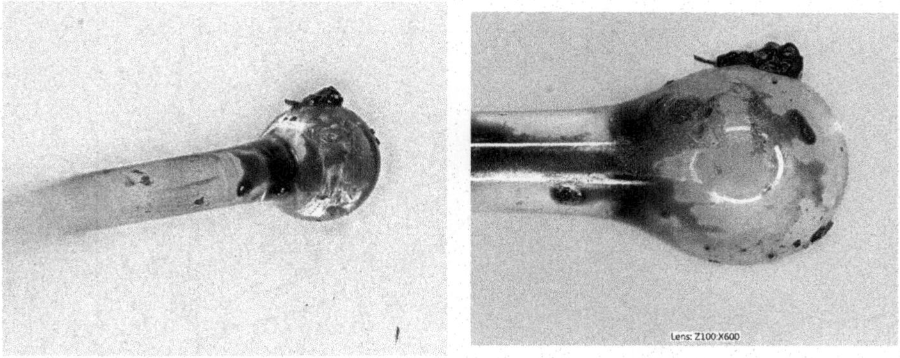

a) b)

Figure 9. Improperly deposited coating causes its run-off and lack of uniformity: a) view along the fiber, b) view of the microsphere.

Moreover, the selection of the coating material has to be thoroughly thought-out. Firstly, attention must be paid to the compatibility of the coating material with the material of the fiber. Deposition of the coating on an uneven surface and the application of an unsuitable material may not be possible or might even damage the microsphere (Figure 10).

Figure 10. Optical microsphere after application of the coating; the method used for applying the coating irreparably damaged the structure of the sensor.

Secondly, the coating should be selected with due consideration to its compatibility with the measured medium. Improper material choice results in possible contamination of the environment in which the sensor is used, as well as the inadvertent degradation of the coating, ultimately rendering the sensor unfit for further operation.

4. Tuning the metrological parameters of fiber-optic sensors with microspheres

A research has been conducted to determine the optimum diameter of the microsphere for interferometric measurements. Following the formulas presented in [27], theoretical calculation of various sizes of the microsphere show that an increase in diameter leads to a lower intensity of the reflected signal (Figure 11). However, considering the size of the standard optical fiber that has a diameter of 120 µm, a fully formed spherical shape of the sensor-head can be achieved with spheres of diameter well above 200 µm (as can be observed in Figure 6).

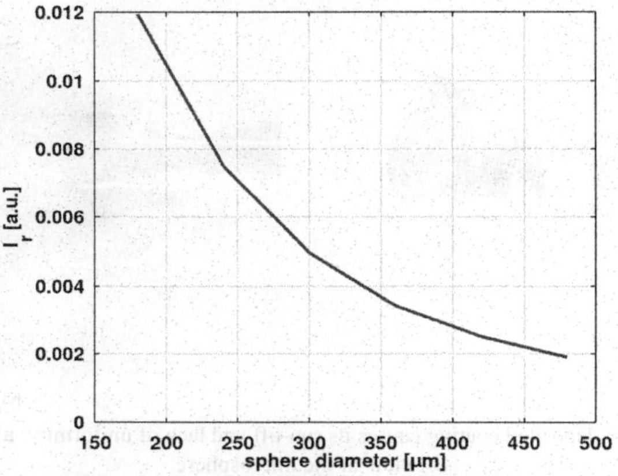

Figure 11. Results of numerical simulations: reflected signal intensity vs. sphere diameter for a sensor illuminated by an SLD-like source with a central wavelength of 1290 nm.

In this chapter, a microsphere-based fiber-optic sensor with a diameter of 240.7 µm has been investigated for the measurement of refractive index. During the investigation, the surface of the microsphere was coated with Zinc oxide (ZnO) using ALD method. Unlike other methods, Atomic Layer Deposition allows for uniform deposition of coating on the fibers, regardless of the structure. Homogeneity of the coating, along with its comparison with an uncoated microsphere, has been presented through SEM images in Figure 12 (Reproduced with permission from [28], Copyright 2019, MDPI.)

Thin ALD ZnO layer has a thickness in the order of only hundreds of nanometers; however, it substantially influences the optical parameters of the structure, thus in turn influencing the metrological properties of the fiber-optic sensor. Careful selection of both the deposition method and its parameters are crucial for obtaining the optimal sensitivity and measurement range of a fiber-optic sensor [27]. Metrological parameters of microsphere-based fiber-optic sensors with ALD ZnO coating can be optimized by modifying the thickness of the coating.

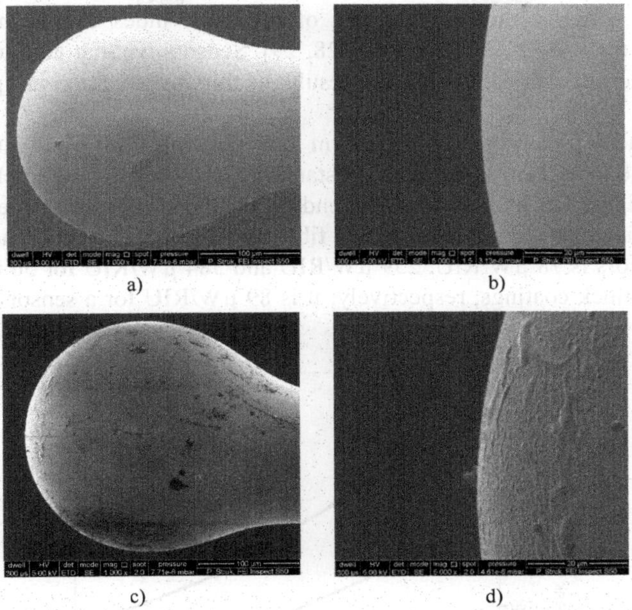

Figure 12. SEM images of microsphere-based fiber-optic sensors: a) uncoated microsphere, b) surface topography of an uncoated microsphere, c) microsphere with a 200 nm ALD ZnO coating, d) surface topography of the microsphere with a 200 nm ALD ZnO coating.

Furthermore, deposition of ALD ZnO coating on the surface of the optical microsphere allows for broadening the range and the sensitivity of refractive index measurements. When the refractive index of a measured medium is similar to the refractive index of an optical fiber core (1.43), the value of the intensity of the obtained signal is close to 0. By using thin ALD coatings with a higher refractive index (such as with ZnO), the transition point can be shifted to a higher value, therefore allowing for carrying out such measurements. The highlighted range of 1.4–1.5 in Figure 13 shows the calculation of the increase in the reflected signal intensity between sensors with an uncoated microsphere and those with a microsphere having 200 nm ZnO ALD coating.

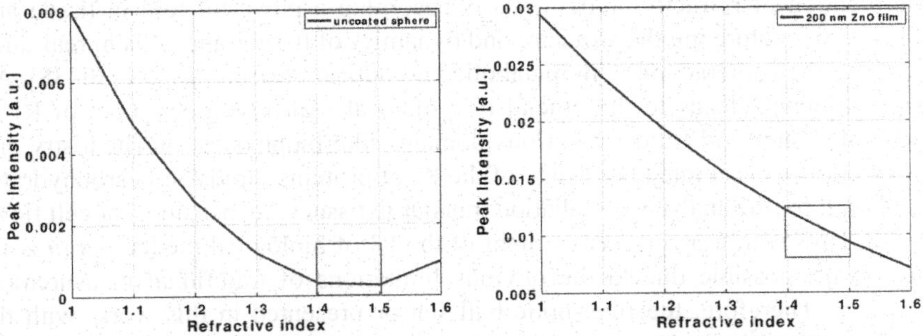

Figure 13. Coatings with a high refractive index shift the transition point, allowing for the measurement of refractive index of media with a refractive index of 1.4–1.5.

Sensors used for the measurement of refractive index use 50 nm, 100 nm and 200 nm thick ZnO ALD coatings [28, 29]. Sensors without additional coating have also been examined herein. The results of this investigation are presented in Figure 14.

Experimental measurements confirm that the value of peak intensity in a 1.4–1.5 range of refractive index is substantially higher when the thickness of the coating is higher. In other words, depending on the thickness of the ALD ZnO coating, the intensity of the investigated fiber-optic sensor changes. The sensitivity of these sensors is 78 μW/RIU, 259 μW/RIU and 384 μW/RIU for 50 nm, 100 nm and 200 nm thick coatings, respectively; it is 89 μW/RIU for a sensor without any coating.

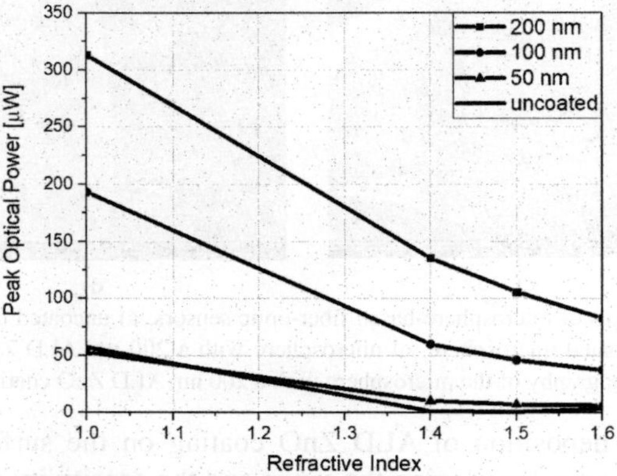

Figure 14. Dependence of the absolute values of signal intensity on refractive index, measured using microsphere-based fiber-optic sensors with ALD ZnO coatings.

Parameters of microsphere-based fiber-optic sensors can be altered by modifying their structure – their geometrical parameters (form, dimensions by selecting a predetermined fusion current supplied to the electrodes, fiber prefuse time and the type of fiber) and deposition of thin ALD coatings on their surface (e.g., ZnO).

5. Potential biological applications

Quantification of refractive index (RI) is becoming a powerful tool in tissue and cell biology for studying the structure and dynamics of tissues and cells in real time, along with their interaction with metabolic and optical clearing agents [30–38]. All of this ultimately leads to the crucial role of local real-time assessment of RI of biological tissues and fluids in various domains like biophysics, biochemistry and biomedicine. An accurate knowledge of the RI of proteins, lipids and carbohydrates is important for the analysis of radiation transfer in tissues, suspensions, or cell flows [30–38]. However, a precise assessment of the RI of biological media *in situ* is far from always possible due to the obvious limitations of traditional measurement techniques. Therefore, the fiber-optic ball sensor presented in this work, with the size of the sensitive part comparable to the size of cells can be efficiently used for monitoring the pathological conditions of cells, tissues and organs, and prove to be a significantly useful tool in this field where the RIs of proteins, lipids and carbohydrates constituting tissue and cells have an inherernt relation with the existing types and states of these molecular structures in health and disease [30, 35, 38].

Current trends in the study of biological solutions and cells, including new imaging methods using microfluidic platforms [34, 36] and flow cytometry [37], provide grounds for using fiber-optic sensors with high spatial and temporal resolution for RI measurements, which is necessary for a deeper understanding of pathological processes.

Thus, the physiological or pathological state of tissues and cells can be investigated, monitored and interpreted based on the spatio-temporal variations in RI. In particular, RI measurements provide information on the morphology of individual cells and their structure, the concentration and the type of substances contained in the cells, the cell cycle, the dry matter content of living cells, and apoptosis or cell necrosis; it is hence possible to monitor functional dynamics and metabolic processes in tissues and organs at cellular and sub-cellular levels [30, 32, 34, 35, 37, 38].

Modern optical technologies for imaging tissues and cells based on measurements of the spatial distribution of RI values include tomographic phase microscopy, digital holographic microscopy, bright field tomographic microscopy, integrated chip technique, optical trapping technique, etc. [30, 34]; these are technically complex devices, which do not allow for *in situ* measurements. This lends credibility to the developed fiber-optic biosensor discussed in this chapter, which shows promise for such studies on cerebral cortex, oral mucosa and other cavity organs of humans and other animals, as well as for express diagnostics of liquid biopsies, while being easily integrated into the microfluidic diagnostic system [36].

For instance, RI mapping can aid in automated cancer screening. In general, malignant cells have a higher average RI than their normal counterparts. RI mapping can effectively identify fundamental precancerous and cancerous changes such as increased nuclear size, irregular cell and nuclear shape, and increased chromatin concentration. RI mapping also shows promise for differentiating between circulating tumor cells (CTCs) in the bloodstream and normal blood cells, which may help in the early detection of metastases, which present a major obstacle in the optimization of cancer treatment [30, 34, 37, 38].

6. Conclusion

A novel implementation of optical microspheres integrated with fiber-optic sensors allows one to detect, in real time, the damage to the structure of the sensor head, which may occur during measurement execution. Optimization of the metrological properties of fiber-optic sensors can be achieved by modifications in optical microspheres and thin ALD ZnO coating deposited on their surface.

Ultimately, microsphere-based sensors are ideal for long-term and remote measurements of parameters such as refractive index due to their ability to constantly monitor the integrity of the sensor head.

Acknowledgment

The authors acknowledge the financial support of the DS Programs of the Faculty of Electronics, Telecommunications and Informatics of the Gdańsk University of Technology.

VVT was supported by the grant from the Government of the Russian Federation #075-15-2019-1885.

References

[1] Tripathi, S.M., W.J. Bock and P. Mikulic. 2017. A wide-range temperature immune refractive-index sensor using concatenated long-period-fiber-gratings. Sensors and Actuators B: Chemical 243: 1109–1114. Doi: 10.1016/j.snb.2016.12.012.

[2] Chiavaioli, F., C.A.J. Gouveia, P.A.S. Jorge and F. Baldini. 2017. Towards a uniform metrological assessment of grating-based optical fiber sensors: From refractometers to biosensors. Biosensors (Basel) 7. Doi: 10.3390/bios7020023.

[3] Islam, M.R., M.M. Ali, M.-H. Lai, K.-S. Lim and H. Ahmad. 2014. Chronology of Fabry-Perot interferometer fiber-optic sensors and their applications: A review. Sensors (Basel) 14: 7451–7488. Doi: 10.3390/s140407451.

[4] Hromadka, J., N.N. Mohd Hazlan, F.U. Hernandez, R. Correia, A. Norris, S.P. Morgan and S. Korposh. 2019. Simultaneous *in situ* temperature and relative humidity monitoring in mechanical ventilators using an array of functionalised optical fibre long period grating sensors. Sensors and Actuators B: Chemical 286: 306–314. Doi: 10.1016/j.snb.2019.01.124.

[5] Matsunaga, T., Y. Haga, S. Osaki, Y. Shimizu, S. Tupin, M. Ohta, M. Shojima, H. Yoshida and N. Tsuruoka. 2017. Multipoint pressure measurement in blood vessel model for evaluation of intravascular treatment of cerebral aneurysm using fiber-optic pressure sensors. In Proceedings of the 2017 IEEE International Conference on Cyborg and Bionic Systems (CBS); IEEE: Beijing, pp. 136–139.

[6] Chen, J., Y. Yang, X. Pan, Z. Wang, Z. Zhang and H. Ge. 2019. A new measurement method for resonance frequency of fiber optic interferometric vibration detector. In Proceedings of the 2019 4th International Conference on Measurement, Information and Control (ICMIC); IEEE: Harbin, China, pp. 25–27.

[7] Wang, X. and O.S. Wolfbeis. 2020. Fiber-optic chemical sensors and biosensors (2015–2019). Anal. Chem. 92: 397–430. Doi: 10.1021/acs.analchem.9b04708.

[8] Pérez, M.A., O.G. Arias and J.R. 2013. Optical fiber sensors for chemical and biological measurements. Current Developments in Optical Fiber Technology. Doi: 10.5772/52741.

[9] Li, M., T. Dubaniewicz, H. Dougherty and J. Addis. 2018. Evaluation of fiber optic methane sensor using a smoke chamber. International Journal of Mining Science and Technology 28: 969–974. Doi: 10.1016/j.ijmst.2018.05.010.

[10] Alves, H.P., J.F. Nascimento, E. Fontana, I.J.S. Coelho and J.F. Martins-Filho. 2018. Transition layer and surface roughness effects on the response of metal-based fiber-optic corrosion sensors. J. Lightwave Technol. 36: 2597–2605. Doi: 10.1109/JLT.2018.2817517.

[11] Xue, Z., J.-Q. Shi, Y. Yamauchi and S. Durucan. 2018. Fiber optic sensing for geomechanical monitoring: (1)-distributed strain measurements of two sandstones under hydrostatic confining and pore pressure conditions. Applied Sciences 8: 2103. Doi: 10.3390/app8112103.

[12] Kou, J., J. Feng, L. Ye, F. Xu and Y. Lu. 2010. Miniaturized fiber taper reflective interferometer for high temperature measurement. Opt. Express 18: 14245. Doi: 10.1364/OE.18.014245.

[13] Arumona, A.E., A. Garhwal, P. Youplao, I.S. Amiri, K. Ray, S. Punthawanunt and P. Yupapin. 2020. Electron cloud spectroscopy using micro-ring Fabry–Perot sensor embedded gold grating. IEEE Sensors J. 20: 10564–10571. Doi: 10.1109/JSEN.2020.2994240.

[14] Ma, J., J. Ju, L. Jin, W. Jin and D. Wang. 2011. Fiber-tip micro-cavity for temperature and transverse load sensing. Opt. Express 19: 12418. Doi: 10.1364/OE.19.012418.

[15] Shankar, R. and M. Lončar. 2014. Silicon photonic devices for mid-infrared applications. Nanophotonics 3: 329–341. Doi: 10.1515/nanoph-2013-0027.

[16] Ferreira, M.S., P. Roriz, S.O. Silva, J.L. Santos and O. Frazão. 2013. Next generation of Fabry-Perot sensors for high-temperature. Optical Fiber Technology 19: 833–837. Doi: 10.1016/j.yofte.2013.07.006.

[17] Eryürek, M., Z. Tasdemir, Y. Karadag, S. Anand, N. Kilinc, B.E. Alaca and A. Kiraz. 2017. Integrated humidity sensor based on SU-8 polymer microdisk microresonator. Sensors and Actuators B: Chemical 242: 1115–1120. Doi: 10.1016/j.snb.2016.09.136.

[18] Bhattacharya, S., A.V. Veluthandath, C.C. Huang, G.S. Murugan and P.B. Bisht. 2020. Effect of coating few-layer WS 2 on the Raman spectra and whispering gallery modes of a microbottle resonator. J. Opt. 22: 105003. Doi: 10.1088/2040-8986/abad50.

[19] Liang, L., M. Li, N. Liu, H. Sun, Q. Rong and M. Hu. 2018. A high-sensitivity optical fiber relative humidity sensor based on microsphere WGM resonator. Optical Fiber Technology 45: 415–418. Doi: 10.1016/j.yofte.2018.07.023.

[20] Mahmood, A., V. Kavungal, S.S. Ahmed, G. Farrell and Y. Semenova. 2015. Magnetic-field sensor based on whispering-gallery modes in a photonic crystal fiber infiltrated with magnetic fluid. Opt. Lett., OL 40: 4983–4986. Doi: 10.1364/OL.40.004983.

[21] Xu, X., W. Chen, G. Zhao, Y. Li, C. Lu and L. Yang. 2018. Wireless whispering-gallery-mode sensor for thermal sensing and aerial mapping. Light: Science & Applications 7: 62. Doi: 10.1038/s41377-018-0063-4.

[22] Fujii, S. and T. Tanabe. 2020. Dispersion engineering and measurement of whispering gallery mode microresonator for Kerr frequency comb generation. Nanophotonics 9: 1087–1104. Doi: 10.1515/nanoph-2019-0497.

[23] Cai, L., J. Pan, Y. Zhao, J. Wang and S. Xiao. 2020. Whispering gallery mode optical microresonators: Structures and sensing applications. Phys. Status Solidi A 217: 1900825. Doi: 10.1002/pssa.201900825.

[24] Dong, B., J. Hao, T. Zhang and J.L. Lim. 2012. High sensitive fiber-optic liquid refractive index tip sensor based on a simple inline hollow glass micro-sphere. Sensors and Actuators B: Chemical 171-172: 405–408. Doi: 10.1016/j.snb.2012.05.001.

[25] Tang, S., B. Zhang, Z. Li, J. Dai, G. Wang and M. Yang. 2015. Self-compensated microstructure fiber optic sensor to detect high hydrogen concentration. Opt. Express 23: 22826. Doi: 10.1364/OE.23.022826.

[26] Zhang, Y., T. Zhou, B. Han, A. Zhang and Y. Zhao. 2018. Optical bio-chemical sensors based on whispering gallery mode resonators. Nanoscale 10: 13832–13856. Doi: 10.1039/C8NR03709D.

[27] Hirsch, M. 2018. Fiber optic microsphere with ZnO thin film for potential application in refractive index sensor—theoretical study. Photonics Letters of Poland 10: 85–87. Doi: 10.4302/plp.v10i3.835.

[28] Listewnik, P., M. Hirsch, P. Struk, M. Weber, M. Bechelany and M. Jędrzejewska-Szczerska. 2019. Preparation and characterization of microsphere ZnO ALD coating dedicated for the fiber-optic refractive index sensor. Nanomaterials 9: 306. Doi: 10.3390/nano9020306.

[29] Hirsch, M., P. Listewnik, P. Struk, M. Weber, M. Bechelany and M. Szczerska. 2019. ZnO coated fiber optic microsphere sensor for the enhanced refractive index sensing. Sensors and Actuators A: Physical 298: 111594. Doi: 10.1016/j.sna.2019.111594.

[30] Wang, Z., K. Tangella, A. Balla and G. Popescu. 2011. Tissue refractive index as marker of disease, J. Biomed. Opt. 16(11): 116017. Doi: 10.1117/1.3656732.

[31] Yanina, I. Yu., E.N. Lazareva and V.V. Tuchin. 2018. Refractive index of adipose tissue and lipid droplet measured in wide spectral and temperature ranges. Applied Optics 57: 4839. https://doi.org/10.1364/AO.57.004839.

[32] Lazareva, E.N. and V.V. Tuchin. 2018. Measurement of refractive index of hemoglobin in the visible/NIR spectral range. J. Biomed. Opt. 23: 035004. Doi: 10.1117/1.JBO.23.3.035004.

[33] Bashkatov, A.N., K.V. Berezin, K.N. Dvoretskiy, M.L. Chernavina, E.A. Genina, V.D. Genin, V.I. Kochubey, E.N. Lazareva, A.B. Pravdin, M.E. Shvachkina, P.A. Timoshina, D.K. Tuchina, D.D. Yakovlev, D.A. Yakovlev, I. Yanina, O.S. Yu.; Zhernovaya and V.V. Tuchin. 2018. Measurement of tissue optical properties in the context of tissue optical clearing. J. Biomed. Opt. 23: 091416. Doi: 10.1117/1.JBO.23.9.091416.

[34] Gul, B., S. Asharf, S. Khan, H. Nisar and I. Ahmad. 2020. Cell refractive index: Models, insights, applications and future perspectives. Photodiagnosis and Photodynamic Therapy. Doi: https://doi.org/10.1016/j.pdpdt.2020.102096.
[35] Li, X., B. Xie, M. Wu, J. Zhao, Z. Xu and L. Liu. 2020. Visible-to-Near-infrared optical properties of protein, lipid and carbohydrate in both solid and solution state at room temperature. Journal of Quantitative Spectroscopy & Radiative Transfer. Doi: https://doi.org/10.1016/j.jqsrt.2020.107410.
[36] Credi, C., C. Dallari, E. Lenci, A. Trabocchi, S. Nocentini, D. Wiersma, R. Cicchi and F.S. Pavone. 2020. 3D printing of multifunctional optofluidic systems for high-sensitive detection of pathological biomarkers in liquid biopsies. Proc. SPIE 11361: 113610A. https://doi.org/10.1117/12.2553685.
[37] Tuchin, V.V. 2014. *In vivo* optical flow cytometry and cell imaging. Rivista Del Nuovo Cimento 37: 375. Doi: 10.1393/ncr/i2014-10102-x.
[38] Oliveira, L. and V.V. Tuchin. 2019. The Optical Clearing Method: A New Tool for Clinical Practice and Biomedical Engineering, Basel: Springer Nature Switzerland AG, 177 p. https://www.springer.com/gp/book/9783030330545.

8

Nanosensors for Diagnostics and Conservation of Works of Art

Georgia-Paraskevi Nikoleli

1. Introduction

Nanosensors have high sensitivity and selectivity, and use electrical, optical and acoustic transduction schemes to detect low limits of analytes. The exceptional properties that nanomaterials possess are: large surface area to volume ratio, use of charge and reactive sites, and physical structure and potentiality to detect low limits of analytes; these characteristics provide an advantage towards ideal sensing properties. High-sensitivity is managed mainly through signal amplification. With the use of such nanosensors, their sensitivity and selectivity can provide various advantages in biomedical analysis, such as earlier diagnosis of diseases and biotoxin threats, and subsequently create significant improvements in clinical, environmental and industrial outcomes. The emerging discipline of nanotechnology at the boundary of life sciences and chemistry offers a wide range of prospects within a number of fields like fabrication and characterization of nanomaterials, supramolecular chemistry, targeted drug supply and early detection of disease-related biomarkers.

Till date, research has focused on developing non-invasive or micro-invasive techniques to characterize components and degradation products. However, these techniques usually require advanced instrumentation and highly specialized personnel for data acquisition and processing. Moreover, although conservators and curators recognize the importance of scientific analyses in restoration campaigns, the time and financial costs involved are prohibitive. An alternative approach is to develop easy-to-use analytical kits, which would enable conservators to rapidly obtain information to guide their restoration strategy.

Department of Conservation of Antiquities and Works of Art, School of Applied Arts & Culture, University of West Attica, 12243 Egaleo, Greece.
Email: gnikoleli@uniwa.gr

Based on point-of-care testing (POCT) approach, miniaturized nanosensors which can be used by conservators to identify proteins in artworks have been developed [1, 2]. Thanks to the miniaturization of instrumentation and delocalization of analyses, conservators thus become akin to a primary care physician, performing anamnesis (that is, cultural context, characterization of materials), diagnosis (evaluation of conservation state, identification of degradation processes), prognosis, and therapy (restoration).

Proteinaceous materials have been used as fixatives or binders in paintings since ancient times; yet, even now, their characterization is a challenge for diagnostic studies. Immunological approaches have proved to be particularly powerful, for they are simple, highly specific, can be implemented with relatively inexpensive instruments, can recognize even denaturized proteins, and require minimal sample processing. In fact, although originally developed for medicine, immunoassays have been used to detect proteins in paintings for over 45 years now. Immunofluorescence microscopy (IFM) and chemiluminescence (CL) detection have been used for single or multiplexed localization of proteins in artwork samples [2, 3]. However, these techniques do not offer the advantages of nanosensing devices, have high costs, are time consuming, and will therefore not be included in this chapter.

The analysis of materials composing works of art has been a fundamental issue for long. Moreover, the techniques should essentially be non-destructive and require the use of a very small amount of sample. The development of portable devices that give quick and reliable chemical information to the restorer during the restoration work has prompted the evolution of an efficient scientific approach. Analytical diagnostics on a work of art is fundamental to restoration and conservation purposes or for certifying the authenticity of an ancient artefact. Analyzing the chemicals of which the work of art is composed is thus crucial. A layer of complexity is added by the microscopic structure of the cross section of a paint, which is in actuality composed of five or even more superimposed layers, each with its own typical composition and requiring a specific analytical approach to be identified and classified. Some of these layers are made of composite materials: in particular, the pictorial layer is made by pigments (inorganic oxides and salts and/or organic dyes or pigments) bound and dispersed in the binder, the latter being a glue of animal or vegetal origin for tempera paints or a drying oil for oil paintings. Moreover, fillers, emulsifiers, preservatives and/or other additives are also present, depending on factors like period of production, painter, following restoration intervention, etc.

The point-of-care testing concept has been used to design and develop portable and cheap bioanalytical systems that can be used on-site by conservators. These systems deploy lateral flow immunoassays to simultaneously detect two proteins (ovalbumin and collagen) in artworks. For an in-depth study on the application of these portable biosensors, both chemiluminescent and colorimetric detections have been developed and compared in terms of sensitivity and feasibility. The chemiluminescent system displays the best analytical performance (that is, two orders of magnitude lower limits of detection than the colorimetric system). To simplify its use for this specific application, disposable cartridges have been designed ad hoc. These results highlight the enormous potential of these inexpensive, easy-to-use and minimally invasive diagnostic tools for conservators in the cultural heritage field.

This chapter focuses on the recent advances and prospects of the application of two classes of nanosensors—surface enhanced Raman scattering (SERS) and nanostructured electrochemical (EC) nanosensors, both characterized by high selectivity and sensitivity, in the domain of cultural heritage. Both of these techniques require the use of a metal surface which acts as a transducer for EC nanosensors or, when nanostructured, as a signal amplifier in the case of SERS. It is noteworthy that both these techniques use metal substrates and this, in the near future, would lead to the development of nanosensors capable to provide both spectroscopic and electrochemical information.

2. SERS

As discussed in the introduction, a painting is a multilayer matrix and the pictorial layer itself is a complex system composed of pigments dispersed in a binding medium. The nature of the binder determines the working behavior of the paint and the final characteristics of the work of art, not only from an aesthetic point of view, but also from the perspectives of durability and conservation. Western artists have been using a variety of binders, including animal glue, casein, eggs, drying oils, resins, natural gums and, more recently, synthetic materials such as acrylic and vinyl polymers. The detection of binders has long presented a key issue in the study of cultural heritage materials. Vibrational spectroscopic techniques (e.g., IR and Raman spectroscopy) have been widely used for preliminary identification of binders. However, their lack of specificity in identifying the biological provenance of these macromolecules warrants the use of chromatography (e.g., GC/MS and HPLC) for a more precise diagnosis. Among the methods based on optical detection, the chemiluminescent approach gives the best results in terms of sensitivity. The use of enzyme-labeled antibodies in the presence of a chemiluminescent substrate (e.g., horseradish peroxidase (HRP) reacts with luminol when hydrogen peroxide is added, generating an emission of photons) allows for the simultaneous detection of ovalbumin and casein on cross sections of real samples. A promising alternative to this approach is represented by SERS, whose potential has not been exploited to its maximum yet. NPs can be bound to specific antibodies for the target proteins and to a Raman-probe by means of functional thiols. In this way, the so-called SERS nano-tags can be employed even directly on cross sections. Once bound to the surface, the nano-tags are mapped using the SERS technique, revealing the distribution of the protein in the cross-section [3, 4].

Interesting avenues in this area can be opened by combining SERS detection with lateral flow immunoassay (LFIA) to substitute typical fluorescence detection methods. A recently developed lateral flow immunoassay demonstrated the possibility to simultaneously detect the presence of two proteins (ovalbumin and collagen) in artworks; it was based on the use of disposable cartridges and a chemiluminescent detection system [5].

The extraordinary positive results achieved by means of SERS make it a worthy substitute detection technique in such an approach. Furthermore, a SERS substrate immunochemically functionalized and labeled with a Raman-probe can prove to be a promising biosensor to be employed on site, taking advantage of the strong Raman enhancement produced by anisotropic Ag nanostars [6], which are able to create a larger number of SERS hot spots than spherical NPs.

3. Electrochemical immunosensors for pictorial binders and paper substrates

One of the oldest painting techniques is the tempera technique, based on the use of water dispersible binders which, upon drying, become insoluble. The ancient tempera technique was described by Cennino Cennini in his XVth century book "*Il Libro dell'Arte*" and is based simply on mixing finely grounded pigments with egg yolk; however, from Renaissance onward, tempera recipes became more complex and different binders came to be in use, such as whole eggs or egg whites, singly or in combination with flour or animal glue or oily components [7]. From an analytical perspective, identifying proteins from different animal sources is not an easy task, typically requiring the use of spectroscopic and chromatographic or hyphenated mass spectrometric techniques.

As also mentioned in the previous section dealing with SERS sensors, an alternative approach to detection is offered by the use of immunosensors coupled with optical detection methods. The first study in this direction was aimed at developing a sensor suitable for the detection of the glycoprotein immunoglobulin IgY, the main immunoglobulin present in chicken egg yolk [7]. This approach is summarized in Figure 1.

IgY is captured by the polycarbonate surface (A), after blocking with bovine serum albumin; the electrode is incubated with anti-IgY labeled with HRP (anti-IgY-HRP) (B) and, finally, the presence of the label HRP is detected by adding the enzyme substrate H_2O_2 and a mediator (C). Also, other molecules in the sample can spontaneously bind to the polycarbonate; however, only IgY is recognized by the HRP-labeled anti-IgY, which finally generates the electrocatalytic signal, detected by cyclic voltammetry. This allows the development of a high precision diagnostic scheme able to qualitatively distinguish egg-yolk tempera (which contains IgY) from other kinds of tempera (not containing IgY) and oil or acrylic paints. In the experiment, the capability of the sensor to identify egg yolk tempera was successfully tested on 24 samples, ranging from paint mockups to artworks from XVIII–XX centuries, and validated by a chemiometrical approach [7].

A similar approach has been used for the detection of ovalbumin (OVA) from egg white (or albumin) [8], which is also used as a binder in works of art—for egg white or whole-egg tempera or for preparing the photosensitive emulsion of XIXth century albumen photographic prints for instance. Ovalbumin is extracted from microsamples by ultrasonication in aqueous buffer; then, it is incubated and bound directly to the polycarbonate surface of the NEE, thus reducing the number of analytical steps and the volume of reagents required to a minimum. The protein

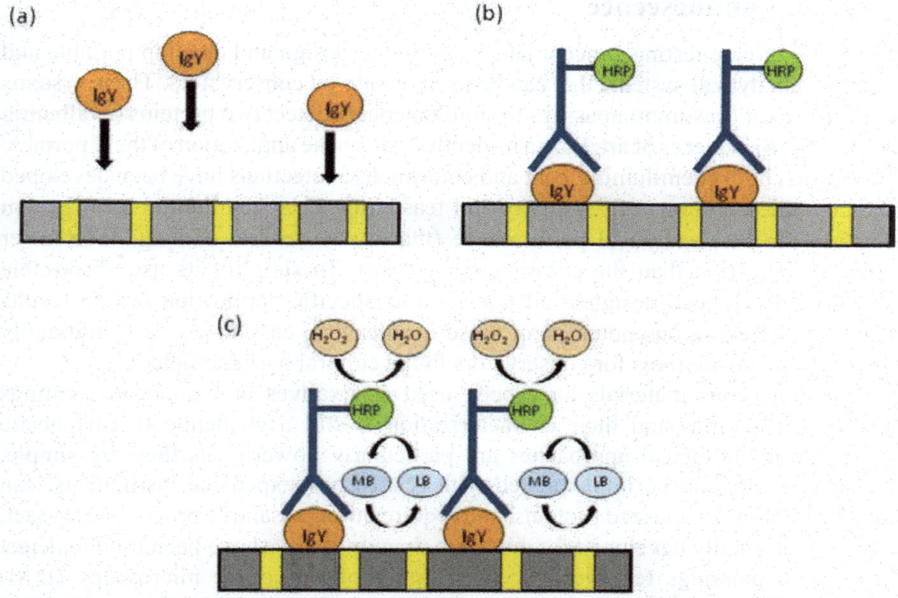

Figure 1. Schematic representation of the analytical protocol used to detect immunoglobulin IgY extracted from tempera paints. MB and LB are the oxidized and reduced forms of the redox mediator methylene blue and HRP represents horseradish peroxidase (used as an enzymatic label for the anti-IgY antibody); (a) capture on the polycarbonate of the NEE of immunoglobulin IgY extracted by ultrasonication from a paint microfragment; (b) molecular recognition by anti-IgY labeled with HRP; (c) generation of electrocatalytic cycle and detection by cyclic voltammetry, by adding the enzyme substrate H_2O_2 and the mediator MB (Reprinted with permission from [7]).

captured on the polycarbonate membrane is reacted selectively with anti-OVA antibody labelled with glucose oxidase (GOx). The addition of the GOx substrate (i.e., glucose) and a ferrocenyl redox mediator provides an electrocatalytic oxidation current, strictly dependent on the OVA concentration present in the extracts from historical photographic prints or ancient tempera paintings. It has also been demonstrated that the combined use of IgY and OVA NEE-based sensors allows to distinguish egg yolk from whole egg tempera in real samples [8].

The qualitative identification of substances, as well as their location within a complex paint stratigraphy (e.g., a cross section), is one of the most challenging issues in the characterization of painting materials. This requirement led G. Sciutto et al. [9] to develop a new detection approach for the immunochemical localization of ovalbumin in paint cross-sections based on the use of scanning electrochemical microscopy (SECM). To identify ovalbumin in paint samples, immunochemical analyses were performed using an anti-ovalbumin primary antibody and a secondary antibody labelled with HRP. SECM measurements were performed in feedback mode using benzoquinone (BQ)/hydroquinone (H_2Q) redox couple as the SECM probe. The results obtained were promising, foreseeing a wider application of SECM in cultural heritage researches and allowing localized identification of the binder in different layers of a cross-section from real paint samples.

4. Chemiluminescence

The point-of-care testing concept has been used to design and develop portable and cheap bioanalytical systems that can be used on-site by conservators. These systems employ lateral flow immunoassays to simultaneously detect two proteins (ovalbumin and collagen) in pieces of art. For an in-depth study on the application of these portable biosensors, both chemiluminescent and colorimetric detections have been developed and compared in terms of sensitivity and feasibility. The chemiluminescent system displays the best analytical performance (that is, two orders of magnitude lower limits of detection than the colorimetric system). To simplify its use, disposable cartridges have been designed ad hoc for this specific application. These results highlight the enormous potential of these inexpensive, easy-to-use, and minimally invasive diagnostic tools for conservators in the cultural heritage field.

Proteinaceous materials have been used as fixatives or binders in paintings since ancient times and their characterization is still a challenge for diagnostic studies. Immunological approaches are particularly powerful as they are simple, highly specific, can be implemented with relatively inexpensive instruments, can recognize even denaturized proteins, and require minimal sample processing. In fact, although originally developed for medicine, immunoassays have been used to detect proteins in paintings for over 45 years. Immunofluorescence microscopy (IFM) and chemiluminescent (CL) detection have been used for single or multiplexed localization of proteins in art samples. Previous studies have supported the suitability of immunochemical analyses for aged and historical samples, showing that aging does not hamper recognition of proteins by immunological methods [10–12]. Past research has also confirmed the absence of detectable interferences owing to paint materials [10, 12]. In particular, the effect of metal ions in pigments was excluded probably due to a limited release of such ions in the solution during the incubation steps [10].

Recently, a chemiluminescence immunochemical portable imaging device which can be used to detect ovalbumin in paintings has been developed [13]. The analytical prototype used a silanized glass slide with an array of immobilized antibody spots and a polydimethylsiloxane (PDMS) fluidic element.

A design of new portable devices that can simultaneously detect two proteins, using the lateral flow immunoassay (LFIA) technology, was recently described in the literature [10]. This is one of the most commercially successful POCT techniques, owing to its ease of use and quickness. LFIA technology uses cellulose-based materials assembled in a strip format, which acts as an inexpensive and easy-to-use platform for performing immunological assays [14]. The strip is preloaded with appropriate reagents and the capillary force makes the sample and reagents flow across it without the need of external pumps. Immunoreagents are immobilized in specific areas on the nitrocellulose membrane, namely on one or more test (T)-line(s), in which the analyte(s) are bound and detected, and on a control (C)-line, in which the excess labeled probe is captured to confirm the validity of the analysis.

The simplest LFIAs employ colorimetric detection, often based on gold nanoparticles (AuNPs) as labels, which may suffer from poor detectability and limited quantitative performances [15]. As an alternative, chemiluminescent (CL)-based LFIAs provide high detectability and a wide linear range for quantitative analyses [16]. In this case, the detection probe is often labeled with an enzyme which catalyzes a CL reaction in the presence of a proper CL substrate. However, CL detection involves the addition of a series of reagents, and thus requires a more complex analytical procedure.

5. Conclusions and prospects

This review on the development and application of SERS and electrochemical sensors for diagnostics, conservation and restoration of works of art indicates the great potentialities offered by such approaches. Of course, sensors complement the analytical information obtained by consolidated analytical methods based on the use of complex and expensive instrumentation. Yet, sensors are capable of providing reliable information at a competitive cost and can be used by non-specialized personnel in raw samples, with no or minimal use of reagents and/or standards. These features make sensors particularly apt for performing on-site analysis, to be used by conservation scientists alongside restorers or conservators, during the restoration intervention. It is worth stressing that only in a few cases can sensors be used in a completely non-destructive manner. However, the most recent works of research have shown that both SERS and electrochemical sensors can be successfully applied to analyze cross-sections of works of art. Note that such a sampling, although nowadays miniaturized to a microsampling level, is almost always necessary for a full characterization of a piece of art due to its complexity and multilayered structure. The application of SERS and electrochemical techniques to characterize these multilayered samples has demonstrated their ability to provide valuable chemical information which implements significantly the information obtained by optical and electronic microscopies and spectroscopic techniques – the techniques most widely applied on cross-sections.

References

[1] Nikoleli, G.-P. 2020. Biosensors for bacteria for artwork. Int. J. Adv. Res. in Chem. Sci. 7(3): 1–7.
[2] Palmieri, C.L., M. Vagnini and L. Pitzurra. 2016. Immunochemical methods applied to art-historical materials: Identification and localization of proteins by ELISA and IFM. Top. Cur. Chem. 374 (1): Article No. 5.
[3] Arslanoglu, J., S. Zaleski and J. Loike. 2010. An improved method of protein localization in artworks through SERS nanotag-complexed antibodies. Anal. Bioanal. 399: 2997–3010.
[4] Sciutto, L., C. Lilli, S. Lofrumento, M. Prati, M. Ricci, A. Gobbo, E. Roda, M. Castellucci, M. Meneghetti and R. Mazzeo. Alternative SERRS probes for the immunochemical localization of ovalbumin in paintings: An advanced mapping detection approach. Analyst 138(16): 4532–4541.
[5] Sciutto, G., M. Zangheri, L. Anfossi, M. Guardigli, S. Prati, M. Mirasoli, F.D. Nardo, C. Baggiani and A. Roda. 2018. Miniaturized biosensors to preserve and monitor cultural heritage: From medical to conservation diagnosis. Angew. Chem. 130(25): 7507–7511.
[6] García-Leis, A., J.V. García-Ramos and S. Sánchez-Cortés. 2013. Silver nanostars with high SERS performance. J. Phys. Chem. C 117: 7791–7795.
[7] Bottari, F., P. Oliveri and P. Ugo. 2014. Electrochemical immunosensor based on ensemble of nanoelectrodes for immunoglobulin IgY detection: Application to identify hen's egg yolk in tempera paintings. Biosens. Bioelectron. 52: 403–410.

[8] Gaetani, C., G. Gheno, M. Borroni, K.D. Wael, L.M. Moretto and P. Ugo. 2019. Nanoelectrode ensemble immunosensing for the electrochemical identification of ovalbumin in works of art. Electrochim. Acta 312: 72–79.
[9] Sciutto, G., S. Prati, R. Mazzeo, M. Zangheri, A. Roda, L. Bardini, G. Valenti, S. Rapino and M. Marcaccio. 2014. Localization of proteins in paint cross-sections by scanning electrochemical microscopy as an alternative immunochemical detection technique. Anal. Chim. Acta 831: 31–37.
[10] Sciutto, G., M. Zangheri, S. Prati, M. Guardigli, M. Mirasoli, R. Mazzeo and A. Roda. 2016. Immunochemical micro Imaging analyses for the detection of proteins in artworks. Top. Curr. Chem. 374: 31–59.
[11] Palmieri, M., M. Vagnini, L. Pitzurra, P. Rocchi, B.G. Brunetti, A. Sgamellotti and L. Cartechini. 2011. Development of an analytical protocol for a fast, sensitive and specific protein recognition in paintings by enzyme-linked immunosorbent assay (ELISA). Anal. Bioanal. Chem. 399: 3011–3023.
[12] Lee, H.Y., N. Atlasevich, C. Granzotto, J. Schultz, J. Loieke and J. Arslanoglu. 2015. Development and application of an ELISA method for the analysis of protein-based binding media of artworks. Anal. Methods 7: 187–196.
[13] Wong, C. and H.Y. Tse. 2009. Lateral Flow Immunoassay, Humana Press, New York.
[14] Huang, X., Z.P. Aguilar, H. Xu, W. Lai and Y. Xiong. 2016. Membrane-based lateral flow immunochromatographic strip with nanoparticles as reporters for detection: A review. Biosens. Bioelectron. 75: 166–180.
[15] Zangheri, M., L. Cevenini, L. Anfossi, C. Baggiani, P. Simoni, F. Di Nardo and A. Roda. 2015. A simple and compact smartphone accessory for quantitative chemiluminescence-based lateral flow immunoassay for salivary cortisol detection. Biosens. Bioelectron. 64: 63–68.
[16] Madariaga, J.M. 2015. Analytical chemistry in the field of cultural heritage. Anal. Methods 7: 4848–4876.

9

Applications of Biosensors in Animal Biotechnology

Georgia-Paraskevi Nikoleli,[1,*] *Marianna-Thalia Nikolelis*[2] and *Vasillios N. Psychoyios*[2]

1. Introduction

One of the novel detection methods that are used in analytical chemistry involve an analytical device for the detection of an analyte, combining a biological component with a physicochemical detector [1, 2]. These devices are called biosensors. In truth, they are systems that typically consist of a bio-recognition site, a biotransducer component, and an electronic system which includes a signal amplifier, a processor and a display. Figure 1 provides the schematic of a typical lipid membrane-based biosensor.

Transducers and electronics can be combined. One such example can be found in CMOS-based microsensor systems [3, 4]. The recognition component, often called bioreceptor, use biomolecules from organisms or receptors that are modeled after interaction with biological systems. This interaction is usually measured by the biotransducer, the output of which is a measurable signal proportional to the presence of the target analyte in the sample. The general aim of a biosensor design is to enable quick and convenient testing; therefore, within the domain biosensors, there is a class – the so-called point of concern or point of care class – which works at the site where the sample is procured [5–8].

[1] Laboratory of Statistical Modelling and Educational Technology in Public and Environmental Health, Department of Public and Community Health, Faculty of Public Health, University of West Attica, 122 43 Egaleo, Athens, Greece.

[2] Laboratory of Inorganic & Analytical Chemistry, School of Chemical Engineering, Department 1, Chemical Sciences National Technical University of Athens, Athens, Greece.

* Corresponding author: gnikoleli@uniwa.gr

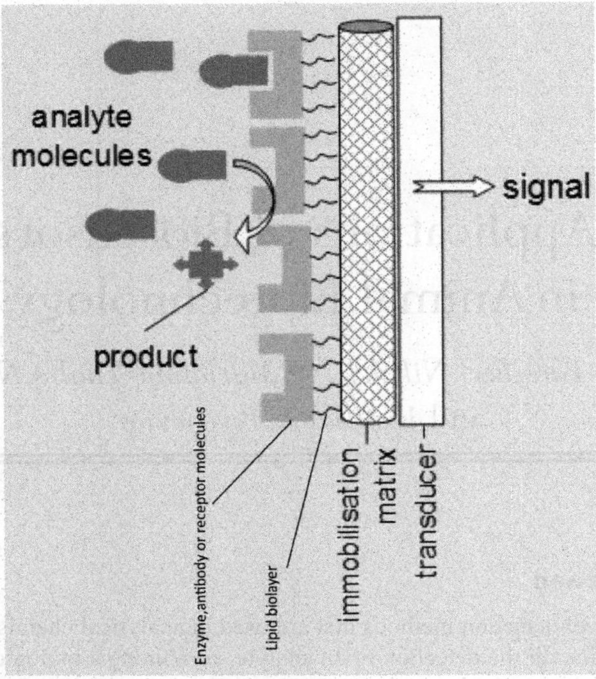

Figure 1. Schematic of a lipid membrane-based biosensor (the lipid bilayer serves as an intermediate transduction element and exists only in a lipid membrane-based biosensor) [Reprinted from: Nikoleli, G.-P., D.P. Nikolelis, C.G. Siontorou, S. Karapetis and M.-T. Nikolelis. 2018. Application of biosensors based on lipid membranes for the rapid detection of toxins. Biosensors, MPDI 8(3): 61].

Nanotechnology deals with the generation and alteration of materials in the nano-scale (10^{-9} m). Nanomaterials are materials with dimensions between 1–100 nm. The size constraints of these materials makes them very special as they have most of their constituent atoms located at or near their surface and have all vital physicochemical properties significantly different from materials of the bulk scale. Nanomaterial-based biosensors or "nano-biosensors" – biosensors made from nanomaterials – represent the integration of material science, molecular engineering, chemistry and biotechnology, and have improved the sensitivity and specificity for biomolecule detection; they have a great potential for applications in fields such as molecular recognition, food analysis, environment monitoring, biomedical and clinical analysis, and pathogen diagnosis [9]. Hence, nanomaterials can play a very important role in the future of biosensor technology.

Several nanomaterials have been discovered and many of those have been researched for their electronic and mechanical properties that can be improved in biological signaling and transduction mechanisms. Notable examples include nanotubes, nanowires, nanorods, nanoparticles, and thin films made of nanocrystalline matters [10].

Cytosensors are biosensors that use living cells as the measuring element. They are typically used to detect substances according to the effect of the substance on the functional activities of living cells. A primary impetus to cytosensors research was provided due to their prospects in building instruments that are triggered by broad ranges of biological or chemical substances. Such broadly specific cytosensors can find many applications in areas where the question is not as to what is in a sample, but rather whether anything dangerous is present in the sample. The cytosensors in this study identify bio-active agents by detecting changes in the morphology and/or physiology of a living cell.

This work reviews the status of various nanostructure devices that are used as biosensors in animal biotechnology. The chapter provides the state of the art of design and microfabrication of prototype nanosensing devices for the rapid detection of viruses, pathogens, antibiotics, infectious agents, toxins, etc., in animals, so as to prevent diseases and deaths. These devices also are integrated inside an animal body to provide useful information on the animal's medical condition and the challenges that lie ahead.

2. Animal biosensors

Sensors and wearable technologies can be implanted on animals to detect their sweat constituents [11–13] and pH [14], prevent diseases, and detect analytes such as viruses and pathogens [15–18], antibiotics, infectious diseases and toxins. Wearable sensors help farmers detect the disease early, and thereby prevent the death of animals. Farmers can also cull diseased animals in time, hence preventing the spread of diseases in cattle herds, through prediction.

Apart from collecting useful data regarding animal health, general farm monitoring can also be made easier and more reliable through the use of biosensors integrated with cellphones and hand-held devices, as opposed to conventional methods like writing notes, keeping a farm diary or using a simple equipment without data-sharing functions. A number of systems have been developed on cellphones and hand-held devices to reduce the effort of recording data manually [19]. Solar-powered receivers mounted on livestock can collect data that is transmitted to a central server. The final data can easily be viewed on a custom dashboard or an office computer, making this technology very convenient for farmers.

Commercially available biosensor collars are now being used on cows for the detection of estrus period [20, 21]. Herein, an innovative robotic grazing system uses electronic leg bands that interact with sensors mounted on the animal to record data on its feeding and milking behavior and pattern. It is a big challenge to provide good quality safe meat to meet the increasing global demand for meat and poultry products. With rising demand comes growing concerns of animal health [22]. Devices that can be integrated inside the body of an animal, patched under its skin, or that remain in its stomach, provide the animal owner with useful information regarding their behavior and medical conditions.

The abovementioned electronic devices have applications in the medical treatment of animals, detection on their heating and cooling needs, iontophoretic drug delivery, and even conservation of wild species. The ability to quickly, accurately and reliably detect the presence or absence of biomarkers or specific chemicals can be a matter of life or death for farmed animals. Monitoring glucose or proteins or enzymes in the bloodstream, testing for harmful compounds such as metals or antibiotic residues in animals and early warning of biological and chemical agents in the livestock requires sensitive and reliable sensing devices. While the demand for real-time detection of diseases using sensors and devices is ever more urgent, the capability of several relevant enabling technologies to build such sensing devices is also unprecedented. Bionanotechnology and microelectronics have made the fabrication of transistors smaller than 100 nm and integration of several hundreds of them into a functional circuit on a small chip possible. Rapid progress in nanofabrication has also offered novel enabling technologies.

A variety of methods are used to detect the level of antibiotics in a living body to avoid health hazards. Biosensors are the most prevalent method herein. The mechanism and configuration of biosensors is very simple and easy to understand, and provides fast and accurate detection of antibiotics. Biosensors work with the help of a recognition element and a transducing device. The recognition element works on the mechanism of affinity-pairing, such as enzyme substrate and antibody-antigen receptors. The transducer detects any contact between such pairs by producing a detectable electrical signal in response to biological activity, which is later analyzed.

Biosensors, for their applications in animal health management, are an emerging market that is quickly gaining recognition globally. Worldwide, numerous sensors being produced for animal health management are at various stages of commercialization. Some technologies that are used for reporting the health status accurately and for disease diagnosis are applicable only for humans, with a few modifications (or in the testing phase) for animals. Now, these innovative technologies are being considered for their future use in livestock development and welfare. However, there is a need to integrate all the available sensors and create an efficient online monitoring system so that animal health status can be monitored in real time, without delay.

The commercialization of biosensors for applications in animal biotechnology is currently limited mainly because the biological sensing element is affected by different factors, including environmental factors and type of molecules. Moreover, the size of the transducer can also affect the efficiency and functioning of a biosensor. The biological element can be affected by the temperature, pH, ionic strength, etc., of the electrolyte solution. These factors affect the sensitivity of response. The size of the transducer also affects the response time; it is well known that thinner transducers have a faster response. Also, the smaller the surface, the more pronounced is its effect on the response time.

The main methods used in the detection of antibiotics in animals are Surface Plasmon Resonance technology (SPR), Surface-enhanced Raman Spectroscopy or Surface-enhanced Raman Scattering (SERS), Enzyme-Linked Immunosorbent Assay (ELISA), Liquid Chromatography–Mass Spectrometry (LC-MS), High-Performance Liquid Chromatography (HPLC) and other types of chromatography and spectroscopy.

Each of the aforementioned methods has its advantages and limitations in the analysis of antibiotics. The main advantage of chromatography techniques is that they are straightforward and yield reliable results; however, the main disadvantage is that they are time consuming and have an extensive preparation procedure. The main advantages of biosensor technology in comparison with traditional analytical methods are fast detection (in the order of minutes) and response (in the order of seconds), high sensitivity (typically nM; even higher sensitivity with nanoparticles – in the order of pM and better), high selectivity, and easy preparation and operation of the assay. Furthermore, most of these devices are reusable and have low-cost assays (typically less than 10 EUR/sensor).

Methods like SPR make the future of biosensors very promising. About 20 commercial standard SPR platforms are currently present in the market. The application of gold nanoparticles for detecting antibiotics in food residues has caught the attention of the food industry. Recently, nanoparticles have been shown to amplify SPR signals. Gold nanoparticles, upon combination with screen-printed electrodes can exhibit versatile and suitable electrochemical properties [23–26]. The physical and chemical properties of these particles, such as efficient mass transport and enhanced surface area, are desirable due to their high surface to volume ratio.

Owing to a fast binding rate with the biomolecules and low toxicity, gold nanoparticles have found their application in antibiotic detection. Using Solid Phase Extraction (SPE) in combination with Self Assembled Monolayer (SAM) cysteine (Cys) on gold nanoparticles also raises the selectivity of this technique and boosts sensitivity by increasing the efficiency [27]. Impedimetric immunosensors constitute another highly sensitive and fast way to detect biomarkers and analytes of interest. Integrating a soluble gel-derived silica-based material with these immunosensors ensures the encapsulation of biorecognition elements with adequate mechanical stability [28, 29].

The most recent development in biosensing technology is the introduction of an SPR device for portable detection of antibiotics. This approach is significantly practical and yields highly sensitive results towards antibiotic detection in chicken muscle/blood serum in slaughterhouses. However, there is still a huge potential of growth for the technology to detect larger families of antibiotics more easily and quickly, and apply SPR to the detection of catalase in milk samples [30, 31].

3. Transduction Systems

Biosensors can be classified based on the transduction methods they employ. Transduction can be accomplished through a large variety of methods that can be categorized into three main classes: (1) optical detection methods, (2) electrochemical detection methods, and (3) mass-based detection methods. These methods have been discussed in subsequent sections.

3.1 Spectroscopic detection methods

This method offers the largest number of possible sub-categories as optical biosensors can be used with various types of spectroscopies like absorption, fluorescence, phosphorescence, Raman, surface-enhanced Raman scattering (SERS), refraction, and dispersion spectrometry. The signal intensity associated with optical detection is usually directly proportional to the number of a specific chromophore within a certain path length and can be directly dependent on the concentration of the chromophore. The linear relationship between signal intensity and concentration of a species is known as Beer–Lambert's law. The cases in which discrepancies from this linear relationship occur are usually a result of secondary effects such as self-absorbance, and equilibrium conditions. Every spectroscopy has different properties in terms of amplitude, energy, polarization, decay time and phase. Amplitude is the most commonly measured parameter of the electromagnetic spectrum, for it can be correlated with the concentration of the analyte of interest.

The energy of the electromagnetic radiation measured usually provides information about changes in the environment of the analyte. Generally, the information comes from molecular vibration – Raman or infrared absorption spectroscopy (IR) – or from the formation of new energy levels. When the interaction of a free molecule with a fixed surface is investigated in polarization measurements is usually differ from emitted light; but, when a molecule gets bound to a fixed surface, the emitted light often remains polarized.

The decay time of a specific emission signal (fluorescence or phosphorescence), being highly dependent on the excited state of the molecules and their environment, can also be used to derive information about molecular interactions. Another property that can be measured is the phase of the emitted radiation. When an electromagnetic radiation interacts with a surface, the speed or phase of that radiation is altered. This difference is based on the refractive index of the analyte.

When the analyte changes, the refractive index may change and the phase of the impinging radiation may be altered. This property of electromagnetic radiation has been successfully exploited in commercial applications using surface plasmon resonance sensors.

3.2 Fluorescence

Fluorescence is one of the most sensitive spectroscopic techniques that makes biosensors suited for the detection of very low concentrations of analytes. Particularly when a biosensor is coupled with a high-power light source (such as a laser beam), it can yield very high signal-to-noise ratios (S/N). Single-molecule detection using laser-induced fluorescence has been reported in many studies. Its inherent high sensitivity has traditionally made this technique usable for the optical detection of trace-level analytes. For high-quantum-yield fluorophores, the effective fluorescence cross sections can be as high as 10^{-16} cm^2/molecule.

A typical optical setup for a fluorescence biosensor using laser as the light source is a one-instrument system consisting of an optical fiber with antibodies immobilized at the sensor tip. Excitation light from a laser is sent through a beam splitter onto the incidence end of the optical fiber. The laser beam is transmitted inside the fiber onto the sensor tip, where it excites the analyte molecules bound to the antibodies. The excited antigen, fluorescence and retransmitted to the incidence end of the fiber, directed by the beam splitter onto the entrance slit of a monochromator, and recorded by a photomultiplier. This fluoroimmunosensor (FIS) has been used to detect the carcinogen benzo[a]pyrene (BaP) [32, 33].

Fluorescence detection is also suitable for time- or phase-resolved measurements. For this purpose, the excitation laser beam is modulated with an acousto-optic modulation system. The function of the generator provides the waveforms to drive the modulator. Laser light is delivered to the sample through an optical fiber and fluorescence is collected by the same fiber. The fluorescence from the sensing probe is collimated by appropriate optics and focused onto the entrance slit of a monochromator equipped with a photomultiplier. A lock-in amplifier synchronized with the function generator is used to measure phase-resolved signals.

3.3 *Surface plasmon resonance*

Since the first application of the surface plasmon resonance (SPR) phenomenon for sensing almost two decades ago, this method has made great strides in terms of instrumentation development and applications [34]. SPR biosensor technology has been commercially available and has now become a useful tool for characterizing and quantifying biomolecular interactions.

SPR makes it possible to monitor the binding process as a function of time by following the increase in refractive index that occurs when interacting partners bind to their ligand immobilized on the surface of an SPR sensor substrate [35]. The most important advantage of this technique is that it does not require the labeling of reactants. This simplifies the data collection process. Biosensor binding data is also useful in the selection of peptides used in diagnostic solid-phase immunoassays. Consequently, very small changes in binding affinity can be measured with good precision. This is a prerequisite for analyzing the functional effect and thermodynamic implications of limited structural changes in interacting molecules. For example, the on-rate (ka) and off-rate (kd) kinetic constants of the interaction between a protein and an antibody can be readily measured and the equilibrium affinity constant, K, can be calculated from the ratio ka/kd = K [35].

The transduction principle involved in surface plasmon resonance sensors is based on the arrangement of a dielectric-metal-dielectric sandwich; hence, when light impinges a metal surface, a wave is excited within the plasma, formed by the conduction of electrons in the metal [36, 37]. A surface plasmon is a surface charge density wave occurring at a metal surface. When plasmon resonance is induced on the surface of a metal conductor due to the impact of light and the critical wavelength and angle, the effect, observed as a minimum in the intensity of light, is reflected off the metal surface. The critical angle is naturally very sensitive to the dielectric constant of the medium immediately adjacent to the metal and therefore lends itself to exploitation for bioassays. For example, the metal can be deposited upon illumination with a wide band of frequencies and the absence of reflected light at the frequencies at which the resonance matching conditions are met can be observed.

Because of the intrinsic dependence on the refractive index of the surface, surface plasmon resonance can be used as a sensor transducer for indicating when alterations at the surface happened. The binding event involves antibody-antigen recognition or DNA hybridization at the SPR sensor surface. These are the most common SPR applications. SPR can detect small variations in the refractive index at the metal-coated interface caused by changes in a few monolayers above the surface.

In biosensor devices, surface plasmon resonance is detected as a very sharp decrease in light reflectance when the angle of incidence is varied. The resonance angle is very sensitive to variations in the refractive index of the analyte. The electric field probes the analyte within only a few hundreds of nanometers from the metal surface. The conditions of the resonance are very sensitive to variations in thin films on the surface. Changes in the refractive index are about 10^{-5}, and can be thus easily detected.

The surface plasmon wave penetrates in both directions of normal to the interface of the incident angle or frequency resonance that is observed due to the refractive index of the dielectric of the interface. Liedberg and co-workers have shown that surface plasmon resonance can be used as the basis of a genuine reagentless immunosensor if large analytes can be monitored in the antibody that is immobilized on the metal [36]. When a large antibody binds, it displaces the solution with protein or other bioreceptors. The effective refractive index of the dielectric adjacent to the metal is thus changed in proportion to the amount of analyte bound, and the surface plasmon resonance is shifted accordingly. Flanagan and Pantell have shown that the amount of analyte bound can be directly related to the resonance shift even when the resonance curve is distorted by scattering caused by surface roughness, thus relieving one of the constraints of precise control of metalization, which would have been unattractive in the mass production of inexpensive sensors [37].

SPR biosensors can give qualitative information on macromolecular assembly processes under a variety of conditions. Such information can be obtained in a manner similar to conventional solid phase assays. The main advantage of SPR biosensors here is that the formation and breakdown of complexes can be monitored in real time. This offers a possibility of determining the mechanism and kinetic rate. These constants are associated with a binding event. Hence, this information is essential for understanding how biological systems work at the molecular level.

However, accurate interpretation of biosensor data is not always straightforward. A few software programs can interpolate SPR data and provide an estimated binding constant [38]. For example, the interactions between adenylate kinase (AK) and a monoclonal antibody against AK have been examined with an optical biosensor, and the sensograms were fitted to four models using numerical integration algorithms [39].

3.4 Electrochemical detection

Electrochemical detection is another possible means of transduction that has been extensively used in biosensors [40–44]. This technique is complementary to optical detection methods such as fluorescence. Since many analytes of interest are not strongly fluorescent, they can be tagged by a molecule with a fluorescent label that is often labor-intensive. Electrochemical transduction can be very useful in such cases as it can combine the sensitivity of electrochemical measurements with the selectivity provided by bioreception. Detection limits comparable to those of fluorescence biosensors are often achievable by electrochemical sensors as well.

The associated electrochemical techniques are based on the principle that the potential, current or charge in an electrochemical cell serves as the analytical signal. These analytical techniques measure potential, charge or current to determine an analyte's concentration or to characterize an analyte's chemical reactivity. Collectively, we call this area of analytical chemistry electrochemistry as it originated from the study of movement of electrons in an oxidation-reduction reaction. Despite the difference in instrumentation, all electrochemical techniques share several common features. Electrochemical methods of analysis include three main categories: (1) coulometry, (2) potentiometry, and (3) voltammetry. Coulometry is based on exhaustive electrolysis of the analyte. In other words, the analyte is completely oxidized or reduced at the working electrode, or it reacts completely with a reagent generated at the working electrode. There are two forms of coulometry: controlled-potential coulometry (in which a constant potential is applied to the electrochemical cell) and controlled-current coulometry (in which a constant current is passed through the electrochemical cell). In potentiometry, the potential of an electrochemical cell under static conditions is measured. Because no or negligible level of current flows through the electrochemical cell, its composition remains unchanged. For this reason, potentiometry is a useful quantitative method. The first quantitative potentiometric applications appeared soon after the formulation of the Nernst equation, which relates an electrochemical cell's potential to the concentration of electroactive species in the cell. In the case of voltammetry, a time-dependent potential is applied to an electrochemical cell and the resulting current is measured as a function of that potential. The resulting plot of current versus applied potential is called a voltammogram; it is the electrochemical equivalent of a spectrum in spectroscopy, providing quantitative and qualitative information about the species involved in the oxidation or reduction reaction. Voltammetry includes polarography and amperometry.

Polarography is a type of voltammetry where the working electrode is a dropping mercury electrode (DME) or a static mercury drop electrode (SMDE); these electrodes are useful for their wide cathodic ranges and renewable surfaces. Polarography was invented in 1922 by the Czech chemist, Jaroslav Heyrovský, for which he won the Nobel prize in 1959. Amperometry is the term indicating the entirety of electrochemical techniques in which current is measured as a function of an independent variable which is, typically, time or electrode potential. Chronoamperometry is the technique in which the current is measured at a fixed potential, at different times since the start of polarisation. It is typically carried out in an unstirred solution and at a fixed electrode, that is, under experimental conditions avoiding convective mass transfer to the electrode. On the other hand, voltammetry is a subclass of amperometry, in which the current is measured by varying the potential applied to the electrode. According to the waveform that describes the way how the potential is varied as a function of time, different voltammetric techniques are defined. Recently, confusion arose regarding the correct use of many terms appropriately in electrochemistry/ electroanalysis, often owing to the diffusion of electroanalytical techniques in fields where they constitute an instrument for use, not being the 'core business' of the study. Though electrochemists are pleased about this, they welcome the usage the terms properly in order to avoid grave misunderstandings. Electrochemical detection is usually based on the chemical potential of a particular species in the analyte and can be measured by comparison with a reference electrode.

Electrochemical response is dependent on the activity of the analyte species and not on their concentration. However, for dilute solutions of low ionic strength, the thermodynamic parameter activity approaches the physical parameter concentration (in molar terms).

4. Detection of toxins

Biosensor devices allow effective toxin and antigen detection, which has many beneficial applications in health care of animals. Microporous calcium alginate gels, low-energy food products, foams and emulsions can be produced through this technology as well. Nanoparticle synthesis and formation of plastic microfluidic chips for applications in food safety have also increased recently. Microfluidic devices are being used for the culture and manipulation of embryos in assisted reproduction of cattle [45].

The interactions of aflatoxin MI (AFM1) with bilayer lipid membranes (BLMs) composed of egg phosphatidylcholine (PC) have been investigated by researchers [46]. These interactions were found to be electrochemically transduced by BLMs in the form of a transient current signal of duration in the order of seconds and appeared within 7 seconds of exposure of the membranes to this toxin. BLMs composed of PC demonstrated the maximum sensitivity. The mechanism of signal generation was also explored. This mechanism takes place due to alterations in the membrane surface electrostatics (that is, reorganization of the electrochemical double layer of membranes), which results in a charging current signal due to the adsorption of AFM1. The magnitude of the transient current signal is related to the concentration of AFM1 in the bulk solution in the concentration range of 2–15 nM. The application of the electrochemical transduction system for the determination of aflatoxin MI in skimmed milk was also studied.

The rapid, sensitive and selective electrochemical flow injection monitoring of AFM1 using stabilized filter-supported BLMs has been reported in the literature [47]. In the referred study, injections of AFM1 were made into the flowing streams of a carrier electrolyte solution (KCl) and a transient current signal with duration in the order of seconds reproducibly appeared less than 10 seconds after the exposure of the lipid membranes to the toxin. The magnitude of this signal was related linearly to the concentration of AFM1, with detection limits being in subnanomolar levels. The mechanism of signal generation was explored by physicochemical (differential scanning calorimetric) experiments and the results showed phase structure alterations in the lipid membranes.

The development of a one-shot electrochemical sensor for AFM1 has been also reported [48]. The interactions of AFM1 with these Langmuir-Blodgett films were found to be electrochemically transduced as a transient current signal with a duration of seconds and appeared within 7 seconds after the exposure of the membranes to AFM1. The magnitude of the transient current signal was related to the concentration of AFM1 in a range of 2–15 nM. The present preliminary work included novel applications of BLM-based films in real samples and explored the matrix effects in electrochemical transduction for the determination of AFM1 in skimmed milk.

The interactions of AFM1 with self-assembled metal-supported bilayer lipid membranes and its effects on DNA hybridization have been investigated previously [49]. Therein, alterations of electrochemical signals due to DNA hybridization were used for the rapid detection of this toxin. The interactions of AFM1 with these membranes (composed of egg PC) provided ion current increases which reproducibly appeared within 8–10 seconds after the lipid membranes were exposed to the toxin. The magnitudes of the current signals were related to the toxin concentration, which could be determined within the range of 1.9–20.9 nM. Furthermore, AFM1 was found to affect the kinetics and time of signal generation due to DNA hybridization, which was electrochemically monitored using s-BLMs. The "receptor" oligomer was single-stranded deoxyribonucleic acid (ssDNA) thymidylic acid icosanucleotide that terminated with a C16 alkyl chain to assist incorporation into s-BLMs (dT20-C16). The target oligomer was deoxyadenylic acid icosanucleotide (dA20). The "receptor" was incorporated into s-BLMs and complementary dA 20 (cDNA) was injected into the stirred bulk electrolyte solution. The electrochemical ion current across s-BLMs was found to increase due to the presence of ssDNA and decrease due to the formation of double-stranded DNA (dsDNA). The toxin reduced the initial rate of signal alterations and increased the time of equilibrium. This provided a means for rapid (less than 1 minute) and sensitive (detection limit of 0.5 nM) detection of AFM1, based on measurements of the initial rate of hybridization.

A multi-assay flow through sensor has been proposed in literature for the detection of AFM1 in cheese [50]. In the referred case, stabilized filter-supported BLMs were used as detectors. Single-stranded dT20-C16 was incorporated into the membranes. The incorporation of dT20-C16 in BLMs lowered the detection limit for this toxin by one to four orders of magnitude, compared to the detection limit obtained in the absence of DNA. This biosensor provided the possibility to continuously monitor this toxin at concentrations that approached those that could be of interest as set by the US Food and Drug Administration (FDA) and most European countries. Injections of AFM1 were made into flowing streams of a 0.1 M KC1 electrolyte solution, and a transient current signal with duration in the order of seconds appeared 12 seconds after the exposure of the BLMs to the toxin. The magnitude of this signal was related linearly to the concentration of the toxin, with detection limits being at sub-nM concentration levels. The effect of interferents like proteins and lipids was also explored. The interferences from proteins could be eliminated by adjustment of the flow rate of the carrier electrolyte solution. The technique was applied for the rapid flow determination of aflatoxin MI in cheese samples with a rate of at least 4 samples per minute. Repetitive cycles of injection of AFM1 showed no signal degradation during each cycle for experiments that attempted over 30 cycles of detection.

In one study, lipid films were deposited on the sensing surface in an optical biosensor instrument [51]. The membranes were mixtures of biologically occurring lipids. Eight surfaces were prepared, some of which contained various glycolipids as minor components. One of the surfaces was supplemented with membrane proteins. The binding of six protein toxins (cholera toxin, cholera toxin B sub-unit, diphtheria toxin, ricin, ricin B sub-unit and staphylococcal enterotoxin B) and of bovine serum albumin at pH 7.4 and pH 5.2 respectively to each of the sensor surfaces was investigated. Each of the seven proteins gave a distinct binding pattern. The assay was rapid and simple, with no need of reagents. The lipid sensor surface was readily regenerated after binding and was very stable. The concept with mixed lipid layers and assays at different pHs gives numerous combinations and can be applied for developing a sensor for protein toxins.

The rabbit pyrogen test and limulus amoebocyte lysate (LAL) [52, 53] have been used to detect endotoxins, and are considered conventional methods. The United States Food and Drug Administration has approved these methods as the standard methods. The rabbit pyrogen test was first developed in 1920. For the test, a test solution is injected into a rabbit's body and one waits for some time to witness any change in the body temperature to detect an endotoxin. However, an animal rights group has opposed the use of this method for preventing the killing of rabbits. Moreover, the method is time-consuming and expensive. Now this method is losing favor, and only the LAL test is used, which was first discovered 67 years ago in 1950, by Dr. Frederik Bang [54], after he first observed that horseshoe-crab blood forms clots when exposed to endotoxins.

For the LAL test, endotoxin-contaminated food or water is mixed with horseshoe-crab blood to derive the amoebocyte extraction; then, one waits for a response from the endotoxins. The FDA has approved four tests: chromogenic assay, colorimetric [55] (lower protein), gel-clot and turbidity metric (spectrophotometric) tests. The particular reaction of amoebocyte/endotoxin characterizes these methods. The method of gel clots is based on the occurrence or non-occurrence of gel formation in the sample; when endotoxins are present in the sample, gelation occurs due to the coagulation of proteins. Turbidity occurs due to the sharp division of an endotoxin-sensitive substrate, and the turbidimetric methods use this turbidity to detect endotoxins. Another technique, known as the chromogenic technique, relies on the change in color during the division of a complex into a peptide and a chromogen [56].

The concentration of Lipopolysaccharide (LPS) is expressed as EU/mL or EU/mg, where EU stands for endotoxin unit for biological activity in LPS. Suppose that in one EU, 10^{-15} g of LPS is contributed by gram-negative bacteria. This implies that at most 10^5 bacteria can be generated thereafter. The response of an LAL test is quick and it takes approximately 30 minutes to get the result. The detection limit is considerably low and the technique is highly sensitive compared to other detection methods. However, a major disadvantage is that it requires expert personnel to complete all the complex steps to avoid any external interference. Another disadvantage of the LAL test is that the testing kits are expensive for some sampling tests.

5. Detecting influenza virus using FRET

Influenza virus is highly contagious among birds. It can spread through saliva, nasal secretions and other excretions from infected birds [57].

A homogenized and uniform fluorescence-quenching-based assay has been recently developed for specifically detecting the influenza virus surface antigen hemagglutinins (HAs) [58]. This assay is expected to revolutionize the process of detecting viruses and can be a sensitive diagnostic tool for influenza virus detection. The assay consists of two nanoprobes: glycan-conjugated highly luminescent quantum dots (Gly-QDs), and HA-specific antibody-modified gold nanoparticle (Ab-Au NPs). These nanoprobes are brought together when exposed to strain-specific HA. This happens due to a very specific binding event between the HA and the two nanoprobes, which forms a sandwich complex. QDs fluorescence intensity is hence diminished as a non-radiative energy transfer occurs from QDs to Au NPs [44]. The fluorescence changes, and a resulting correlation between the target's HA concentrations is easily observed. Moreover, the origin of the HAs can be easily detected and, due to the specific interaction between HA and glycan with sialic acid residues, we can distinguish between human (H1) and avian (H5) viral sub-types.

Normally, the influenza virus is detected using the so-called "gold standard" viral culture and RT-PCR, which is both time-intensive and labor-intensive. Immunoassays like ELISA are less specific and less sensitive, which makes them inconvenient to be used for this purpose. However, fluorescence resonance energy transfer (FRET) is sensitive enough to detect small changes due to interactions between the biomolecules. It is a very powerful method in which non-radiative energy is transferred from an excited donor to an acceptor at ground state (quencher) placed almost 1 to 10 nm away [59].

A homogenous fluorescence-quenching-based assay for specific and sensitive detection of influenza virus a hemagglutinin antigen [60, 61]. Illegal feed additives can also be detected using fluorescence quenching and immunochromatography; determination of ractopamine and clenbuterol in porcine muscle and swine urine, for instance [60, 61].

Direct contact with infected poultry can infect healthy birds with the avian influenza virus. Even sharing water and feed can result in contamination. The disease manifests itself in two forms, with low and high extremes of virulence. The low pathogenic form is less virulent and shows mild symptoms, like lesser egg production. The highly pathogenic form manifests itself by rapidly spreading through flocks and affecting multiple internal organs. A system called Sensor Networks assists in collecting data from a desired location and transmitting it to a computer. Sensor Networks contain different sensors which perform extraordinary signal, alarm and decision-making functions. The sensor system uses an open-source technology, Arduino – an electronics prototype that can be integrated with various sensors to monitor parameters like body temperature and movement of the animals. The alarm is sent to an observing station, which is used to monitor whether the observed value is less or greater than the normal value. The detection of avian influenza is thus assisted by monitoring the level of H5H1 virus, which causes mortality in chickens [62].

6. Detection of bacteria using SERS

SERS, a label-free biosensing method for bacterial detection, provides information about the chemical structure of analytes. It uses the intrinsic vibrational fingerprint of analytes to detect molecules. Moreover, the performance of Raman spectroscopy remains unaffected by the surrounding water, allowing easy detection of bacteria [17].

Raman spectroscopy has found another application in the label-free modes of analysis of chemical and biological components. The detection of disease-causing bacteria in drinking water through label-free near-infrared surface-enhanced Raman scattering/spectroscopy (NIR-SERS) method is a recent development, which provides a diagnostic platform. This analytical method is a rapid way for the successful label-free identification of pathogenic bacteria in health-care applications. The *in situ* synthesis of silver nanoparticles (Ag NPs) within bacterial cell suspensions provides a means for the detection of food-borne bacteria.

This method does not require a preparatory phase and is label-free. To enhance the assay's sensitivity, Triton X-100 is used to pre-treat the bacterial cells. However, probing the fingerprints of bacteria through this method becomes difficult due to poor selectivity as it is a simple mixing process. This remarkable technology not only helps detect pathogenic bacteria, but also enables one to distinguish between different types of these bacteria, such as Methicillin-resistant, *Pseudomonas aeruginosa*, *Escherichia coli*, *Staphylococcus aureus* (MRSA) and *Listeria* spp.

This method also helps differentiating between two species—*L. monocytogenes* and *L. innocua*—through the comparison of the SERS spectra and Raman frequencies. Similarly, it is also possible to differentiate between two MRSA strains from clinical isolates [63]. Even the most intricate details, like the molecular composition of a sample, can be revealed using this spectroscopic technique at a micrometer scale. Metallic nanoparticles (NPs) in the diameter range of 104 to 106 nm have been used with SERS via SPR. This method ensures that there is homogenous contact of the constituents of the bacterial cells with nanoparticles, and provides an intense spectrum with better selectivity [63]. Another study explored a novel microfluidic platform employing methodologies for chemometric data analysis, including a combination of principle component analysis and linear discriminant analysis, as well as silver nanoparticles. Distinguishing between eight key food-borne pathogens (*E. coli*, *L. monocytogenes*, *L. innocua*, *S. typhimirium*, *S. enteritis*, *Pseudomonas aeruginosa*, MRSA 35 and MRSA 86), which significantly affect the food industry, has been made possible through this method.

Surface-enhanced Raman spectroscopy has made the imaging and detection of gram-negative and gram-positive bacteria successful as well. The distinction between these two types of bacteria depends on the difference in the scattering intensity; it is higher for gram-positive bacteria, compared to that of gram-negative bacteria. Another method for the detection of bacteria suggests the synthesis of magnetic–plasmonic Fe_3O_4–Au core–shell nanoparticles to concentrate bacterial cells. This is accomplished through the application of an external point magnetic field and SERS [17]. Biosensors can also be used to detect live bacteria in drinking water using Ag nanoparticles. This novel technology has also detected anthrax spores on nanosphere substrates. Even multi-drug-resistant strains of bacteria can be detected using complex nanohybrid systems. These systems are developed by combining antibody-conjugated gold nanoparticles with single-walled carbon nanotubes.

7. Pathogen detection methods

In conventional methods, the detection of pathogens mostly depends on the identification of precise microbiological and biochemical constituents [64]. There are three types of conventional methods: (1) immunology-based method, (2) count method of culturing and colony, and (3) polymerase chain reaction method (PCR) [64–66]. The culturing technique gives accurate results due to its high selectivity and sensitivity [67, 68]. Various microorganisms grow on food samples and, depending on the growth of the microorganisms, the specific pathogens are identified. The process requires selective plating, pre-enrichment, selective enrichment, and identifications, taking a few days to yield results. This detection method is monotonous and lengthy.

The polymerase chain reaction (PCR) method is very popular for detection of pathogens [69, 70]. Specific bacteria based on their nucleic acid sequence [71–73], protozoa [74, 75], and viruses [76, 77] are targeted when PCR is used for pathogen detection. Different PCR methods are available for pathogen detection, like reverse transcript PCR (RT-PCR) [78, 79], real-time PCR [80] and multiplex PCR [81].

The immunological detection technique is also used for the detection of pathogens [82, 83]. The antigen-antibody bindings are utilized widely in immunological detection for pathogens from gram-negative bacteria. This method has been successfully used to detect *Salmonella* and *E. coli* [84, 85]. Enzyme immunoassay (EIA) [86, 87], enzyme-linked fluorescent assay (ELFA) [88, 89], enzyme-linked immunosorbent assay (ELISA) [90, 91], flow injection immunoassay [92], and other immunological methods [93, 94] are mostly used for immunological detection. They require less time for assay preparation, compared to a culturing technique. However, real-time pathogen detection is not possible with this method [95].

Strategies like elemental doping and photo-irradiation are deployed to vary the physico-chemical functionalities of GO. Biosensing applications can be improved to a great extent by incorporating a durable single hybrid nanostructure on the GO surface, with suitable optical and biocompatible capabilities. Reactivity of the oxygen functional groups present on the edges of GO govern the chemical functionalization of materials on the GO's surface [96]. The detection of pathogens from the skins, oral cavity, feces of farm animals and also in the environment of a barn or pen can be efficiently quantified using hybrid nanoparticle-based biosensors.

HNPs-GO electrodes, known for their remarkable electrochemical immune-sensing properties, are being currently employed to detect Listeria monocytogenes (Lm), a major foodborne pathogen. Lm is a gram-positive bacterium that causes listeriosis, which is widespread and has a high mortality rate. This alarming situation calls for an efficient system to detect Lm in food products. Nanosheets of graphene oxide (GO) coated with hybrid nanoparticles of a silver–ruthenium bipyridine complex (Ag@ $[Ru(bpy)_3]^{2+}$) core and chitosan shell have significant immunosensing properties.

The oxygenated groups of GO and the amine groups on the surface of hybrid nanoparticles play a significant role in unique immunosensing application. Contamination in milk and other food products can be easily detected using monoclonal antibodies and HNPs-GO immunosensors. Hence, intelligent and specific optimization of the bio-recognition elements on HNPs-GO electrodes can provide a bright future in the food processing industry. Having good fabrication, thin layering and tunable oxygen functional groups make two-dimensional Graphene Oxide (GO) and reduced GO (rGO) preferred choices. Surface treatment of the active components on GO nanosheets changes the crystallite size and the properties associated with it.

8. Detection of infectious agents

Globalization has led to the rapid and unhindered distribution of animal products all over the world, posing great threats to humans. Transboundary animal diseases (TADs), including foot-and-mouth disease and classical swine fever, can spread very quickly across borders and countries. These diseases affect animal trade and have a devastating impact on animal husbandry. Some animal diseases cross the species barrier and can affect humans, causing zoonotic infections.

Therefore, appropriate methods must be used for the diagnosis of such diseases. This would help in devising special precautionary measures like vaccinations and quarantine. In direct detection methods, infectious agents can easily be detected in samples collected from animals. Classical methods for detection of microbes include identification by culture techniques and immunofluorescence. Molecular techniques include Polymerase Chain Reaction (PCR) and loop-mediated isothermal amplification (LAMP) for the detection of infectious agents. Occurrence of infections in hosts can be diagnosed though indirect methods as well; for instance, by identifying the antibodies against various infectious agents.

The advantage of PCR is its high specificity. PCR is a highly sensitive assay that can diagnose infectious agents at the molecular level. Since every microorganism has its own unique genome, PCR enables amplification of the genetic material, including DNA and RNA. The real-time PCR technique has many variants: FRET-based assays, TaqMan assays, etc. SYBR Green is a cost-effective method avoiding the use of probes. PCR techniques are thus affordable and are put to use in light-weight portable devices for on-site detection of infection. Use of novel isothermal amplification methods further facilitates the on-site diagnosis of infections in animals. For example, loop-mediated isothermal amplification (LAMP) runs at a single temperature level and gives results readable by the naked eye. Enzymes used in PCR include polymerases, reverse transcriptases, nucleases, etc. Use of thermostable polymerases allows amplification through thermocycling. The PCR product can be visualized through agarose gel electrophoresis using fluorescent dyes. Electrophoresis allows estimation of the amplicon length and maintains the specificity. But using gel-based PCR can sometimes be laborious and does not allow the quantification of the initial viral load. Due to lack of specificity, problems like false positive detection arise. This problem has been solved with real-time PCR, which allows a closed-tube assay with minimum risk of cross contamination. In this assay, the product is monitored during the reaction of the DNA-binding moieties that bind to the amplified DNA and emit fluorescence without the requirement of the gel formation step.

The cycle number at which the fluorescence reaches its threshold level depends on the initial viral load. Communication of laboratory results with the health authorities in a rapid manner can lead to the successful eradication of infectious diseases from animals; rapid two-way communication between laboratories and practitioners would ensure the success of the control program. Three final PCR technologies aiding the detection of infectious agents include MAP, the Field Effective sensor, and Vantix.

Johne's disease (JD) is a major gastrointestinal disease of cattle; it is caused by Mycobacterium Avium subspecies Paratuberculosis (MAP). This cattle disease causes premature culling and reduced milk production. Nevertheless, it can be controlled using conductometric biosensors which combine immunomigration technology with electronic signal detection [97].

9. Sweat analyzers

Analyzing sweat can relay useful information about an individual animal's health [98–103]. Wearable sweat analyzers have not yet been made commercial, mainly because of the size constraints of the equipment. However, low-cost robust designs have been developed in laboratories [104, 105]. Recent developments made in sweat analyzers aim to restrict the size of the system so it is wearable and easy to handle. Researchers have worked on real-time sweat monitoring of sodium by disposable potentiometric strips integrated with microfluidic chips; this system is connected to a mini wireless system to detect sodium levels in sweat [106]. Monitoring a number of electrolytes simultaneously is however more useful; hence, the system developed by Gao et al. detects the levels of sodium, potassium, lactose, glucose, and skin temperature simultaneously. Therein, integrated Bluetooth technology enables sharing and monitoring of the measured data [107]. Biomonitoring of sweat in animals has great potential for animal health because of its non-invasive nature. The amount of metals can also be detected by sweat analyzers [108]. If such a technology is introduced on farms, changes in animal health can be monitored in a novel fashion to significantly prevent health and economic loss.

10. Farm monitoring

Sensor systems for measuring fat and protein content in milk are frequently used on farms nowadays. The sensor system used differs according to the milking system used on each farm. These sensors provide health and fertility data on the cattle monitored.

Reproductive performance in dairy herds can be analyzed through estrus detection. Sensor systems have been reported to detect roughly 80–85% of cows in estrus. It is not fully clear whether the use of sensor systems also benefits the health and production of milk in cows. In older studies, it was proved that higher estrus detection resulted in a shorter calving interval, consequently leading to increased milk production. Furthermore, high somatic cell count has been linked to lower milk production. Usage of automatic milking systems has also proved to increase milk production. Different statistical analyses can be used to study the role of sensors for mastitis and estrus detection in dairy cows. Apart from these, fat, protein, temperature and milk temperature sensors can also be used [109–111] to enhance the animal production systems. Studies have suggested that dogs can be trained, through positive reinforcement and an optimized training protocol, to differentiate between vaginal mucus samples from cows which are in estrus and those from cows in diestrus [112].

Better reproduction rates in cattle can be obtained if hormone levels are measured efficiently. Using Electrochemical Impedance Spectroscopy (EIS) techniques, which can detect progesterone in purified water, is one option. Planar capacitive sensors use silicon substrate thin film microelectromechanical-based semiconductor device fabrication technology. This sensor can evaluate conductivity, permeability and dielectric properties of the reproductive hormone progesterone and quantify its concentration in purified water [113]. Integrated wireless sensors for online health monitoring systems have been designed and investigated as monitoring systems [114–118]. Proportional Integral Derivative (PID) control technique has also been applied to reduce labor and yield higher profits in poultry feeding management [119].

11. Saliva analyzer

Biological fluids of living beings, like tears, sweat and saliva, can be used for monitoring health and detecting pathological conditions. Similarly, breath and interstitial fluids of the body can also be used for this purpose. Non-invasive monitoring of uric acid in saliva can be done using a mouth guard with an integrated screen-printed electrode system. The uricase enzyme is utilized in this system, which uses electronics (potentiostat, Bluetooth and microcontroller). Generally, biosensors require a lot of power. However, this platform is capable of transmitting information to laptops and smartphones, where the information can be processed and stored. This mouth guard biosensor is highly selective and stable for uric acid detection in saliva, since it covers a large range of concentrations. This real-time biosensor is a wearable monitor which can be used in different health applications [99]. Analyzing saliva can be done through non-invasive and readily available methods. It is extremely useful in analyzing the mouth conditions and Gastrophageal Reflux Diseases.

12. Conclusion

Animal health is a serious global issue that demands apt scientific techniques. For this purpose, innovative approaches like the use of biosensors for animal health management have gained recognition. These sensors are at various steps of commercialization, but are making their way into practice and application in the domain of animal health.

Some technologies for getting an accurate health status and for disease diagnosis are applicable only for humans. With modifications and testing in animal models, these innovative technologies are now being considered for their future use in livestock development and welfare as well. Precision livestock farming techniques, which include a wide span of technologies, are being applied, along with advanced technologies like microfluidics, sound analyzers, image detection techniques, sweat and saliva sensing, and serodiagnosis.

However, there is a need to integrate all the available sensors and create an efficient online monitoring system, so that animal health can be monitored in real time, without delay. Looking at an optimistic future of different wearable technologies for animals, including nano biosensors and advanced molecular biology diagnostic techniques for the detection of various infectious diseases of cattle, a large-scale adoption of the modern techniques discussed here is likely.

References

[1] Turner, Anthony, P.F. and G.S. Wilson. 1989 Biosensors Fundamentals and Applications.

[2] Bănică, F.-G. 2012. Chemical Sensors and Biosensors.

[3] Hierlemann, A., O. Brand, C. Hagleitner and H. Baltes. 2003. Microfabrication techniques for chemical/biosensors. Proc. IEEE 91: 839–863 .

[4] Hierlemann, A. and H. Baltes. 2003. CMOS-based chemical microsensors. Analyst 128: 15–28.

[5] Chandra, P. 2016. Nanobiosensors for personalized and onsite biomedical diagnosis. The Institution of Engineering and Technology. (1st ed). UK, Cryodon.

[6] Mahato, K., A. Srivastava and P. Chandra. 2017. Paper based diagnostics for personalized health care: Emerging technologies and commercial aspects. Biosensors and Bioelectronics 96: 246–259.

[7] Mahato, K., P.K. Maurya and P. Chandra. 2018. Fundamentals and commercial aspects of nanobiosnsors in point-of-care clinical diagnostics. 3 Biotech. 8: 149.

[8] Chandra, P., T.Y. Nee and S.P. Singh. 2017. Next generation point-of-care biomedical sensors technologies for cancer diagnosis. Springer Singapore.

[9] Baranwal, A., A. Srivastava, P. Kumar, V.K. Bajpai, P.K. Maurya and P. Chandra. 2018. Prospets of nanostructure materials and their composites as antimicrobial agents. Frontiers Microbiology 9: 422.

[10] Jianrong, C., M. Yuqing, H. Nongyue et al. 2004. Nanotechnology and biosensors. Biotechnol. Adv. 22: 505–518.

[11] Glennon, T., C. O'Quigley, M. McCaul et al. 2016. SWEATCH: A wearable platform for harvesting and analysing sweat sodium content. Electroanalysis 28: 1283–1289.

[12] Heikenfeld, J. 2016. Bioanalytical devices: Technological leap for sweat sensing. Nature 529: 475–476.

[13] Garcia, S.O., Y.V. Ulyanova, R. Figueroa-Teran et al. 2016. Wearable sensor system powered by a biofuel cell for detection of lactate levels in sweat. ECS J. Solid State Sci. Technol. 5: M3075–M3081.

[14] Kim, J., T.N. Cho, G. Valdés-Ramírez and J. Wang. 2016. A wearable fingernail chemical sensing platform: PH sensing at your fingertips. Talanta 150: 622–628.

[15] Mungroo, N.A. and S. Neethirajan. 2014. Biosensors for the detection of antibiotics in poultry industry—A Review. Biosensors 4: 472–493.

[16] Vidic, J., M. Manzano, C.M. Chang and N. Jaffrezic-Renault. 2017. Advanced biosensors for detection of pathogens related to livestock and poultry. Vet. Res. 48: 11.

[17] Mungroo, N.A., G. Oliveira and S. Neethirajan. 2016. SERS based point-of-care detection of food-borne pathogens. Microchim. Acta 183: 697–707.

[18] Posthuma-Trumpie, G.A., J. Korf and A. Van Amerongen. 2009. Lateral flow (immuno)assay: Its strengths, weaknesses, opportunities and threats. A literature survey. Anal. Bioanal. Chem. 393: 569–582.

[19] Li, Y., R. Vyas, A. Rida et al. 2008. Wearable RFID-enabled sensor nodes for biomedical applications. *In*: Proceedings—Electronic Components and Technology Conference, pp. 2156–2159.
[20] Andersson, L.M., H. Okada, Y. Zhang et al. 2015. Wearable wireless sensor for estrus detection in cows by conductivity and temperature measurements. *In*: 2015 IEEE SENSORS—Proceedings.
[21] Andersson, L.M., H. Okada, R. Miura et al. 2016. Wearable wireless estrus detection sensor for cows. Comput. Electron. Agric. 127: 101–108.
[22] Ivanov, S., K. Bhargava and W. Donnelly. 2015. Precision farming: Sensor analytics. IEEE Intell. Syst. 30: 76–80.
[23] http://www.dropsens.com/en/screen_printed_electrodes_pag.html.
[24] Gerd-Uwe Flechsig, G.-U. 2018. New electrode materials and devices for thermoelectrochemical studies and applications. Current Opinion in Electrochemistry 10: 54–60.
[25] Zhang, Y., G. Wang, L. Yang, F. Wang and A. Liu. 2018. Recent advances in gold nanostructures based biosensing and bioimaging. Coordination Chemistry Reviews 370: 1–21.
[26] Kim, S.H. 2018. Nanoporous gold: Preparation and applications to catalysis and sensors. Curr. Appl. Phys. 18(7): 810–818.
[27] Galal, A., N.F. Atta and E.H. El-Ads. 2012. Probing cysteine self-assembled monolayers over gold nanoparticles—towards selective electrochemical sensors. Talanta 93: 264–73.
[28] Wang, J. and P.V. Pamidi. 1997. Sol-gel-derived gold composite electrodes. Anal. Chem. 1, 69(21): 4490–4494.
[29] Wang, J., P.V.A. Pamidi and K.R. Rogers. 1998. Sol-gel-derived thick-film amperometric immunosensors. Anal. Chem. 70(6): 1171–1175.
[30] Ashley, J. and S.F.Y. Li. 2013. An aptamer based surface plasmon resonance biosensor for the detection of bovine catalase in milk. Biosens. {&} Bioelectron 48: 126–131.
[31] Meng, K., W. Sun, P. Zhao et al. 2014. Development of colloidal gold-based immunochromatographic assay for rapid detection of Mycoplasma suis in porcine plasma. Biosens. Bioelectron 55: 396–399.
[32] Khansili, N., G. Rattu and P.M. Krishna. 2018. Label-free optical biosensors for food and biological sensor applications. Sens. Actuators (Chemical) 265: 35–49.
[33] Vo-Dinh, T. 2014. Nanosensing at the single cell level, Spectrochim Acta Part B, At Spectrosc. Author manuscript; available in PMC 2014 May 15. Published in final edited form as: Spectrochim Acta Part B At Spectrosc. 2008 Feb; 63(2): 95–103.
[34] Homola, J., S.S. Yee and G. Gauglitz. 1999. Surface plasmon resonance sensors: Review. Sensors Actuators B Chem. 54: 3–15.
[35] Van Regenmortel, M.H., D. Altschuh, J. Chatellier et al. 1997. Uses of biosensors in the study of viral antigens. Immunol. Invest. 26: 67–82.
[36] Liedberg, B., C. Nylander and I. Lunström. 1983. Surface plasmon resonance for gas detection and biosensing. Sensors and Actuators 4: 299–304.
[37] Flanagan, M. and R. Pantell. 1984. Surface plasmon resonance and immunosensors. Electron Lett. 20: 968–970.
[38] Morton, T.a and D.G. Myszka. 1998. Kinetic analysis of macromolecular interactions using surface plasmon resonance biosensors. Methods Enzymol. 295: 268–94.
[39] Luo, J., J. Zhou, W. Zou and P. Shen. 2001. Antibody—Antigen interactions measured by surface plasmon resonance: Global fitting of numerical integration algorithms. J. Biochem. 130: 553–559.
[40] Gyurcsányi, R.E., Z. Vágföldi, K. Tóth and G. Nagy. 1999. Fast response potentiometric acetylcholine biosensor. Electroanalysis 11: 712–718.
[41] Dobay, R., G. Harsanyi and C. Visy. 1999. Conducting polymer based electrochemical sensors on thick film substrate. Electroanalysis 11: 804–808.
[42] Coche-Guérente, L., V. Desprez, J.-P. Diard and P. Labbé. 1999. Amplification of amperometric biosensor responses by electrochemical substrate recycling: Part I. Theoretical treatment of the catechol–polyphenol oxidase system. J. Electroanal. Chem. 470: 53–60.
[43] Dall'Orto, V.C., C. Danilowicz, I. Rezzano et al. 1999. Comparison between three amperometric sensors for phenol determination in olive oil samples. Anal. Lett. 32: 1981–1990.
[44] Karyakin, A.A., M. Vuki, L.V. Lukachova et al. 1999. Processible polyaniline as an advanced potentiometric pH transducer. Application to biosensors. Anal. Chem. 71: 2534–2540.

[45] Neethirajan, S., I. Kobayashi, M. Nakajima et al. 2011. Microfluidics for food, agriculture and biosystems industries. Lab Chip 11: 1574.

[46] Andreou, V.G., D.P. Nikolelis and B. Tarus. 1997. Electrochemical investigation of transduction of interactions of aflatoxin M1 with bilayer lipid membranes (BLMs). Anal. Chim. Acta 350(1-2): 121–127.

[47] Andreou, V.G. and D.P. Nikolelis. 1998. Flow injection monitoring of aflatoxin M1 in milk and milk preparations using filter-supported bilayer lipid membranes. Anal. Chem. 70: 2366–2371.

[48] Andreou, V.G. and D.P. Nikolelis. 1997. Electrochemical transduction of interactions of aflatoxin M_1 with bilayer lipid membranes (BLMs) for the construction of one-shot sensors. Sensors and Actuators, B 41(1-3): 213–216.

[49] Siontorou, C.G., D.P. Nikolelis, A. Miernik and U.J. Krull. 1998. Rapid methods for detection of Aflatoxin M1 based on electrochemical transduction by self-assembled metal-supported bilayer lipid membranes (s-BLMs) and on interferences with transduction of DNA hybridization. Electr. Acta 43(23): 3611–3617.

[50] Siontorou, C.G., V.G. Andreou, D.P. Nikolelis and U.J. Krull. 2000. Flow injection monitoring of aflatoxin MI in cheese using filter-supported bilayer lipid membranes with incorporated DNA. Electroanalysis 12(10): 747–751.

[51] Puu, G. 2001. An approach for analysis of protein toxins based on thin films of lipid mixtures in an optical biosensor. Anal. Chem. 73(1): 72–79.

[52] Ding, J.L. and B. Ho. 2010. Endotoxin Detection—From Limulus Amebocyte Lysate to Recombinant Factor C. Volume 53 Springer; Berlin, Germany.

[53] Thorne, P.S., S.S. Perry, R. Saito, P.T. O'Shaughnessy, J. Mehaffy, N. Metwali, T. Keefe, K.J. Donham and S.J. Reynolds. 2010. Evaluation of the limulus amebocyte lysate and recombinant factor c assays for assessment of airborne endotoxin. Appl. Environ. Microbiol. 76: 4988–4995.

[54] Bang, F.B. and J.L. Frost. 1953. The toxic effect of a marine bacterium on limulus and the formation of blood clots. Mar. Biol. Lab. 105: 361–362.

[55] Alhogail, S., G.A. Suaifan and M. Zourob. 2016. Rapid colorimetric sensing platform for the detection of listeria monocytogenes foodborne pathogen. Biosens. Bioelectron. 86: 1061–1066.

[56] Kotanen, C.N., F.G. Moussy, S. Carrara and A. Guiseppi-Elie. 2012. Implantable enzyme amperometric biosensors. Biosens. Bioelectron. 35: 14–26.

[57] Chemburu, S., E. Wilkins and I. Abdel-Hamid. 2005. Detection of pathogenic bacteria in food samples using highly-dispersed carbon particles. Biosens. Bioelectron. 21: 491–499.

[58] Chen, L. and S. Neethirajan. 2015. A homogenous fluorescence quenching based assay for specific and sensitive detection of influenza virus A. Sensors (Basel) 15(4): 8852–8865.

[59] Hung, H.C., C.L. Liu, J.T. Hsu, J.T. Horng, M.Y. Fang, S.Y. Wu, S.H. Ueng, M.Y. Wang, C.W. Yaw and M.H. Hou. 2012. Development of an anti-influenza drug screening assay targeting nucleoproteins with tryptophan fluorescence quenching. Anal. Chem. 7, 84(15): 6391–6399.

[60] Jing, X., B. Bai, C. Zhang et al. 2015. Rapid and sensitive determination of clenbuterol in porcine muscle and swine urine using a fluorescent probe. Spectrochim. Acta—Part A Mol. Biomol. Spectrosc. 136: 714–718.

[61] Zhang, W., X. He, P. Liu et al. 2016. Rapid Determination of ractopamine in porcine urine by a fluorescence immunochromatography assay. Anal. Lett. 49: 2165–2176.

[62] Dovas, C.I., M. Papanastassopoulou, M.P. Georgiadis et al. 2010. Detection and quantification of infectious avian influenza a (H5N1) virus in environmental water by using real-time reverse transcription-PCR. Appl. Environ. Microbiol. 76: 2165–2174.

[63] Zhao, H. and W. Su. 2010. Cooperative wireless multicast: Performance analysis and power/ location optimization. IEEE Trans. Wirel. Commun. 9: 2088–2100.

[64] Velusamy, V., K. Arshak, O. Korostynska, K. Oliwa and C. Adley. 2010. An overview of foodborne pathogen detection: In the perspective of biosensors. Biotechnol. Adv. 28: 232–254.

[65] Lazcka, O., F.J. Del Campo and F.X. Munoz. 2007. Pathogen detection: A perspective of traditional methods and biosensors. Biosens. Bioelectron. 22: 1205–1217.

[66] Leonard, P., S. Hearty, J. Brennan, L. Dunne, J. Quinn, T. Chakraborty and R. O'Kennedy. 2003. Advances in biosensors for detection of pathogens in food and water. Enzyme Microb. Technol. 32: 3–13.

[67] Lee, K.M., M. Runyon, T.J. Herrman et al. 2015. Review of Salmonella detection and identification methods: Aspects of rapid emergency response and food safety. Food Control 47: 264–276.
[68] Leoni, E. and P.P. Legnani. 2001. Comparison of selective procedures for isolation and enumeration of Legionella species from hot water systems. J. Appl. Microbiol. 90: 27–33.
[69] Ratnam, S., S.B. March, R. Ahmed et al. 1988. Characterization of *Escherichia coli* serotype O157:H7. J. Clin. Microbiol. 26: 2006–2012.
[70] Fratamico, P.M. 2003. Comparison of culture, polymerase chain reaction (PCR), taqman salmonella, and transia card salmonella assays for detection of *salmonella* spp. in naturally-contaminated ground chicken, ground turkey, and ground beef. Mol. Cell. Probes 17: 215–221.
[71] Oh, S.J., B.H. Park, J.H. Jung, G. Choi, D.C. Lee and T.S. Seo. 2016. Centrifugal loop-mediated isothermal amplification microdevice for rapid, multiplex and colorimetric foodborne pathogen detection. Biosens. Bioelectron. 75: 293–300.
[72] Jensen, M.A., J.A. Webster and N. Straus. 1993. Rapid identification of bacteria on the basis of polymerase chain reaction-amplified ribosomal DNA spacer polymorphisms. Appl. Environ. Microbiol. 59: 945–952.
[73] Belgrader, P., W. Benett, D. Hadley, J. Richards, P. Stratton, R. Mariella and F. Milanovich. 1999. Infectious disease—PCR detection of bacteria in seven minutes. Science 284: 449–450.
[74] Naravaneni, R. and K. Jamil. 2005. Rapid detection of food-borne pathogens by using molecular techniques. J. Med. Microbiol. 54: 51–54.
[75] Russell, S., S. Frasca, I. Sunila and R.A. French. 2004. Application of a multiplex PCR for the detection of protozoan pathogens of the eastern oyster crassostrea virginica in field samples. Dis. Aquat. Org. 59: 85–91.
[76] Lee, S.H., M. Joung, S. Yoon, K. Choi, W.Y. Park and J.R. Yu. 2010. Multiplex PCR detection of waterborne intestinal protozoa: Microsporidia, cyclospora, and cryptosporidium. Korean J. Parasitol. 48: 297–301.
[77] Traore, O., C. Arnal, B. Mignotte, A. Maul, H. Laveran, S. Billaudel and L. Schwartzbrod. 1998. Reverse transcriptase PCR detection of astrovirus, hepatitis a virus, and poliovirus in experimentally contaminated mussels: Comparison of several extraction and concentration methods. Appl. Environ. Microbiol. 64: 3118–3122.
[78] Morales-Rayas, R., P.F.G. Wolffs and M.W. Griffiths. 2010. Simultaneous separation and detection of hepatitis a virus and norovirus in produce. Int. J. Food Microbiol. 139: 48–55.
[79] Yaron, S. and K.R. Matthews. 2002. A reverse transcriptase-polymerase chain reaction assay for detection of viable *Escherichia coli* O157: H7: Investigation of specific target genes. J. Appl. Microbiol. 92: 633–640.
[80] Choi, S.H. and S.B. Lee. 2004. Development of reverse transcriptase-polymerase chain reaction of fima gene to detect viable salmonella in milk. J. Anim. Sci. Technol. 46: 841–848.
[81] Mukhopadhyay, A.K. and U. Mukhopadhyay. 2006. Novel multiplex PCR approaches for the simultaneous detection of human pathogens: *Escherichia coli* 0157:H7 and *Listeria monocytogenes*. J. Microbiol. Methods. 68(1): 193–200.
[82] Iqbal, S.S., M.W. Mayo, J.G. Bruno, B.V. Bronk, C.A. Batt and J.P. Chambers. 2000. A review of molecular recognition technologies for detection of biological threat agents. Biosens. Bioelectron. 15: 549–578.
[83] Gracias, K.S. and J.L. McKillip. 2004. A review of conventional detection and enumeration methods for pathogenic bacteria in food. Can. J. Microbiol. 50: 883–890.
[84] Chen, C.-S. and R.A. Durst. 2006. Simultaneous detection of Escherichia coli O157:H7, Salmonella spp. and Listeria monocytogenes with an array-based immunosorbent assay using universal protein G-liposomal nanovesicles. Talanta 69: 232–238.
[85] Magliulo, M., P. Simoni, M. Guardigli et al. 2007. A rapid multiplexed chemiluminescent immunoassay for the detection of Escherichia coli O157:H7, Yersinia enterocolitica, Salmonella typhimurium, and Listeria monocytogenes pathogen bacteria. J. Agric. Food Chem. 55: 4933–4939.
[86] Qadri, A., S. Ghosh, K. Prakash et al. 1990. Sandwich enzyme immunoassays for detection of Salmonella typhi. J. Immunoassay 11: 251–270.

[87] Chapman, P.A., A.T. Cerdan Malo, C.A. Siddons and M. Harkin. 1997. Use of commercial enzyme immunoassays and immunomagnetic separation systems for detecting Escherichia coli O157 in bovine fecal samples. Appl. Environ. Microbiol. 63: 2549–2553.

[88] Rozand, C. and P.C.H. Feng. 2009. Specificity analysis of a novel phage-derived ligand in an enzyme-linked fluorescent assay for the detection of Escherichia coli O157:H7. J. Food Prot. 72: 1078–81.

[89] Maria De Giusti, Daniela Tufl, Caterina Aurigemma, Angela Del Cimmuto, Federisa Trinti, Alice Mannocci et al. 2011. Italian of public detection of Escherichia coli O157 in raw and cooked meat: Comparison of conventional direct culture method and Enzyme Linked Fluorescent Assay (ELFA). Ital. J. Public Health 8: 28.

[90] Vázquez, F., E.A. González, J.I. Garabal et al. 1996. Development and evaluation of an ELISA to detect Escherichia coli K88 (F4) fimbrial antibody levels. J. Med. Microbiol. 44: 453–463.

[91] Song, C., C. Liu, S. Wu et al. 2016. Development of a lateral flow colloidal gold immunoassay strip for the simultaneous detection of Shigella boydii and Escherichia coli O157: H7 in bread, milk and jelly samples. Food Control. 59: 345–351.

[92] Abdel-Hamid, I., D. Ivnitski, P. Atanasov and E. Wilkins. 1999. Highly sensitive flow-injection immunoassay system for rapid detection of bacteria. In: Anal. Chim. Acta 399: 99–108.

[93] Valdivieso-Garcia, A., A. Desruisseau, E. Ricche et al. 2003. Evaluation of a 24-hour bioluminescent enzyme immunoassay for the rapid detection of Salmonella in chicken carcass rinses. J. Food Prot. 66: 1996–2004.

[94] Rasooly, A. and R.S. Rasooly. 1998. Detection and analysis of Staphylococcal enterotoxin A in food by Western immunoblotting. Int. J. Food Microbiol. 41: 205–212.

[95] Meng, J.H. and M.P. Doyle. 2002. Introduction. Microbiological food safety. Microbes Infect. 4: 395–397.

[96] Veerapandian, M. and S. Neethirajan. 2015. Graphene oxide chemically decorated with Ag–Ru/chitosan nanoparticles: Fabrication, electrode processing and immunosensing properties. RSC Adv. 5: 75015–75024.

[97] Okafor, C., D. Grooms, E. Alocilja and S. Bolin. 2014. Comparison between a conductometric biosensor and ELISA in the evaluation of Johne???s disease. Sensors (Switzerland) 14: 19128–19137.

[98] Bandodkar, A.J. and J. Wang. 2014. Non-invasive wearable electrochemical sensors: A review. Trends Biotechnol. 32(7): 363–371.

[99] Kim, J., S. Imani, W.R. de Araujo et al. 2015. Wearable salivary uric acid mouthguard biosensor with integrated wireless electronics. Biosens. Bioelectron. 74: 1061–1068.

[100] Gao, W., S. Emaminejad, H.Y.Y. Nyein et al. 2016. Fully integrated wearable sensor arrays for multiplexed *in situ* perspiration analysis. Nature 529(7587): 509–514.

[101] Sonner, Z., E. Wilder, J. Heikenfeld et al. 2015. The microfluidics of the eccrine sweat gland, including biomarker partitioning, transport, and biosensing implications. Biomicrofluidics 9(3): 031301.

[102] Kennedy, G.A. 2011. U. S. Patent, No. 7,964, 409, Washington, D.C.: U.S. Patent and Trademark Office.

[103] Matzeu, G., L. Florea and D. Diamond. 2015. Advances in wearable chemical sensor design for monitoring biological fluids. Sensors Actuators, B Chem. 211: 403–418.

[104] Modali, A., S.R.K. Vanjari and D. Dendukuri. 2016. Wearable woven electrochemical biosensor patch for non-invasive diagnostics. Electroanalysis 28(6): 1276–1282.

[105] Bandodkar, A.J., W. Jia and J. Wang. 2015. Tattoo-based wearable electrochemical devices: Areview. Electroanalysis 27(3): 562–572.

[106] Matzeu, G., C. O'Quigley, E. McNamara et al. 2016. An integrated sensing and wireless communications platform for sensing sodium in sweat. Anal. Methods 8(1): 64–71.

[107] Gao, W., S. Emaminejad, H.Y.Y. Nyein et al. 2016. Fully integrated wearable sensor arrays for multiplexed *in situ* perspiration analysis. Nature 529(7587): 509–514.

[108] Kim, J., W.R. De Araujo, I.A. Samek et al. 2015. Wearable temporary tattoo sensor for real-time trace metal monitoring in human sweat. Electrochem. Commun. 51: 41–45.

[109] Steeneveld, W., J.C.M. Vernooij and H. Hogeveen. 2015. Effect of sensor systems for cow management on milk production, somatic cell count, and reproduction. J. Dairy Sci. 98(6): 3896–905.

[110] Yunkwang, O.h., Youngmi Lee, J. Heath and Moonil Kim. 2015. Applications of animal biosensors: A review. IEEE Sens. J. 15: 637–645.

[111] Staric, K.D., B. Cvetkovic, A.U. Levicnik and J. Staric. 2015. One health concept of measuring and monitoring wellbeing. ICT Innovations 2015 Web Proceedings, 303–312.

[112] Johnen, D., W. Heuwieser and C. Fischer-Tenhagen. 2015. How to train a dog to detect cows in heat-Training and success. Appl. Anim. Behav. Sci. 171: 39–46.

[113] Zia, A.I., A.R. Mohd Syaifudin, S.C. Mukhopadhyay et al. 2012. Sensor and instrumentation for progesterone detection. *In*: 2012 IEEE I2MTC—International Instrumentation and Measurement Technology Conference, Proceedings(I2MTC), 2012 IEEE International, IEEE (2012, May), pp. 1220–1225.

[114] Nagl, L., R. Schmitz, S. Warren et al. 2003. Wearable sensor system for wireless state-of-health determination in cattle. *In*: Proceedings of the 25th Annual International Conference of the IEEE EMBS, Cancun, Mexico, Vol. 4(2003, September): 3012–3015.

[115] Martínez-Avilés, M., E. Fernández-Carrión, J.M. López García-Baones and J.M. Sánchez-Vizcaíno. 2015. Early detection of infection in pigs through an online monitoring system. Transbound Emerg. Dis. 10.1111/tbed. 1237.

[116] Rutten, C.J., A.G.J. Velthuis, W. Steeneveld and H. Hogeveen. 2013. Invited review: Sensors to support health management on dairy farms. J. Dairy Sci. 96(4): 1928–1952.

[117] Kashiha, M., A. Pluk, C. Bahr et al. 2013. Development of an early warning system for a broiler house using computer vision. Biosyst Eng. 116: 36–45.

[118] Busse, M., W. Schwerdtner, R. Siebert et al. 2015. Analysis of animal monitoring technologies in Germany from an innovation system perspective. Agric. Syst. 138: 55–65.

[119] Mikail, O.O., T.A. Folorunso, M.A. Akogbe and A. Adejumo. 2015. Design of a mobile poultry liquid, feed dispensing system using PID control technique. Conference: International Multidisciplinary and Interdisciplinary Conference on Science.

10

Electrochemical Nano-aptamer-based Assays for the Detection of Mycotoxins

Sondes Ben Aissa,[1,2] *Rupesh K. Mishra*,[3,4] *Noureddine Raouafi*[2] and *Jean Louis Marty*[1,]*

1. Introduction

The advent of nanoscience and recent strides in nanotechnology have enabled the extension of its application to biochemistry, aimed at developing powerful devices for environmental, industrial, and healthcare fields. Biosensors are an excellent example of such a combination that integrates a large part of fundamental and applied research. On the other hand, the mass industrial production of various eatables has rapidly become a part of everyday life. This was accompanied by various global health risks, including the increase of pollutants and toxins in the food industry. Consequently, food contamination has been getting more attention and, nowadays, it represents one of the most important worldwide issues. Food commodities are often exposed to a variety of mycotoxins which are generally produced in inconvenient storage conditions. Over 400 mycotoxins have been identified so far, displaying a wide range of toxic effects for animals as well as humans. According to the Global Health Program launched by non-governmental organizations (IARC, BMGF,

[1] Laboratoire BAE-LBBM, Université de Perpignan Via Domitia, 52 Avenue Paul Alduy, CEDEX 9,66860 Perpignan, France.

[2] Sensors and Biosensors Group, Laboratoire de Chimie Analytique et Electrochimie (LR99ES15), Faculté des Sciences de Tunis, Université de Tunis El Manar, Tunis 2092, Tunisia.

[3] Bindley Bio-Science Center, Lab 222, 1203 W. State St., Purdue University, West Lafayette, IN 47907, USA.

[4] School of Materials Engineering, Purdue University, 701 West Stadium Avenue, West Lafayette, IN 47907, USA.

* Corresponding author: jlmarty@univ-perp.fr

etc.), there is an urgent need to establish an effective translation of the vast body of science through to subsistence and smallholder farmers to tackle this neglected problem. Therefore, multidisciplinary scientists are pursuing research in this domain to push technological progress to develop innovative devices for the monitoring of mycotoxins in food. Biosensing tools are in the core of these scientific investigations, with the aim of detecting mycotoxins at trace levels that meet regulatory requirements, without the need for sophisticated instrumentation. Nonetheless, biosensors are not intended to replace conventional laboratory equipment like chromatographs or mass spectrometers, which are known for their higher sensitivity and accuracy for a broader list of food-borne analytes. The main merits of such tools rely on their compactness, amenability, rapid response, and easy transfer for on-site analysis. To some extent, the use of biosensors outside chemical laboratories favors the electrochemical instrumentation for an affordable and sensitive signal transduction.

In the last few decades, the increased interest in electrochemical biosensors has been particularly assisted by both the invention of new bio-recognition molecules (antibodies, peptides, aptamers, etc.) as well as the recent boom in nanotechnology. The use of nanomaterials has made it possible to synthesize rather complex layered structures with an enhanced surface-to-volume ratio and sufficient permeability of low-molecular charge carriers required for biosensor operation. Additionally, the discovery of aptamers as biomimetic ligands has revolutionized the fabrication of biosensors, including those specific to mycotoxins. The significance of novel materials in the improvement of the state-of-the-art aptasensors for diverse range of applications, industrial or domestic, is well acknowledged by the field of sensing studies.

Figure 1 shows a stark increase in the number of cited scientific research papers dealing with the design of electrochemical aptasensing assays specific to mycotoxins.

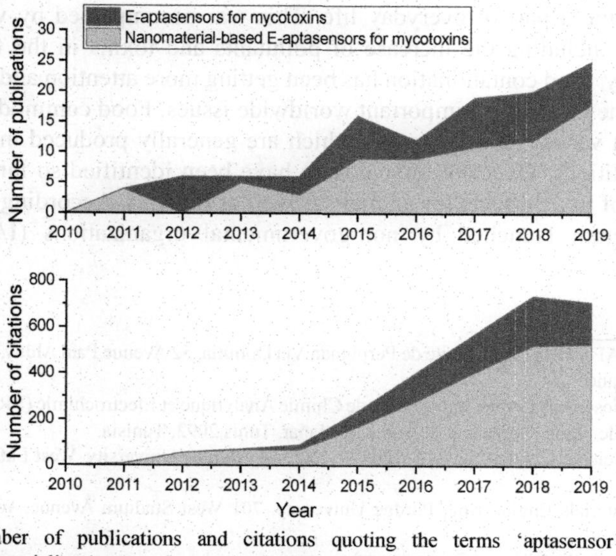

Figure 1. Number of publications and citations quoting the terms 'aptasensor', 'mycotoxin', 'nanomaterial' or 'aptamer', per year, in Web of Science (Accessed in February 2020).

This research interest is supported by the recent boom in nanotechnology, which paved the way for countless nanomaterial/aptamer combinations aimed at enhancing the performance of the detection methods for mycotoxins. The reproducible and consistent sensing uniqueness are regulated by the sensing surface as well as the material. In this chapter, current trends, and figures of merit of electrochemical nano-aptasensors, specific to the analysis of mycotoxins have been perused over.

2. Mycotoxins

Mycotoxins are a chemically diverse group of fungal metabolites that have a wide variety of toxic effects. This section provides an overview of food contamination with mycotoxins and the corresponding recent regulations for food safety control. A brief assessment of the most prominent mycotoxins studied to date, their harmful impacts and their conventional analysis methods have also been discussed alongside. This attempt is not at a complete review; however, at appropriate points, we refer to more extensive accounts.

2.1 General description of mycotoxins

The term *mycotoxin* originated from the Greek word *mukos* (meaning *fungus*), combined with the Latin word *toxicum* (referring to *poison*). It was coined for toxic secondary metabolites produced by certain fungi that can thrive on crops under favorable environmental conditions [1]. Contamination by mycotoxins has proven to be unavoidable and unpredictable. It depends mainly on a variety of weather factors (temperature, precipitation, humidity, and accompanying changes in insect damage and crop stress, etc.), either in the field over the harvesting period, or during shipping, due to mishandling and improper storage [2]. Occurrence of mycotoxins is widespread in agricultural products throughout the world, albeit with varying severity. Unfortunately, it is known that almost no treatment can be efficient to remove mycotoxins from food and feed once contaminated, since most mycotoxins are resistant to heat within the standard range of cooking temperatures [3]. The estimation usually given by the Food and Agriculture Organization (FAO) is that at least one quarter of the world's crops are contaminated to some extent with mycotoxins [4]. However, according to relevant data, Eskola et al. have recently criticized this widely cited 25% FAO estimate, showing that it underestimates the real occurrence of contamination of grains with mycotoxins [5].

Till date, over 300 mycotoxins have been identified and reported. However, only a few regularly contaminate food and animal feed [6]. These consist of aflatoxins, ochratoxins, fumonisins, patulin, zearalenone, and trichothecenes (including nivalenol, deoxynivalenol and T-2 toxin). According to the International Agency for Research on Cancer (IARC), some mycotoxins have proved to be strong carcinogenic agents, such as aflatoxin B1 (AFB1), while others are suspected to have carcinogenic effects. For that purpose, IARC has adopted the following classification of mycotoxins, based on their effects on humans [7]:

- Group 1: Carcinogenic to humans;
- Group 2A: Probably carcinogenic to humans (sufficient evidence for animals);

– Group 2B: Possibly carcinogenic to humans (insufficient evidence for animals);
– Group 3: Not classifiable with respect to its carcinogenicity to humans; and
– Group 4: Probably not carcinogenic to humans.

Of the several hundred mycotoxins identified so far, about a dozen have gained the most attention due to their presence in food and their severe effects on human health. Many national and international public health and governmental authorities around the world, such as the US Food and Drug Administration (FDA), World Health Organization (WHO), Food Agriculture Organization (FAO), the European Food Safety Authority (EFSA), and the Institute of Public Health in Japan (IANPHI), are paying serious attention to mycotoxin contamination in food and feed. Accordingly, they have established strict regulatory guidelines for safe doses of major mycotoxin classes in food and feed; there is still a need for worldwide harmonization of mycotoxin regulations [6].

The chemical structure of most widely occurring mycotoxins, along with their fungal origins, toxic effects and associated EU regulation(s) have been summarized in Table 1. The established maximum limits (MLs) differ depending on the mycotoxin and the targeted foodstuff. In particular, the strictest regulations have been set for aflatoxins and processed food products for infants.

2.2 Risks associated with mycotoxin contamination

Contamination by mycotoxins presents a serious challenge for food safety and poses an acute threat to human and animal health, while also contributing to huge economic losses to the agriculture industry and small-scale farmers.

2.2.1 Health risks

Ingestion of some feed and food-borne mycotoxins leads to a serious health threat called *mycotoxicosis*, which may eventually result in death. The adverse effects of mycotoxins on human health include acute poisoning, cancer, other chronic diseases, and biological effects, including growth impairment and immune deficiency [8]. Kuiper-Goodman [10], a leading figure in the risk assessment field, ranks mycotoxins as the most important chronic dietary risk factor, higher than even synthetic contaminants, plant toxins, food additives, or pesticide residues. However, except for aflatoxins, minimal evidence exists about the direct correlation between the widespread exposure to mycotoxins and their wide range of adverse effects on human health. Thus, more investigations into their entire range of impacts on human health remain to be conducted [8].

Regarding *aflatoxicosis*, advanced studies have linked the exposure to high dietary levels of aflatoxins to human liver cancer, with a particularly elevated risk in people chronically infected with hepatitis B virus [11]. Chronic consumption of aflatoxin-contaminated foods has been reported to cause immunosuppression in humans worldwide. Moreover, aflatoxins have been reported to be responsible for the malabsorption of various nutrients, thus leading to nutritional deficiencies, malnutrition, and stunted growth in children [12]. Nevertheless, more information

Table 1. An overview of major mycotoxins: properties, toxicity and EU limits on food and animal feed levels [1, 7–9].

Mycotoxin	Chemical Structure[a]	Fungal Species[b]	Toxicity Group[c]	Food Matrix	Toxic Effects	EU ML[d] ($\mu g.kg^{-1}$)
Ochratoxins OTA[a], OTB, OTC	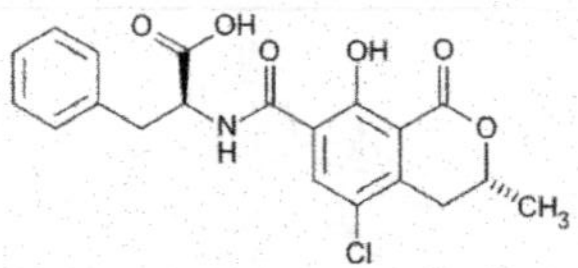	*A. ochraceus* *A. carbonarius* *P. verrucosum*	2B	Corn, barley, oats, rye, wheat, grapes, coffee beans, wine, beer, dried fruits	Nephrotoxic, teratogenic, immunosuppressive, potent teratogen	0.5–10.0
Aflatoxins AFB1[a], AFB2, AFG1, AFG2,	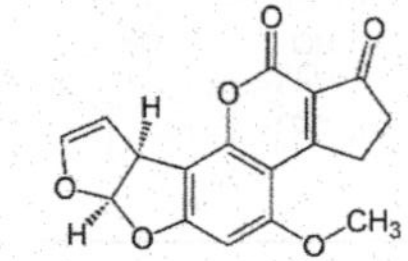	*A. flavus* *A. parasiticus*	1	Nuts, cereals, maize, rice oilseeds	Highly toxic, carcinogenic, immunosuppressive, mutagenic, genotoxic, teratogenic	0.1–12.0 (AFB1) 4–15 (Total sum)
Aflatoxin M AFM1[a], AFM2	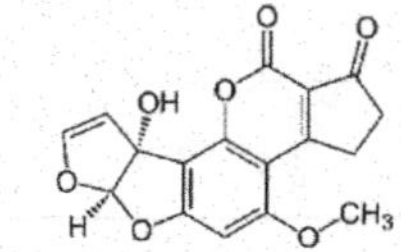	Metabolites of aflatoxin B	1	Milk and derivatives	Carcinogenic, stunting in children mutagenic, genotoxic, teratogenic	0.025–0.050
Fumonisins FB1[a], FB2, FB3	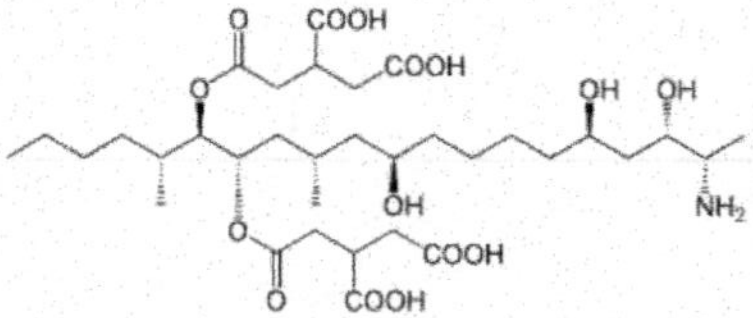	*F. verticillioides* *F. proliferatum*	2B	Soybeans, maize and sorghum	Hepatotoxic and carcinogenic, interferes with sphingolipid metabolism	200–4000
Patulin (PAT)	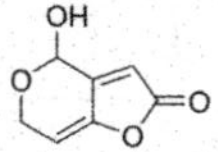	*P. patulum* *P. expansum* *A.* spp	3	Apple and derivatives, pears, cherries, and other fruits	Genotoxic, immunotoxic, neurotoxic, teratogenic	10–50

Table 1 contd. ...

...*Table 1 contd.*

Mycotoxin	Chemical Structure[a]	Fungal Species[b]	Toxicity Group[c]	Food Matrix	Toxic Effects	EU ML[d] ($\mu g.kg^{-1}$)
Zearalenone (ZEA)	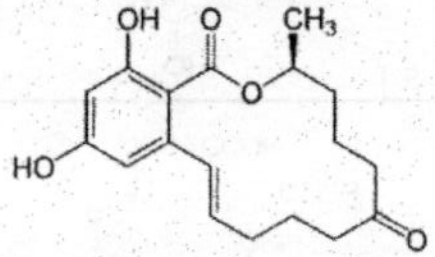	*F. graminearum* *F. culmorum*	2A	Cereal crops such as wheat, maize, barley, and sorghum	Estrogenic	20–350
Citrinin (CTN)	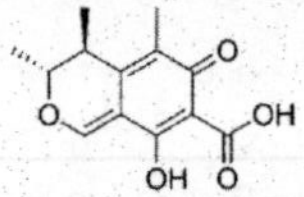	*P. citrinum* *P. camemberti* *A. terreus* *A. niveus*	3	Wheat, oats, rye, corn, barley, rice	Cytotoxic and nephrotoxic	2000 (Rice)
T-2 toxin (T-2)	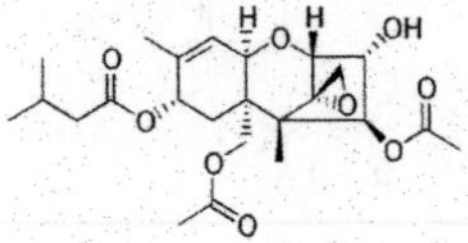	*F. sporotrichioides* *F. poae*	3	Wheat, barley, oats, maize	Cytotoxic, immunosuppressive, potent inhibitors of eukaryotic protein synthesis	15–1000
Nivalenol	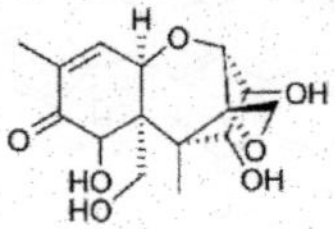	*F. graminearum* *F. culmorum*	3	Oats, barley, maize, wheat, bread, pasta, cereals	Growth retardation, Leukopenia, Intrauterine growth delay	1.2
Deoxynivalenol	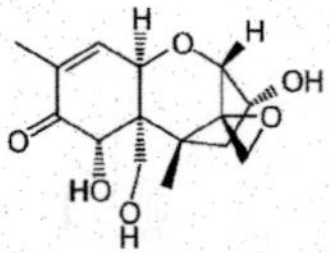	*F. graminearum* *F. culmorum*	3	Grains such as wheat, barley, oats, rye, maize, rice, sorghum	Potent inhibitor of eukaryotic protein synthesis causes nausea, vomiting, and diarrhea	200–1750

[a] Only the main mycotoxin's structure within the corresponding group has been presented.
[b] *A* for *Aspergillus*, *F* for *fusarium* and *P* for *Penicillium*.
[c] Last update [7].
[d] Maximum limits (MLs) vary according to the food matrix.

is required on the potential immune effects of aflatoxins, especially in vulnerable populations.

For ochratoxins, IARC has concluded that there is sufficient evidence that OTA is carcinogenic in experimental animals, but there is inadequate evidence that OTA increases cancer risks in humans (Group 2B) [7]. OTA intake related to its residuals in food commodities causes nephrotoxicity and renal tumors in a variety of animal species. Although human health effects are less well-characterized, it is suspected to induce adverse effects on the fetus in the womb, given its ability to cross the placenta and cause central nervous system malformation and brain damage [13].

Studies on fumonisins indicate its possible role in esophagal cancer and neural tube defects, although no definitive conclusions can be drawn at present [8]. For deoxynivalenol and other trichothecenes, exposure has been linked to acute poisoning outbreaks in large numbers of subjects, where severe gastrointestinal toxicity is the primary symptom [8].

When it comes to zearalenone, its metabolites can bind to estrogen receptors, thus leading to infertility and abortion risks. This effect has been observed in lab animals, but not yet confirmed in humans; hence, ZEA is considered to be of relatively low acute toxicity for humans [14].

Overall, the limited tools available to accurately assess human exposure to mycotoxins and the relative paucity of epidemiological studies need to be addressed. These major limitations can be tackled through the development of validated biomarkers to assess exposure to mycotoxins. The availability of such biomarkers can greatly assist epidemiological studies on mycotoxins and allow a comprehensive evaluation of their effects on human health in order to take adequate public health measures [8].

2.2.2 Economic impacts

Owing to the rejection of contaminated lots with high levels of mycotoxins, the economic impacts of mycotoxins are mostly related to the direct market costs associated with lost trade or reduced incomes. Various studies have attempted to quantify the potential market losses associated with mycotoxins in crops. For instance, in the USA, Vardon and coworkers estimated the total annual losses due to three mycotoxins – aflatoxin, fumonisin, and deoxynivalenol – to range from \$0.5 million to over \$1.5 billion [15]. Moreover, in the next decade, it is expected that the global climate change will deepen economic losses worldwide due to mycotoxin contamination induced by the warmer climate, particularly for aflatoxins [16].

On the other hand, it should be mentioned that the economic impacts of such wide-reaching contaminants also include the human health losses from adverse effects associated with the consumption of mycotoxins [8]. This includes the cost of the regulatory process designed to reduce risks to animal and human health, as well as enforced control of levels of mycotoxins. Consequently, losses related to health vary between developed and developing countries, where different standards for mycotoxins are set.

2.3 Conventional analytical methods

In the past decades, various conventional analytical methods have been devised for the quantitative analysis of some mycotoxins in order to comply with regulations ensuring food safety. Chromatographic methods, including thin-layer chromatography (TLC), gas chromatography (GC) and high-performance liquid chromatography (HPLC), coupled with diode array (DAD), mass spectrometry (MS) or fluorescence (FL) detectors, are the golden methods to strictly control the levels of mycotoxins in food samples. Therefore, these techniques are usually involved in accredited laboratories dedicated to the accurate analysis of mycotoxins. Particularly, the use of HPLC hyphenated to tandem MS (HPLC-MS/MS) has facilitated the multi-analysis of structurally relevant mycotoxins [17]. Overall, typical detection limits of mycotoxins for the above-mentioned conventional methods are usually in the sub-nanomolar concentration range.

These sensitive methods inarguably offer reliable results; however, they involve tedious sample pre-treatment and highly skilled personnel. They generally consist of a lengthy multi-step process involving three basic stages: (a) sample preparation, (b) mycotoxin extraction from the matrix and extract clean-up (usually with mixtures of water and polar organic solvents), and (c) final detection and quantitative determination [18]. In addition to the above, expensive, and bulky instruments also restrict these chromatographic methods from use in *in situ* applications. As a consequence of their constraints, the development of more convenient novel methods is still highly desirable and sought after for rapid monitoring of major mycotoxins at trace concentrations in foodstuff.

In this sense, tremendous progress has been made in the commercialization of various immunological assays, such as enzyme-linked immunosorbent assay (ELISA) and other rapid antibody-based strip test kits, as an alternative to the rapid screening of food contaminants. These tools are based on the specific interaction between antibodies and corresponding mycotoxins and are relatively fast and cost-effective when compared to instrumental techniques, while also maintaining the required sensitivity and specificity [6]. Commonly, antibodies are labeled with enzymes to sensitively detect changes in optical properties or electrochemical reactions following the addition of special substrates. This "golden" concept enables the detection of ultra-low amounts of target analytes. However, ELISA is quite time-consuming and needs multiple addition/washing steps. It also requires specialist plate readers, which renders it unsuitable for field testing [19].

Regarding the analysis of mycotoxins, immunoassaying is complicated by the small size of target analytes. Furthermore, the antibodies are produced by immunization of animals [20], which poses ethical problems and is fiercely advocated against by animal rights organizations [21]. Such small molecules in living cells, commonly known as *haptens*, cannot initiate immune response. Therefore, anti-mycotoxin antibodies are necessarily obtained using conjugates with serum proteins [22]. Moreover, antibodies are costly because they are produced in living animals and involve optimum physiological conditions to avoid denaturation and to preserve their proper sensing ability.

3. DNA aptamers as (bio)receptors for mycotoxin detection

3.1 Generalities on aptamers

Aptamers are artificial single-stranded oligonucleotides (DNA or RNA) selected in a specific way to recognize a target molecule – a small molecule, a protein, a virus, a bacterium or even a living cell. This recognition is established through the formation of stable sequence-dependent tertiary structures. They can be considered as artificial antibodies produced in non-living animals. The name aptamer is derived from the Latin word '*aptus*' (meaning 'to fit') in combination with the Greek word '*meros*' (meaning 'part') [23]. This type of affinity bioreceptors bind to the desired targets by a lock and key mechanism [24], with the advantage of being highly specific within a wide range of targets.

Aptamer ligands usually containing up to 80 nucleotides are isolated from a large random library of synthetic nucleic acids (10^{14}–10^{15} variants) by an iterative process of binding, separation, and amplification, commonly termed as Systematic Evolution of Ligands by EXponential enrichment (SELEX) as summarized in Figure 2. Through this iterative protocol (7 to 15 rounds), non-binding aptamers are discarded via elution, and aptamers binding to the proposed target are expanded and PCR amplified. Initial positive selection rounds are sometimes followed by negative selection. This improves the selectivity of the resulting aptamer candidates. Later, multiple rounds of SELEX are performed with increasing stringency to improve enrichment of the oligonucleotide pool.

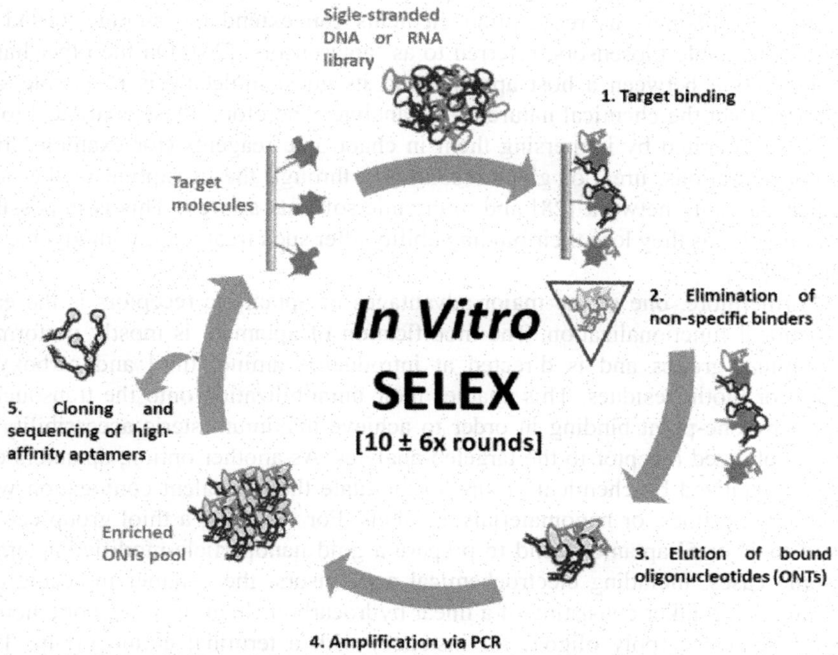

Figure 2. Schematic overview of the common SELEX procedure.

In the last 30 years, the protocol for SELEX has considerably evolved. SELEX takes weeks to months to obtain specific aptamer candidates, with generally low hit rates. Hence, obtaining high-quality aptamers against relevant targets remains a bottleneck for advancing research in this particular field [25]. To shorten the selection time and enhance yields, several customized *in vitro* selection methods derived from standard SELEX have been therefore set up; the relevant examples include Capture-SELEX, Capillary Electrophoresis-SELEX (CE-SELEX), and separation of free nucleotides from bound sequences based on their different affinities towards graphene oxide (GO-SELEX) [26].

Aptamers can be used in electrochemical, colorimetric, fluorometric and mass-sensitive methods. These features make aptamers well suited to real-time screening and on-site analysis, as required for production management of industrial products, including food and beverages [27].

Over the last decade, aptamers, including those specific to mycotoxins, have shown great potential as alternative affinity receptors that can substitute antibodies in biosensors, thanks to their several attractive properties. Aptamers can provide high stability and affinity, along with simplicity, cost effectiveness and excellent batch-to-batch reproducibility [9].

Compared to antibodies, aptamers are more stable towards hydrolysis, pH variations and high temperatures. These receptors meet the requirements of bioethics since their design and *in vitro* production do not require animal immunization. Moreover, once the sequence of aptamer is selected, it can be well reproduced by a routine synthesis method which is more cost-effective than isolation of antibodies. The use of aptamers as recognition elements can extend the storage period of appropriate stable biosensors, referred to as 'aptasensors' [23]. On the other hand, the interaction between a host aptamer and its guest molecule is reversible – it does not affect the chemical nature of the analyte. Therefore, these analytical tools can be regenerated by immersing them in chaotropic reagents (for example, high salt concentrations, urea, or glycine solution), through the disruption of the non-covalent bonding network [28] and in organic solvents as well. This is impossible for antibodies as they lose their binding ability after such treatment, with glycine for example [29].

Furthermore, one of the major advantages of aptameric receptors is the ease of chemical functionalization. The modification of aptamers is mostly performed by terminal groups and is directed at introducing amino, thiol and carboxylic groups, or biotin residues. This enables their immobilization onto the transducing surface by one-point binding in order to achieve maximum steric accessibility of the immobilized receptor to the targeted analytes. As another option, aptamers can be functionalized by chemical groups to facilitate their covalent conjugation with dyes, redox probes, or nanomaterials as labels. For example, a thiol group can be linked to 3' or 5' aptamer's end to prepare a gold nanoparticle-modified aptamer. In some cases, including electrochemical aptasensors, the aptamer molecule can contain a long linker consisting of a linear hydrocarbon chain (C_{12}–C_{16}) or a homo-nucleic sequence (poly oligoT, for instance) with a terminal amino group. This length extension contributes towards minimizing the possible steric limitations of the recognition reaction with the target molecule, near the electrode surface [30].

3.2 Specific aptamers for mycotoxins

Various aptamers for the capture of mycotoxins have been designed since 2008. The first mycotoxin that was targeted by aptasensors was ochratoxin A (OTA) [31]—the most widespread mycotoxin in food. Since selection of aptamers for small molecules poses several challenges [9], the number of aptamers selected for mycotoxin analysis is rather limited, taking into account the variety of potential targets and their binding sites [19]. This is mostly due to the low molecular weight of such organic analytes, usually about few hundred Daltons, which makes the compilation of a primary library and the selection of specific aptamers more challenging than when targeting bigger analytes like proteins. Moreover, there are limited possibilities of interaction between nucleotides and small analytes due to the reduced number of functional groups available in small-sized molecules, especially after further blockage following their immobilization onto solid supports.

Sequence alignments, secondary structure analysis and binding studies are required to identify the final sequence and the characteristics of the selected aptamer after performing SELEX[32]. The dissociation constant, K_D, presents an important affinity metric of aptamers. It can be evaluated by various binding assays. Obviously, the lower the value of K_D, the more stable is the aptamer-target complex formed [29]. Once the sequence affinity is validated, the use of aptamers as bioreceptors proves to be advantageous for the detection of-mycotoxins as food contaminants.

Aptamers can be folded into different well-defined 3D structures as shown in Figure 3. Single-stranded sequences can contain self-complementary fragments to form one or few loops that are unfolded – the so-called pinhole aptamers. Binding to their ligands is then achieved by complementary shape interactions; they can incorporate small molecules into their nucleic acid structure or integrate themselves into the structure of larger molecules [33]. Aptamers can assume a variety of shapes like stems, loops or helices, thanks to their nucleic acid backbone. This explains the versatility of aptamers to bind to very diverse targets. In particular, guanine(G)-rich nucleic acids are able to self-assemble into G-quadruplex four-stranded secondary structures due to the presence of long and contiguous G regions. Electrochemical methods have been successfully used for the rapid detection of the conformational

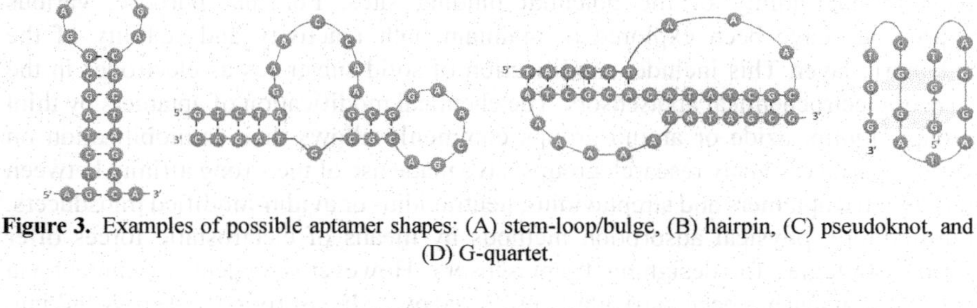

Figure 3. Examples of possible aptamer shapes: (A) stem-loop/bulge, (B) hairpin, (C) pseudoknot, and (D) G-quartet.

changes from single-stranded DNA to G-quadruplex due to their outstanding stiffness [24].

Till date, aptamers directed against ochratoxin A (OTA), aflatoxin B1 (AFB1), versicolorin (a precursor of AFB1), aflatoxin B2 (AFB2), aflatoxin M1 (AFM1), patulin (PAT), fumonisin B1 (FB1), fumonisin B2 (FB2), citrinin (CTN), deoxynivalenol (DON), zearalenone (ZEN) and T-2 toxin have been described and characterized mostly using fluorescence spectroscopy and surface plasmon resonance (SPR) [9]. Aptamers so far synthetized for the capture of mycotoxins are exclusively based on ssDNA sequences, since DNA aptamers are more stable towards nuclease digestion, compared to their RNA analogs. Furthermore, the former shows excellent reproducibility with low-cost mass production. A non-exhaustive list of available aptamers as reported in literature is summarized in Table 2.

More interestingly, the reported dissociation constants corresponding to aptamers of mycotoxins are in the nanomolar range, revealing their high specificity towards such small targets. The numerous advantages of aptamer-based assays open the avenues for the selection of new aptamers and detection kits for other mycotoxins threatening animal and human health [13].

Upon mycotoxin recognition, target binding involves a three-dimensional change of the aptamer structure. This conformational ordering is due to the formation of an analyte-ssDNA complex, resulting from shape-dependent interactions as well as hydrophobic interactions, base-stacking, and intercalations. Such molecular switches, often defined as aptamer folding, alter the steric and electronic density within the surface layer in the case of electrochemical aptasensors. It is thus exploited for the detection of the target. It would be rather interesting to investigate the exact binding site for respective analytes. Unfortunately, in the case of mycotoxin-binding aptamers, the recognizing sites and their detailed interactions are not yet well-known [29]. This needs substantial efforts for structural characterization by Nuclear Magnetic Resonance (NMR), with the support of computational molecular modeling using bioinformatic tools [52].

3.3 Immobilization approaches

The performance of a given biosensor is critically dependent on the functionalization step of the sensor platform. Accordingly, the immobilization protocol of aptamers on the transducer surface plays a pivotal role in the detection. The immobilization strategy should enable the maximum coverage of (bio)receptors without affecting their folding ability or the potential binding sites. For that purpose, various approaches have been explored to maintain both reactivity and stability of the aptameric layer. This includes modification of solid substrates as electrodes in the case of electrochemical aptasensors. The chemical modification of aptamers by thiol groups, biotin, azide or amino groups commonly allows their immobilization on various surfaces. Many research groups have made use of the strong affinity between biotinylated aptamers and streptavidin-, neutravidin- or avidin-modified transducers. Furthermore, physical adsorption methods by means of electrostatic forces offer simple strategies for designing E-aptasensors. However, covalent attachment via covalent linkage has been particularly preferred over physisorption and cross-linking.

Table 2. Updated list of aptamers for mycotoxins, their lengths and dissociation constants (the list summarizes data till 2019).

Mycotoxin	ssDNA aptamer (5' to 3')	Length (mer)	K_D (nM)	Reference
OTA	*GAT CGG GTG TGG GTG GCG TAA AGG GAG CAT CGG ACA*	*36*	**200**	[31]
	GGG AGG ACG AAG CGG AAC CGG GTG TGG GTG CCT TGA TCC AGG GAG TCT CAG AAG ACA CGC CCG ACA	*66*	**96**	[34]
	AGC CTC GTC TGT TCT CCC GGC GCA TGA TCA TTC GGT GGG TAA GGT GGT GGT AAC GTT GGG GAA GAC AAG CAG ACG T	*76*	**110 ± 50**	[35]
AFB1	*TGG GGT TTT GGT GGC GGG TGG TGT ACG GGC GAG GG*	*35*	**NR[1]**	[36]
	CTC GTC TCG TTC TGT CAG TGT TCT TCT GGC TTG GTG GTT GGT GTG GTG GCT TGA TTT GGT AGA CAC GAA GAA GAA GGA AGG A	*80*	**50.4 ± 11.06**	[37]
	GTT GGG CAC GTG TTG TCT CTC TGT GTC TCG TGC CCT TCG CTA GGC CCA CA	*50*	**32**	[38]
AFB2	*AGC AGC ACA GAG GTC AGA TGC TGA CAC CCT GGA CCT TGG GAT TCC GGA AGT TTT CCG GTA CCT ATG CGT GCT ACC GTG AA*	*80*	**9.83 ± 0.99**	[39]
AFM1	*ACT GCT AGA GAT TTT CCA CAT*	*21*	**0.02**	[40]
	GTT GGG CAC GTG TTG TCT CTC TGT GTC TCG TGC CCT TCG CTA GGC CCA CA	*50*	**10**	[41]
	ATC CGT CAC ACC TGC TCT GAC GCT GGG GTC GAC CCG GAG AAA TGC ATT CCC CTG TGG TGT TGG CTC CCG TAT	*72*	**35.6**	[42]
PAT	*GGC CCG CCA ACC CGC ATC ATC TAC ACT GAT ATT TTA CCT T*	*40*	**21.83 ± 5.02**	[43]
FB1	*AAT CGC ATT ACC TTA TAC CAG CTT ATT CAA TTA CGT CTG CAC ATA CCA GCT TAT TCA ATT*	*60*	**100 ± 30**	[44]
	CGA TCT GGA TAT TAT TTT TGA TAC CCC TTT GGG GAG ACA T	*40*	**62 ± 5**	[45]
FB2	*GCA TCA CTA CAG TCA TTA CGC ATC GTG ACG AGG GTG ACT ATG GCG GTG GCG TCT GTG AGC ATC GTG TGA AGT GCT GTC CC*	*80*	**NR[1]**	[46]

Table 2 contd. ...

...Table 2 contd.

Mycotoxin	ssDNA aptamer (5' to 3')	Length (mer)	K_D (nM)	Reference
CTN	*GGC CAG GCG GGG CCT GTT CGC TGG GGC CGT GTC TTC GGC TCG CTC GGT GTT GTT TGG CG*	*59*	**60**	[47]
DON	*GCA TCA CTA CAG TCA TTA CGC ATC GTA GGG GGG ATC GTT AAG GAA GTG CCC GGA GGG GTA TCG TGT GAA GTG CTG TCC*	*78*	**NR[1]**	[48]
ZEN	*AGC AGC ACA GAG GTC AGA TGT CAT CTA TCT ATG GTA CAT TAC TAT CTG TAA TGT GAT ATG CCT ATG CGT GCT ACC GTG AA*	*80*	**41 ± 5**	[49]
	CTT ATT CAA TTA TAC CAG CTT ATT CAA TTA TAC CAG CAC AAT CGT AAT CAG TTA G	*55*	**15.2 ± 3.4**	[50]
T-2	*CAG CTC AGA AGC TTG ATC CTG TAT ATC AAG CAT CGC GTG TTT ACA CAT GCG AGA GGT GAA GAC TCG AAG TCG TGC ATC TG*	*80*	**20.8 ± 3.1**	[51]

[1]NR: not reported.

Despite their simplicity and time-saving attributes, the latter methods show lower reproducibility and sensitivity due to the elution of adsorbed species during washing steps, in addition to the non-specific adsorption of small molecules. Hence, covalent immobilization is frequently used to minimize these hurdles, taking advantage of the ease of chemical modification of aptamer sequences, via 3' or 5' tails. Generally, the 3' end is more suitable for immobilization as it simultaneously confers resistance to nuclease after its binding to the electrode surface [53]. For instance, some commonly used immobilization approaches are: covalent attachment of thiolated aptamer on gold substrates or gold-modified surface, immobilization of amine-terminated aptamer on carboxylic acid group functionalized surface (and vice versa) via amide coupling using carbodiimides chemistry, immobilization of azide-ended aptamer on alkyne-functionalized surface via *click* chemistry, and amine-amine coupling through glutaraldehyde. In some cases, molecular spacers are also used as scaffolds to provide sufficient conformational flexibility and freedom to the aptamer. To avoid non-specific adsorption, free sites of the golden or carbonaceous interface are blocked respectively with hydrophobic mercaptans (w-mercaptohexanol) or some proteins (bovine serum albumin). Surfactants and thin polymeric films can be also used for the same purpose [29].

In another approach, the immobilization of aptamers can be performed by hybridization with partially complementary oligonucleotide immobilized onto the transducer surface beforehand. However, the length of the hybridized portion should be carefully chosen to allow aptamer displacement by small targets such as mycotoxins [54].

Regarding the electrochemical detection formats, most development approaches of aptasensing rely on either redox labels or label-free schemes. In both the formats, the electrochemical signal should be recordable by employing one of the conventional electrochemical techniques. The following section therefore provides a brief overview of the prevailing electrochemical detection methods.

4. Electrochemical detection: An overview

Electrochemical biosensors have been considered as one of the most promising analytical tools for rapid detection of various biochemical species. Electrochemical transduction presents several advantages over optical, piezoelectric or thermal detection, given its high signal-to-noise ratio, relative simplicity, well-developed theory, ease of use, cost-effectiveness and fast response. In addition to these, miniaturization opportunities make it possible to deploy such biosensors in portable devices for on-site analysis. For instance, screen-printed electrodes (SPEs) that use mini-electrode systems with working, reference and auxiliary electrodes have gained popularity in electrochemical aptasensing due to their low cost and speed of mass production using thin film technology [55]. Electrochemical detection is essentially based on changes in the distribution of charged species (electrons or ions) within the working solution, and thus on the alteration of its electrical properties upon interaction between the recognition elements and specific targets [56]. It typically monitors redox reactions that generate a measurable current at a given applied voltage (amperometry), or produce a measurable charge accumulation or potential (potentiometry) or change the electrode's capabilities to transport electrons (conductometry), or modify the electronic resistance and capacitance of the bio-functionalized surface (impedance) [55, 57]. In particular, electrochemical aptasensors (E-aptasensors) represent most of the sensitive techniques used for measuring target concentrations based on aptamer recognition [22]. Therefore, they cover over 60% of all the scientific research papers specifically devoted to the aptasensing of mycotoxins [29]. E-aptasensors track the electrochemical changes that occur when the aptamer captures the analyte of interest via 3D folding. The recorded signal is thereby generated by specific redox reactions that involve either electroactive labels or aptamer supports.

Herein, we describe the principles of three main categories of electrochemical transduction techniques commonly used for the detection of food-borne toxins, using aptamers.

4.1 Voltammetric/Amperometric detection

Voltammetry and amperometry are characterized by applying a potential to a working electrode versus a reference electrode and measuring the current generated by the electrochemical oxidation or reduction of given species generated during the process or intentionally added to the electrolyte. It is the absence of a scanning potential that distinguishes amperometry from voltammetry. Basically, the term voltammetry is used for those techniques in which the potential is scanned over a set potential range. The current response is usually a faradaic peak that is proportional to the concentration of the analyte. However, in amperometry, changes in current are monitored directly with time, while a constant potential is maintained at the working electrode with respect to the reference electrode. Generally, amperometric biosensors have additional selectivity in that the oxidation or reduction potential used for detection is a characteristic of the analyte species in the sample [55].

Different current readout methods have been used in mycotoxin aptasensing; relevant examples include chronoamperometry, cyclic voltammetry, linear sweep voltammetry, differential pulse voltammetry and square wave voltammetry. The latter two tend to be the most used techniques since they are sensitive only to the Faradaic processes of interest.

Abundant literature covering the determination of small mycotoxins by these electrochemical processes is available [58–60]. Interestingly, the introduction of nanomaterials was a key turning point in this field. For instance, Abnous et al. [61] developed a voltammetric (DPV) aptasensor using anti-OTA aptamer assisted by two complementary oligo-strands onto a screen-printed gold electrode (Figure 4). In this study, the second DNA strands were conjugated to single-walled carbon nanotubes (SWCNTs) as electrochemical signal amplifiers. Based on methylene blue (MB) as the redox indicator, the described cage-shaped structure of the aptamer allowed the ON/OFF detection of OTA in grape juice and serum samples with LODs of 58 and 134 pM respectively. Despite its good sensitivity, this assembly strategy seems pretty complicated.

Another work reported by Tang et al. [62] suggested the detection of the same mycotoxin (OTA) using chronoamperometry (CA) as shown in Figure 5. This aptasensing strategy is based on the interesting properties of two-dimensional (2D) MoS_2 nanosheets as an emerging nanomaterial. 2D MoS_2 possesses a considerably larger affinity for ssDNA than for dsDNA. Also, it intrinsically exhibits a peroxidase-like activity to catalyze the hydroquinone/benzoquinone redox reaction. Firstly, the platform design involves a self-assembly of a dsDNA probe (consisting of OTA-aptamer/auxiliary DNA hybrid) on a gold electrode. Then, the introduction of the target OTA unwinds the dsDNA probe through aptamer release, leaving the single-stranded auxiliary DNA on the electrode. Hence, MoS_2 performs like a nano-binder to the electrode interface that amplifies efficiently the subsequent electrochemical signal obtained by the catalysis of hydroquinone. This amperometric nano-aptasensor

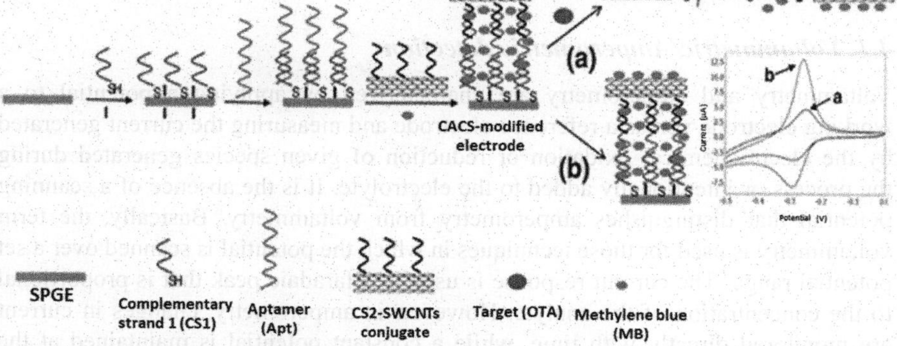

Figure 4. Amperometric aptasensor for ochratoxin A, based on the use of a gold electrode modified with aptamer, complementary DNA and SWCNTs, with Methylene Blue as the redox marker [61].

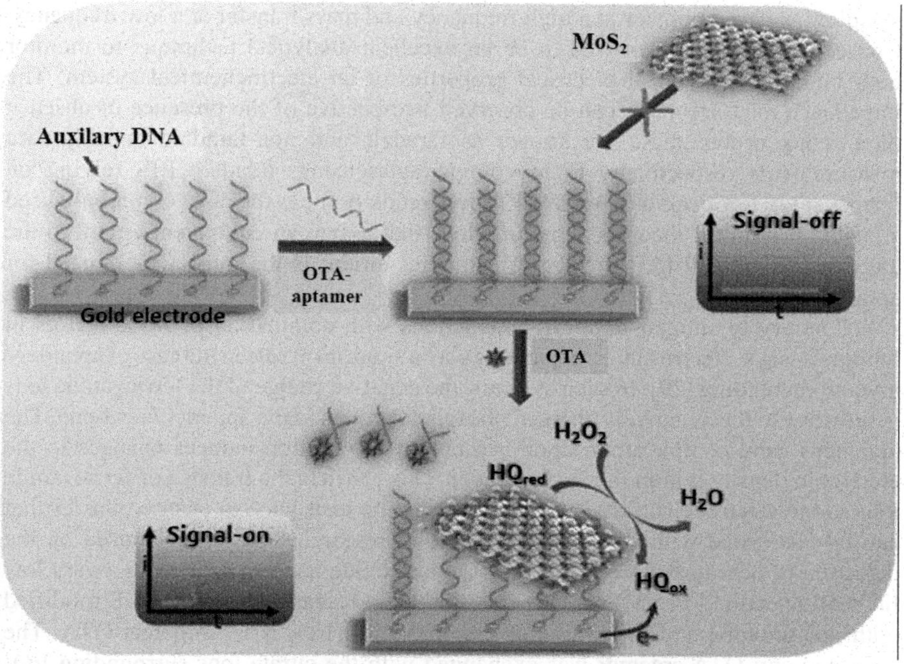

Figure 5. Aptamer-based amperometric detection of Ochratoxin A, using 2D MoS_2 as a nano-binder to catalyze the amplification of the electrochemical signal from the hydroquinone/benzoquinone system [62].

achieved an LOD of 0.57 pM and was used to analyze OTA in spiked red wine samples.

4.2 Potentiometric detection

Potentiometric sensors are based on measuring the potential of an electrochemical cell while drawing negligible current. Practically, potentiometric biosensors measure the changes in the voltage between the working electrode and the reference electrode due to the establishment of an electrostatic interaction. However, their use for the detection of small targets with high sensitivity and good selectivity is still elusive. In 2017, Lv et al. [63] developed a potentiometric aptasensor for small molecules based on the changing of surface charge. As a proof-of-concept, this platform was useful to detect bisphenol A using aptamer-functionalized single-walled carbon nanotubes. Meanwhile, not much work has been done on aptasensors based on potentiometric detection for the analysis of mycotoxins [64].

4.3 Impedimetric detection

E-transduction using electrochemical impedance spectroscopy (EIS) has gained popularity among the biosensor community, thanks to their unique features over the widely used amperometric detection. At a given applied voltage, they are capable of sampling electron transfer at a high frequency and mass transfer at a low frequency. Consequently, EIS has proved to be an excellent analytical technique to monitor both bulk and interfacial electrical properties of an electrochemical system. The impedance measurements can be observed irrespective of the presence or absence of a redox probe; these are known as faradaic and non-faradaic impedimetric measurements respectively. In mycotoxin aptasensing, faradaic EIS relying on ferrocyanide/ferricyanide $[Fe(CN)_6]^{3-/4-}$ free redox probe is the most commonly used strategy. It requires theoretical assumptions that assign an equivalent circuit to the electrochemical cell [65]. In general, Randles circuit is the most widely used for aptasensor modeling.

Contrary to antigen–antibody interactions with domination of steric factors in immunoassays, electrostatic interactions play a significant role in the case of aptamer–mycotoxin binding [29]. In such systems, the negative charge of the ferricyanide ions is sensitive to the negatively charged phosphate groups in the aptamer backbone. The aptamer-ligand complexation upon mycotoxin recognition induces changes in the electronic density within the surface layer. This restricts the transfer of ferricyanide ions to the electrode surface, hence increasing the resistance to charge transfer that can be monitored with EIS. Several works of research have been focused on the detection of ochratoxin A (OTA) using iridium oxide nanoparticles with a very low limit of detection (14 pM) [66]. In one study, the researchers used SPCE modified with polythionine and iridium oxide nanoparticles (IrO_2 NPs) to detect OTA. The aminated anti-OTA aptamer was exchanged with the citrate ions surrounding IrO_2 NPs via electrostatic interactions with the same surface.

However, affinity biosensors using impedance detection have to be carefully designed to avoid non-specific binding of the analyte. To minimize interferences, it is mandatory to block non-reactive surface areas with inert blockers. For instance, protein blockers such as BSA, casein, nonionic surfactant (Tween 20) and many proprietary blended commercial products have been successfully used to prevent non-specific binding [55]. Furthermore, the use of self-assembled monolayers for aptamer immobilization simultaneously enables the suppression of the non-specific binding phenomenon for more selective assays.

It is worth mentioning that dual E-transduction systems, such as the investigation of voltammetric along with impedance response, are also feasible. On these lines, Castillo et al. [67] suggested an AFB1 aptasensor by immobilizing poly (amidoamine) dendrimers of fourth generation (PAMAM G4) on a gold electrode covered with cysteamine (Figure 6). This sensor was assembled in a multi-layer framework that used cyclic voltammetry (CV) and electrochemical impedance spectroscopy (EIS) for acquiring the signal response by means of $[Fe(CN)_6]^{-3/-4}$ as redox indicators. The response obtained was in the AFB1 concentration range of 0.1–10 nM, with an LOD of 0.04 nM. This aptasensor has made it possible to analyze AFB1 in certified peanut samples, without any interference from OTA and AFB2.

On the other hand, the bio-recognition event at a modified surface causes variations in not only the resistive properties, but also the capacitive properties of the impedimetric transducer.

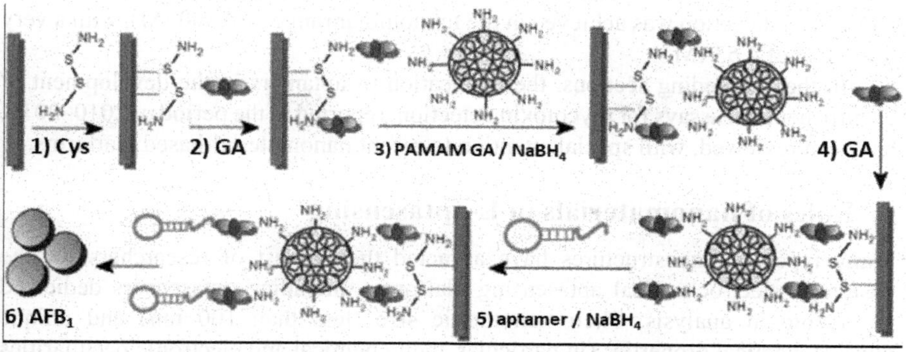

Figure 6. Process of fabrication (schematic) of an aptamer-based biosensor for the detection of AFB1, using PAMAM G4 dendrimers as the linking nanomaterial [67].

In this context, electrochemical capacitance spectroscopy (ECS) is an emerging transduction approach, mainly based on EIS measurements in the non-faradaic regime [68]. Meanwhile, it is solely applicable when the redox probe is confined to the electrode surface, in the close vicinity of bio-receptors. This nano-scale engineering allows us to observe a new pseudo-capacitance called 'redox capacitance (Cr)', which depends on the density-of-states of the confined redox reporter. Overall, the alteration of electronic density through analyte recognition affects the redox fingerprint of the entire bio-film. The experimental data collected from conventional impedance Nyquist spectra can be mathematically converted into Nyquist capacitive spectra [69]. Hence, the treated capacitive signal enables one to derive the redox capacitance value 'Cr' (in μF), which varies in good agreement with target concentration, without the need of any theoretical modeling or equivalent circuit.

Following the ECS approach, Ben Aissa et al. [70] recently designed an aptasensor for the quantification of aflatoxin M1 in milk (Figure 7). Their work consisted of the immobilization of ferrocene moieties as a redox probe, confined to the aptamer layer. Working in a buffer solution, the control of redox capacitance contribution by ferrocene charge/discharge at a fixed potential was found to be highly sensitive for the detection of AFM1. With the use of silicon nanoparticles as capacitive signal

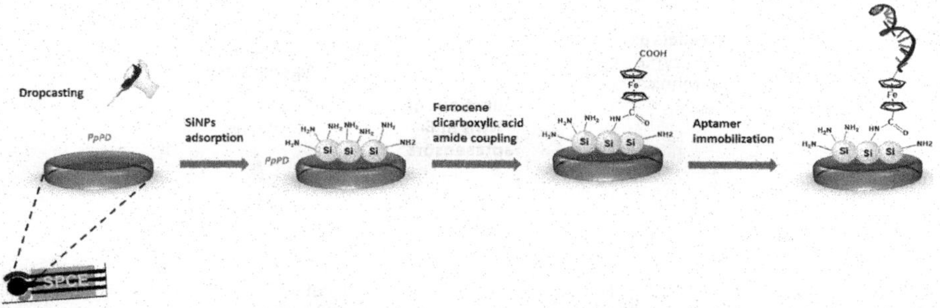

Figure 7. Process of fabrication (schematic) of an AFM1 aptasensor for redox capacitance detection [70].

amplifiers, detection was achieved in the femtomolar range (10–500 fM), with a very low LOD of 4.53 fM.

In the succeeding sections, the application of aptamers in the development of electrochemical assays for mycotoxin detection, reported in the period of 2010–2019, has been reviewed, with special attention to recent nanomaterial-based platforms.

5. Roles of nanomaterials in E-aptasensing

Many types of nanostructures have attracted the interest of researchers for the design of electrochemical aptasensing platforms, including aptasensors dedicated to mycotoxin analysis. Their nanometric size (less than 100 nm) and unique physicochemical properties (in particular, their chemical and electronic peculiarities that differ from the corresponding bulk materials) have promoted their extensive use in the building of aptasensors. Accordingly, scientists handle each nanomaterial differently in order to adapt it for the desired function in a biosensor. For instance, five main roles are depicted in E-aptasensing as summarized in Figure 8. Firstly, nanomaterials such as nanorods, nanowires, nanotubes or nanosheets can be used as immobilization supports to extend the surface area and, consequently, for increasing the biomolecule loading. Also, it has been widely proven that nanostructured platforms allow the amplification of the electrochemical response through electron transfer enhancement, making the assay faster and more sensitive. Moreover, some electroactive nanoparticles can act as redox mediators for label-free analysis or even as electrochemical signal generators [71]. Finally, labelling aptamers with nanoparticles has also been successfully investigated to mimic the catalytic activities of enzymes.

In this section, we explain the possible functions that can be achieved with the integration of nanomaterials in aptameric biosensors. It should be mentioned that

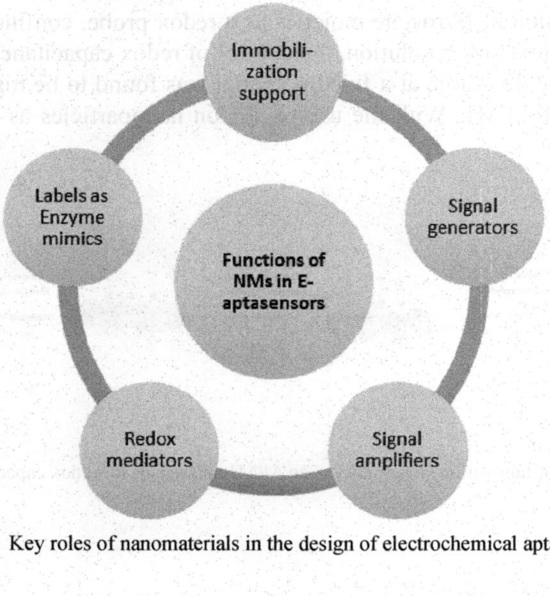

Figure 8. Key roles of nanomaterials in the design of electrochemical aptasensors.

more than one type of nanomaterials can be used in an aptasensor's construction, based on the desired function of the platform. For instance, in 2015, Bulbul et al. [72] used redox active nanoceria particles as both the catalytic label and the redox mediator, and reduced graphene oxide as the immobilization platform in the same assay for sensitive and selective detection of Ochratoxin A. However, since some nano-sized materials can have overlapping intrinsic properties, it can yield certain problems in their particular applications. The reactive nature of nanomaterials (NMs) can be advantageous in one way, but, on the other hand, it may limit other applications of nanomaterials (as immobilization supports, for instance). Therefore, it is crucial to perform control experiments with these NMs and study their reactivity carefully before using them in aptasensors.

5.1 Immobilization supports for aptamers

In the last few years, several attempts at developing aptasensing assays using nanostructured surfaces have been reported. The premise of the use of nanomaterials for immobilization on a transducer surface is to reduce diffusion limits and enhance the surface-to-volume ratio to increase the loading of biomolecules. For instance, AuNPs have been widely used as a substrate in electrochemical sensors [73]. Also, scaffolding aptamers for the detection of mycotoxins is the most extensively reported role of NMs in electrochemical aptasensing designs [74, 75]. This function presents an appealing alternative to conventional polymeric supports such as polysaccharides and hydrophobic materials such as polyvinylchloride.

The introduction of nanomaterials into the design of the transducing platform is generally achieved by their integration within conventional working electrodes in various forms. Contrary to bulk materials, nanomaterials have a strong tendency to uniformly adsorb biomolecules without the risk of bioactivity loss and thereby aid in the immobilization of biomolecules. Since DNA aptamers are negatively charged, the integration of oppositely charged nanomaterials in layer-by-layer biosensor construction facilitates the electrostatic adsorption of such biomolecules and increases the platform physiological stability. More robust immobilization approaches can also be envisaged; for example, covalent attachment using functionalization of both nanostructures' and aptamers' tails, or even by means of affinity-based reactions such as with (strept-, neutr-) avidin-biotin complexes.

5.2 Electrochemical signal generating probes

Various approaches have been explored to obtain the desired (enhanced) electrochemical output signal, in the construction of affinity biosensors. Among these, the use of nano-labels offers great opportunities to generate an amplified electrochemical response. Two main detection schemes can be considered using nanoparticles as labels for aptamers: (i) direct signal generation by the intrinsic electroactivity of nanotags [76], or (ii) indirect signal measurement via the electrolysis of redox substrates [77]. The latter approach refers to enzyme mimicking and has been detailed in Section 5.5. Still, biomolecules labeled with nanomaterials should retain their bioactivity to recognize their counterparts.

Among the many kinds of nanomaterials, gold nanoparticles (AuNPs) are the most frequently used labels in electrochemical aptasensors [78]. Since AuNPs are highly conductive nano-markers, they contribute towards enhancing the loading of electroactive species, resulting in higher signals than conventional redox labels [79]. Thousands of atoms can be released from one nanoparticle, pushing detection limits down to a few thousands of biomolecules.

Quantum dots and semiconductor nanomaterials, such as ZnS, CdS, PbS and CdTe nanoparticles, are also available. In an inspiring piece of work, Wang et al. [80] designed a magneto-controlled aptasensor for the simultaneous electrochemical detection of dual mycotoxins in maize (OTA and FB1), using metal sulfide quantum-dots-coated silica as labels (Figure 9). Herein, CdTe or PbS QDs were partially released from specific aptamers under the action of target recognition, and subsequently from magnetic beads. After magnetic separation, the SWV signals of Cd^{2+} and Pb^{2+} dissolved from the reserved labels in the respective supernatant liquids showed a negative correlation with target quantities. This aptasensor provided a

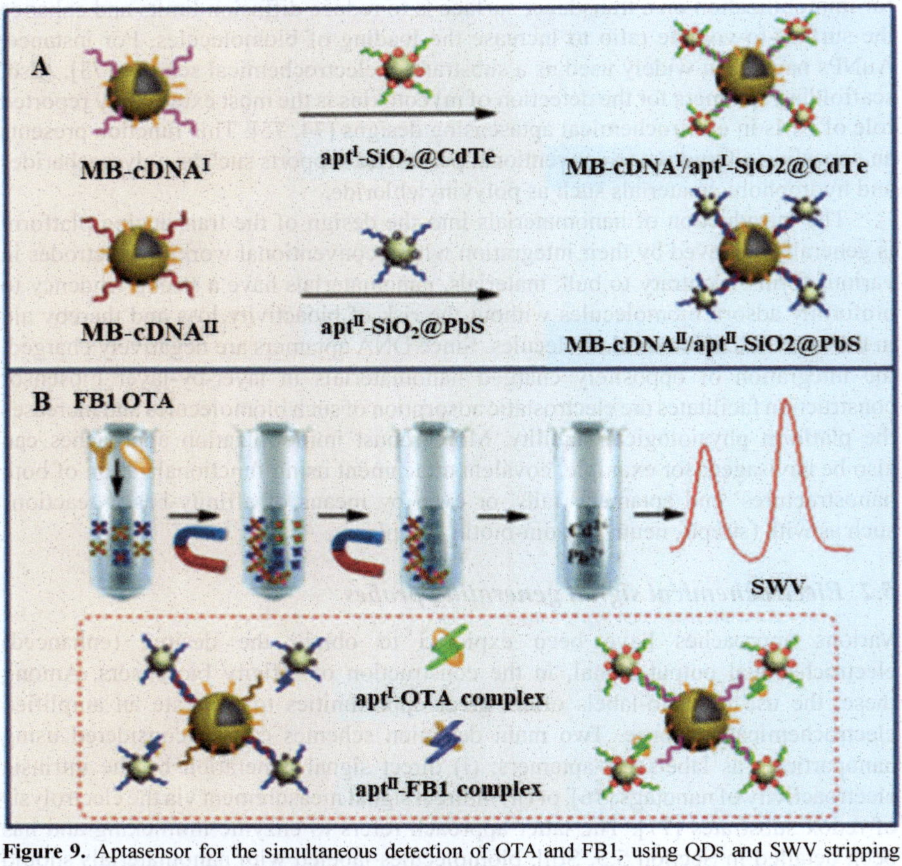

Figure 9. Aptasensor for the simultaneous detection of OTA and FB1, using QDs and SWV stripping measurements: (A) fabrication process (schematic), and (B) working principle [80].

detection range of 10 pg mL^{-1} to 10 ng mL^{-1} for OTA and 50 pg mL^{-1} to 50 ng mL^{-1} for FB1, showing successful application in real maize samples.

Modification with nanoparticles can be performed on not only biomolecules, but also target analytes. This option enables analyte determination after the biorecognition event via a sandwich assay scheme. Such nanomaterial-assisted detection can greatly enhance the output response [81]. Even though such labeling needs more preparative steps than label-free detection, the signal enhancement is higher by several orders of magnitude, which is a convincing argument for using such an approach.

5.3 Redox mediators

It is vital to create an electrical contact between redox-active biomolecules and the electrode surface while developing electrochemical aptasensors. For this purpose, nanomaterials such as conductive metallic, metal oxide and other semiconductor NPs are used as electron shuttles between electroactive species and the electrode surface. In the specific case of aptamer bioreceptors, their spatial 3D conformation can hinder the electron transfer between the mediators and the electrode surface by creating an insulating layer blocking electron-transfer (ET). Conductivity of nanoparticles and their well-defined arrangements can enable the development of nanomaterials-based methods for the construction of sensitive aptasensors with enhanced electron transfer properties. For example, Shi et al. [82] took advantage of the synergistic contribution of ionic liquids and Fe_3O_4 magnetic nanoparticles as the nanocomposite mediators to accelerate ET. The deposition of this nano-mixture onto a screen-printed carbon electrode improved the response speed and sensitivity as previously reported [83].

Unless labelling with a redox probe, most anti-mycotoxin aptamers cannot afford a direct measurable redox signal, except for guanine-rich oligonucleotide sequences such as the anti-OTA aptamer. The inherent electroactivity of these specific aptamers has not been extensively studied in recent literature as guanine oxidation occurs at a relatively high potential (around 1.0 V). In addition to that, it involves a two-step oxidation mechanism at the C8-H position, resulting in a loss of four electrons and four protons. Consequently, despite the potential of conductive NMs as electron shuttles, their use as mediators in aptasensing for mycotoxin detection is still immature.

5.4 Signal amplification

Nanomaterials are frequently deployed as signal amplifiers in electrochemical aptasensing as they can afford ultrasensitive detection with miniaturized platforms. In fact, signal amplification assisted by nanomaterials is possible through three different mechanisms: (1) catalytic reaction, (2) their use as mediators, and (3) employing them as seeds to deposit electrochemically active species and electrocatalysts [77].

As an example of signal amplification through catalytic reaction, Jing et al. reported the use of palladium NPs as a catalyst for H_2O_2, enabling the construction of a sensitive electrochemical aptasensor [84]. More recently, carbon-based nanomaterials have been frequently used as signal amplifiers in electrochemical aptasensing, for they promote and facilitate electron transfer between the biomolecules and the electrode surface [85]. For instance, Goud et al. [36] described an AFB1 aptasensor, using a methylene blue-tagged aptamer serving as a signal moiety, and functional graphene oxide (fGO) acting as a signal amplifier (Figure 10). The fGO was not only used as a platform for the attachment of the aptamer, but also employed to intensify the DPV signal, due to its catalytic property towards MB. The plot of current signal versus analyte concentration showed excellent linearity, ranging from 0.05 to 6.0 ng mL^{-1}, with a very low limit of detection (0.05 ng mL^{-1}). Moreover, the aptasensor was applied to detect AFB1 levels in alcoholic beverages. In this study, fGO exhibited three major merits: (1) better loading of aptamers on the screen-printed carbon electrode, (2) enhanced conductive properties of the electrode surface, and (3) improved electrocatalytic properties of the transducer platform [86].

Furthermore, carbon-based NMs interact with other nanomaterials such as Au NPs, SiO_2, chitosan, etc., facilitating their widespread application in aptamer-based electrochemical bioassays [22]. Hence, several hybrid nanomaterials, such as Au NPs-carboxylic porous carbon (Au NPs-cPC) and Au NPs covalently bound to reduced graphene oxide (Au NPs-rGO), have been employed mainly in impedimetric aptasensors for mycotoxin detection. For instance, Wei et al. [87] reported an ultrasensitive nano-aptasensing platform for OTA measurement, based on Au NPs-cPC nanocomposite, through EIS. The results proved it to be a significant impedimetric amplification strategy for OTA detection, with an extremely low LOD of 1×10^{-8} ng mL^{-1}.

Alternatively, nanomaterial-modified electrode surfaces can be used as seeds to initiate the deposition of metals through a seed-mediated nucleation or

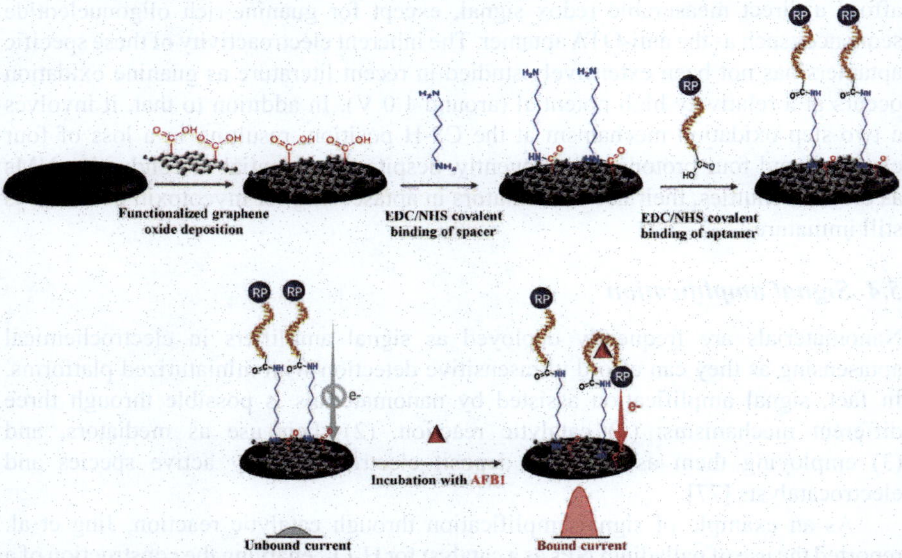

Figure 10. An electrochemical aptasensor based on functionalized-GO-assisted electrocatalytic signal amplification of methylene blue for aflatoxin B1 detection [36].

growth mechanism. The medium consists of metal salts and a reducing agent, and nanoparticles on the electrode surface serve as nucleation sites to facilitate metal deposition; this results in higher loading of electroactive species and subsequently improves the nanoparticle-transducer surface communication [88]. This improvement can be attributed to the fact that the deposited nanoparticles are closer to the electrode surface, compared to the nanoparticle labels [78]. So far, a large number of electrochemical immunosensors based on this method, with developer solutions, have been reported [77, 88]. However, it has not yet been noticeably extended to the design of aptasensors dedicated to mycotoxins.

5.5 Enzyme mimicking: Nanozymes

Most nanomaterials show intrinsic enzyme-like properties that can allow the development of enzyme-free biosensors to detect several target molecules. Compared to natural enzymes, these enzyme-like NMs, commonly called nanozymes, are stable against denaturing, low in cost and highly resistant to high concentrations of substrates [89]. Moreover, nanozymes exhibit environment-dependent properties such as changes in their activity at different pH conditions or with surface modifications. These tunable characteristics can result in changes in their affinities towards substrates, enabling the development of novel aptasensors for many targets [90]. Recently, a surge of nanozymes, including metallic, carbonaceous, and inorganic NPs, have been demonstrated to catalyze some typical enzymatic reactions, mainly mimicking oxidoreductase activities such as those of oxidase, peroxidase, and catalase [91].

It has been reported that peroxidase-like nanozymes, which are highly robust against environmental conditions, are widely used to replace horseradish peroxidase (HRP), providing a cost-effective method for the fabrication of biosensors. Taking advantage of the intrinsic peroxidase-like activity of Au NPs, Cao's study reported that positively charged Au NPs can catalyze the oxidation of TMB by H_2O_2 to produce a blue colored solution [90].

On the other hand, enzyme mimicking applications pose some regeneration problems. In fact, it is difficult to regenerate the surface of NPs and control the reactivity of nanomaterials against certain interfering molecules. Therefore, studies on design and development of selective and specific NPs to overcome matrix interferences are needed to replace enzymes for aptasensing applications.

6. Nanomaterial-based E-aptasensors for mycotoxin analysis

In the last decade, various aptasensors have been proposed for the analysis of mycotoxins by integrating several nanomaterials. Due to excellent characteristics like high surface area-to-volume ratio, these smart nanomaterials acquire inimitable visual/optical, mechanical, electronic, physical, magnetic, chemical and electrochemical properties. In recent times, researchers have synthesized nanomaterials of broad significance for diverse bioassays and bioanalysis. In particular, these materials have found growing applications in biosensor improvement and development, thanks to their exceptional electronic transportation properties. Along with different aptamers' selection, abundant nanomaterials have been synthesized for aptamer-related sensing applications, with the rapid expansion in nanotechnology.

In this section, we lay emphasis on universally used nanomaterials, such as silver and gold nanoparticles, carbon-based nanomaterials, and metal oxide nanoparticles, for sensing applications using electrochemistry. Examples of their applications in recent literature (2010 onwards) have been discussed herein, along with examples of novel design strategies and relevant electroanalytical techniques to highlight the current opportunities, bottlenecks and challenges, and future development trends.

6.1 Metal nanomaterials

The main reason for using noble metal nanoparticles in electrochemical aptasensors is to obtain higher sensitivity by virtue of their electrocatalytic and surface properties.

6.1.1 Silver nanoparticles

Silver nanoparticles (Ag NPs) have attracted growing attention of researchers for a variety of applications. Ag NPs have distinct chemical, physical and biological properties compared to their bulk paternal materials. Some Ag nanomaterials have been used for the detection of fungal toxins, using aptamer-based electrochemical sensors. OTA can be detected using electrochemical aptasensors, using a gold electrode coated with Ag NPs and neutral red by electro-polymerization [92]. In the referred study, aptamers were thiolated and coupled with Ag NPs. Based on their interaction with OTA, conformational adjustment occurred in the aptamer, which triggered an increase in the charge transfer resistance as analyzed by the impedimetric technique. The method was able to achieve a low LOD, down to 0.05 nM. The method was further tested in beer samples [92].

In another approach, Ag nanoparticles were coupled with ssDNA functionalized probes for quantitation of biomolecular targets using chronoamperometric technique. The method showed the detection of the aptamer target OTA in a specific binding event based on induced collision frequency changes and reached a detection limit down to 0.05 nM [93].

6.1.2 Gold nanoparticles

Gold nanoparticles have garnered significant recognition in the field of biosensor improvement over the past decade. This is primarily associated with the capability of Au NPs to offer a well-known coupling of biorecognition elements via their bonding to the metal surface by thiol groups. Additionally, Au NPs authorize direct electron transfer involving redox probes and electrodes, allowing electrochemical sensing without the addition of mediators. Several electrochemical aptasensors involving Au NPs for mycotoxin detection have been described in the literature. An impedimetric aptasensor for OTA has been developed on a gold electrode covered with Au NPs suspended in the dendritic polymer Boltorn H30® [94]. The aptamers were thiolated and covalently coupled to Au NPs through Au-S bonding. The mechanism relied on the conformational switch of the aptamer onto G-quadruplex form upon biosensing of OTA. This was followed by the consolidation of the surface layer and an increase in the charge transfer resistance. The LOD obtained was 0.02 nM. The practicality of the sensor was verified by testing beer samples [94].

In another work, a DPV-based aptasensor was developed for OTA detection. Therein, Au NPs were mixed with porous carbon and subsequnetly employed to modify an aminated Au electrode for the loading of cDNA and to improve electron transfer between the redox probe methylene blue and the sensor surface for sensitivity enhancement (Figure 11). The linear behavior in detection was found in the OTA concentration range of 5.0×10^{-6}–5.0×10^{-4} ng mL^{-1}, with good reproducibility and stability [95].

Researchers have also proposed another rapid and ultrasensitive electrochemical detection method for OTA. Therein, ssDNA with the aptamer was co-immobilized on the electrode and methylene blue was used as a redox probe for OTA detection. The electrochemical signal was improved upon the addition of Au NPs-coupled DNA. In this study, the aptamer changed its conformation, triggered by the interaction with OTA, and an LOD of 30 pg mL^{-1} was achieved [96].

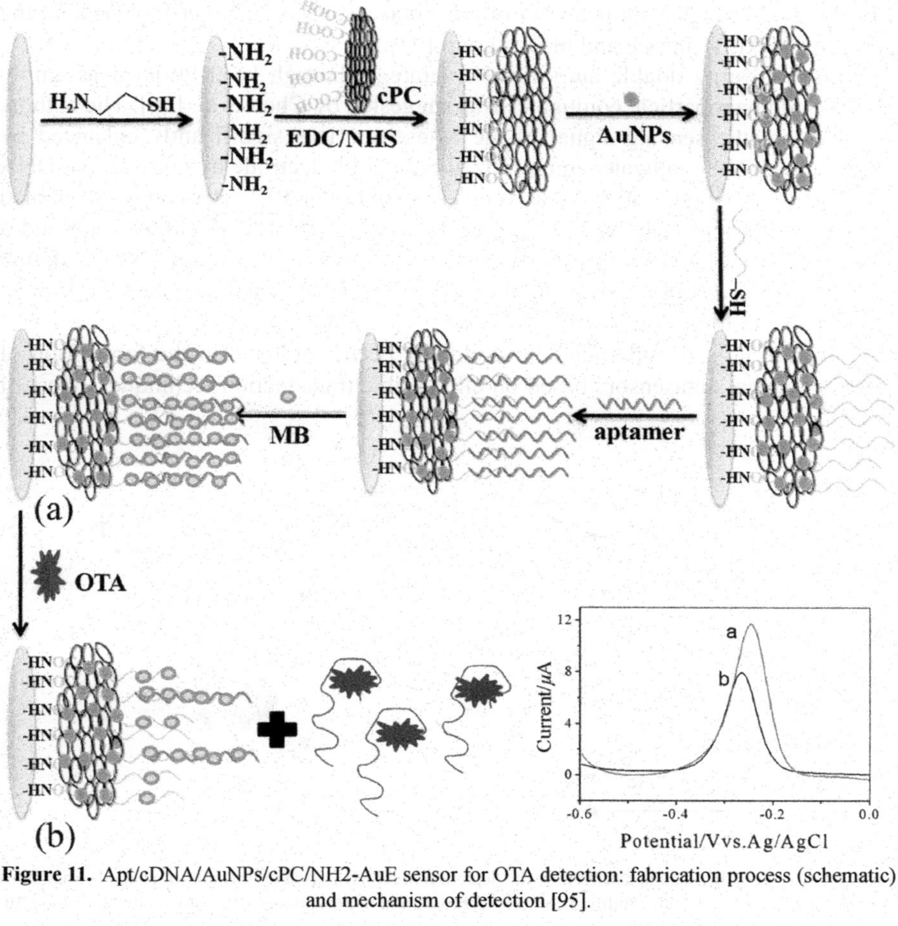

Figure 11. Apt/cDNA/AuNPs/cPC/NH2-AuE sensor for OTA detection: fabrication process (schematic) and mechanism of detection [95].

Novel practices for sensitive and convenient techniques are regularly needed to detect fungal toxins at lower concentration levels. In another strategy, a hairpin-shaped aptamer of AFM1 was deployed to detect AFM1, using Au NPs and a complementary aptamer strand. The detection relied on the conformational changes in the aptamers – the hairpin structure in the absence or presence of AFM1 (Figure 12). The negatively charged Au NPs facilitated AFM1 quantitation, with added advantages of better selectivity and sensitivity. In the absence of AFM1, the hairpin structure was intact and hence less current was recorded; whereas, in the presence of AFM1, disintegration of the aptamers' hairpin structure was suspected, which yielded a higher signal when methylene blue was added as a redox agent; an LOD of 0.9 ng/L was obtained. The method was extended to the detection of AFM1 in milk and serum samples [79].

When it comes to other gold nanomaterials, an aptasensor has been developed for AFB1 detection using gold nanorods as well [97]. The sensing method uses gold nanorods as the transducer, and the aptamer as a biorecognition element. The sensing is done through DPV and the LOD obtained is 0.3 pM in the tested range of 1.0 pM–0.25 nM. The applicability of this aptasensor has been verified through the detection of AFB1 in rice and blood serum [97].

An interesting double amplification strategy through layer-by-layer assembly of dual gold nanoparticle conjugates was suggested by Chen et al. [56]. The authors claimed that the sensing signal of the aptasensor was significantly enhanced by two-round AuNP conjugates immobilized on a gold electrode by consecutive DNA hybridizations (Figure 13). As a result, a signal-on DPV detection was enabled by the second ferrocene-tagged nanoprobe, reaching an LOD of 0.001 ppb and a wide dynamic range of 0.001–500 ppb over 6 orders of magnitude. For validation purposes, a real sample analysis for the detection of OTA in spiked wine samples was performed and found successful.

Overall, gold nanoparticles have already been employed in the fabrication of electrochemical aptasensors based on all possible transduction techniques. Most of

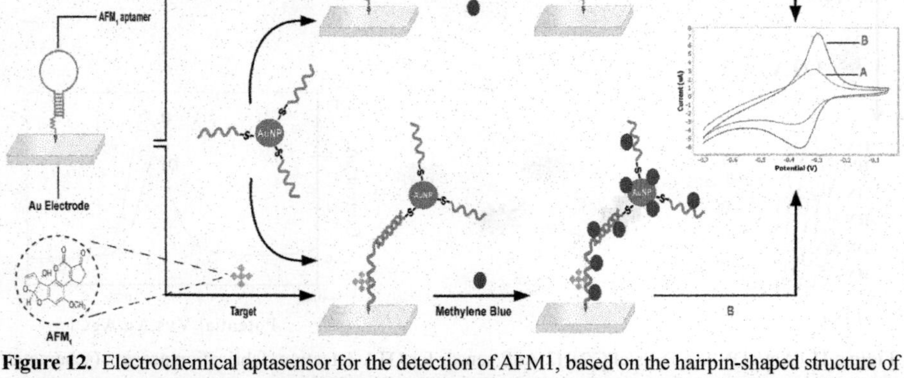

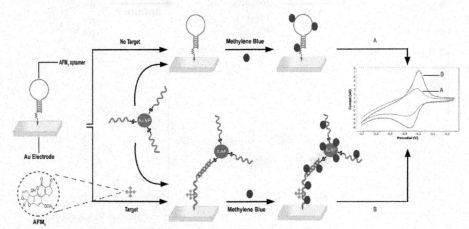

Figure 12. Electrochemical aptasensor for the detection of AFM1, based on the hairpin-shaped structure of AFM1 aptamer, AuNPs, thiol-modified complementary strand of the aptamer (CS), and methylene blue [79].

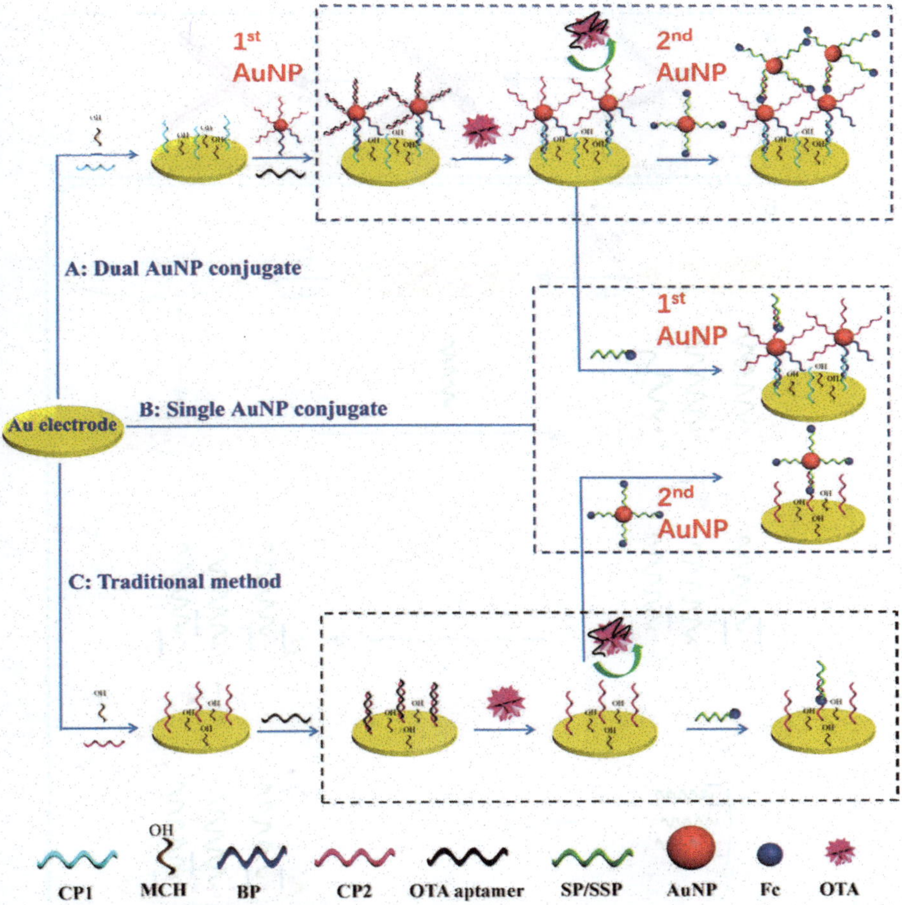

Figure 13. Electrochemical aptasensors for the amplified assay of OTA, based on: (A) dual AuNP conjugates, (B) single AuNP conjugates, and (C) traditional method (no AuNPs) [56].

the electrochemical aptasensors using Au NPs have been used for OTA detection (Table 3).

6.1.3 Other metal nanoparticles

Apart from Au and Ag, some attempts have been made to utilize other metallic nanoparticles in the aptameric analysis mycotoxins, For instance, porous platinum nanotubes have been used for the fabrication of zearalenone (ZEN) aptasensor, deploying the voltammetric technique; however, in the development of this sensor, Au NPs with thionine-coupled GO were also used (Figure 14). Aptamers were thiol-modified and self-assembled on Au NPs previously electrodeposited on the electrode. The obtained LOD for ZEN was 0.17 pg mL^{-1} [98]. Furthermore, Gu et al. recently described an aptasensor assisted by copper nanoclusters for the detection of OTA, wherein metal nanoclusters showed peroxidase-like activity to catalyze hydrogen peroxide reduction [75].

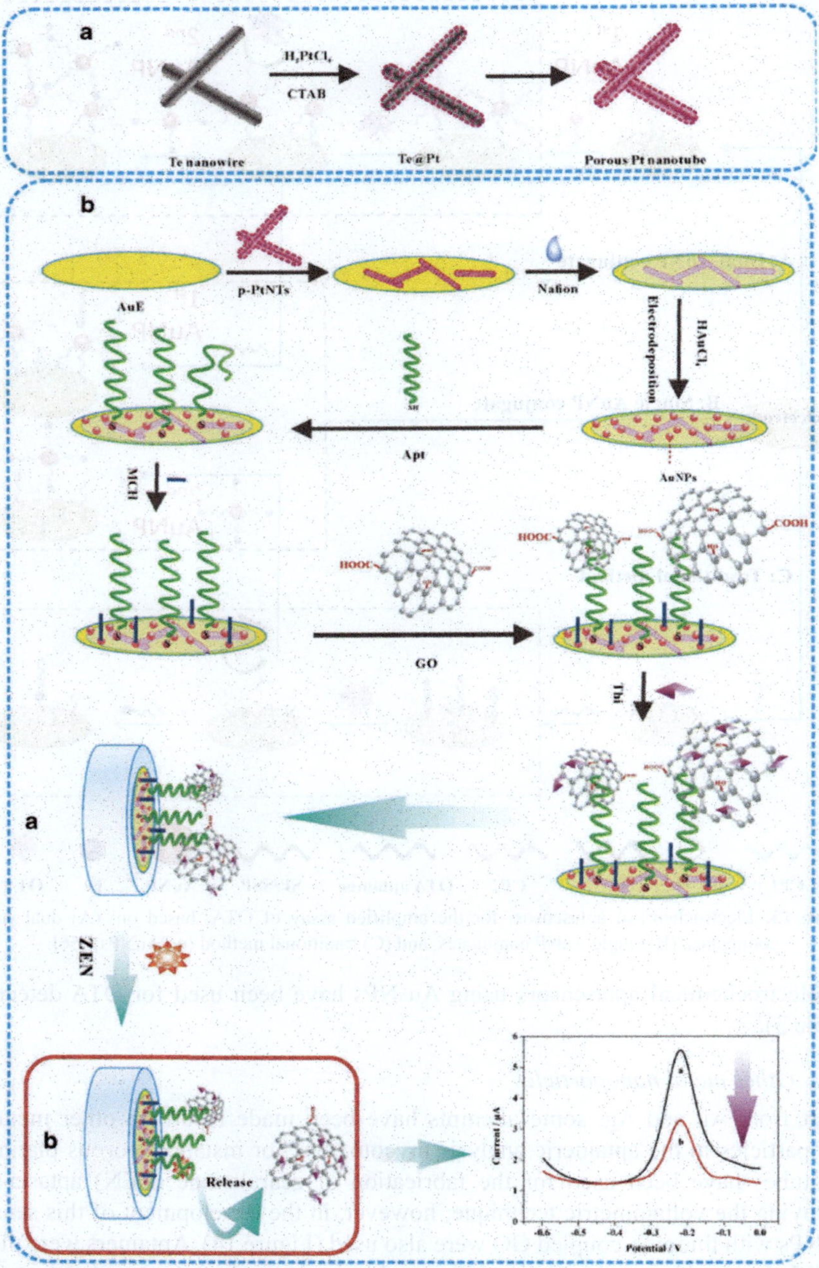

Figure 14. PtNTs-based aptasensor for the detection of OTA: (a) preparation of p-PtNTs, and (b) construction of the aptasensor [98].

Table 3 summarizes recent E-nano-aptasensors for mycotoxins, classified according to the metal nanomaterial used. It provides an overview of their main roles

Table 3. Electrochemical nano-aptasensors for mycotoxin detection, based on metal nanomaterials (2010–2019).

NM	Aptasensor Assembly	Target	Detection Technique	Role of NMs	Conc. Range (ng mL^{-1})	LOD (ng mL^{-1})	Real Applications	Ref.
Gold nanoparticles (Au NPs)	GCE/DNA$_1$/Aptamer/DNA$_2$-**AuNPs**/MB	OTA	CV	cDNA labels for signal enhancement	0.10–20	3.00E-02	Red wine	[96]
	AuE/poly-NR/ Botlorn H30® polymer@**AuNPs**/Aptamer	OTA	EIS	Thiolated aptamer carrier	0.04–40	8.10E-03	Light and dark beer samples	[94]
	AuE/DNA$_1$-MCH/Aptamer/DNA$_3$-**AuNPs**/DNA$_4$/MB	OTA	DPV	DNA label as first-level signal enhancer	0.001–1	0.0003 (± 0.00005)	Red wine	[99]
	GCE/*in situ* **AuNPs**/Aptamer	FB1	EIS	Aptamer loading increase	0.03–32.8	1.40E-03	Maize	[100]
	AuE/Aptamer-MCH/ TS-**AuNPs**-cDNA/ telomerase amplification/ MB	AFB1	SWV	cDNA carrier for sensitivity enhancement	1.0E-7–0.1	6.00E-11	Corn	[101]
	AuE /cPC/**AuNPs**/cDNA/ Aptamer/ MB	OTA	DPV	Aptamer immobilization and signal amplification	5.0E-06–5.0E-04	5.00E-06	Corn	[87]
	AuE/cDNA$_1$/MCH/**AuNPs**@ Aptamer/OTA/**AuNPs**@cDNA$_2$-Fc	OTA	DPV	Dual signal amplification	0.001–500	1.00E-03	Wine samples	[56]
	SPGE/Aptamer/**AuNPs**-cDNA/MB	AFM1	DPV	Signal feasibility and enhancement	0.002–0.6	9.00E-04	Human blood serum, milk	[79]
	AuNPs@CuCoPBA/ Aptamer	OTA	EIS	Aptamer immobilization	50–10	5.20E-06	Watermelon juice	[75]
	AuE-Cys/**AuNPs**/Aptamer/BSA	OTA	EIS	Impedimetric signal enhancement	0.10–10	3.00E-02	Grape commodities (Grape, red wine, white wine, red and purple grape juice)	[102]

Table 3 contd. ...

...Table 3 contd.

NM	Aptasensor Assembly	Target	Detection Technique	Role of NMs	Conc. Range (ng mL^{-1})	LOD (ng mL^{-1})	Real Applications	Ref.
Silver nanoparticles (Ag NPs)	AuE/poly-Neutral Red/**AgNPs**@Aptamer	OTA	EIS	Aptamer carriers	0.12–12.11	2.00E-02	Beer	[94]
	AgNPs/Aptamer coils	OTA	CA	Signal generation through collision's electrochemistry	0.03–4	2.00E-02	–	[93]
Copper nanoclusters (Cu NCs)	AuE-MCH-DNA$_1$ (SAM)/Aptamer/**CuNCs**	OTA	CA	Peroxidase mimicking	–	1.10E-01	Red wine	[103]
Platinum nanotubes (Pt NTs)	AuE/**porous-Pt NTs** /Nafion/**AuNPs**/Apt/MCH/**GO**/Thionine	ZEA	DPV	Signal enhancement	5.0E-04–500	1.70E-04	Maize	[98]

in the assembly of such mycotoxin platforms, the analytical performance obtained, and their potential applications in real sample analysis.

6.2 Aptasensors based on metal-oxides

A great number of nano-biosensors developed so far are based on nanostructured oxides of metals such as zinc, iron, cerium, iridium, titanium, etc. Metal oxide nanomaterials have been found to exhibit interesting electronic, functional, biocompatible, non-toxic and catalytic properties. Various morphologies of these NMs, including nanoparticles, nanorods and nanofibers with controlled shapes, have been obtained using a variety of methods [104].

Ferrimagnetic nanoparticles are particularly interesting and have been used in conjunction with different bioreceptors, thanks to their electron shuttler role and easy recycling by magnetic separation. Nguyen and co-workers fabricated an electrochemical Fe_3O_4/polyaniline-based aptasensor capable of detecting AFM1 in the 6–60 ng/L concentration range, with a detection limit of 1.98 ng/L [40]. This aptasensor was based on the polymerization of Fe_3O_4-incorporated polyaniline (Fe_3O_4/PANi) on an inter-digitized electrode (IDE) as a sensitive film on which the aptamer was immobilized. Magnetic nanoparticles were used as the signal amplification element. AFM1 was detected by following the changes in the electrochemical signal through cyclic and square wave voltammetry.

Zinc oxide nanostructures seem to have gained increasing popularity in the field of biosensing of mycotoxins [105]. In one detection method for PAT, a composite material of ZnO nanorods and chitosan was used to modify a gold electrode. The PAT aptamer was modified by thiol, and self-assembled on the electrode by electrodeposition of Au NPs. The ZnO nanorods increased the stacking with Au NPs and aptamers. When PAT is available in the sample, it forms a complex with the aptamer on the electrode surface and may hamper the transfer of electrons from the electrode to the redox probe hexaferrocyanate, resulting in lower current values. The obtained signal from dosing of PAT can be calculated through the differences in the signal before and after incubation with PAT. The assay showed an LOD of 0.27 pg mL^{-1}. This method was successfully tested for the detection of PAT in juice samples [59].

In the same context, a novel and sensitive electrochemical aptasensor was developed for the same patulin detection through a hierarchical metal organic framework (MOF) and methylene blue as the redox mediator connected to the aptamer as a probe. The aptasensor was arranged through electrodeposited Au NPs on a ZnO-chitosan based electrode. Methylene blue was again used as a redox probe for signal generation. Au NPs coupled with ZnO are likely to load better signal tags and hence achieved signal amplification. The detection was based on structural changes in aptamers resulting from PAT association. The obtained LOD was 1.46×10^{-8} μgmL^{-1}. The developed sensor demonstrated good stability, satisfactory reproducibility and selectivity. It also showed consistent recovery of PAT in spiked apple juice [106].

Additional examples of aptasensing strategies for mycotoxins, based on metal oxides, have been presented in Table 4. It should be mentioned that despite their

Table 4. Electrochemical nano-aptasensors for mycotoxin detection, based on metal oxide nanomaterials (2010–2019).

NM	**Aptasensor Assembly**	**Target**	**Detection Technique**	**Role of NMs**	**Conc. Range ($ng\ mL^{-1}$)**	**LOD ($ng\ mL^{-1}$)**	**Real Applications**	**Ref.**
Fe_3O_4 NPs	Pt (interdigitated electrode array IDA)/ **Fe_3O_4 NPs**-PANI/ glutaraldehyde/ Aptamer	AFM1	CV & SWV	Signal amplification	0.006–0.06	1.98E-03	Milk	[40]
	SPCE/PDMS/Fe3O4@AuNPs-Apt/ MCH	AFB1	EIS	Aptamer carrier	0.020–50	1.50E-02	Peanut	[108]
IrO_2 NPs	SPCE/polythionine/**IrO_2NPs**/Aptamer	OTA	EIS	Aptamer carrier	0.01–100 nM	5.65E-03	White wine	[66]
ZnO (nanorods and nanoflowers)	AuE/**ZnONRs**-Chitosan/*in situ* **AuNPs**/Aptamer/MCH	PAT	DPV	loading increase of AuNPs and aptamers	0.0005–50	2.70E-04	Apple juice	[59]
	MB@MOF-Apt/MCH/cDNA/AuNPs/ CS-**ZnO NFs**/AuE	PAT	DPV	Signal amplification through tags loading	5.0E-05–500	1.46E-05	Apple juice	[106]

particular physicochemical properties suitable to small targets [107], manganese dioxide nanosheets (MnO_2) have been surprisingly yet not explored in the electrochemical analysis of mycotoxins; optical detection schemes using the same nucleic bioreceptors are rather preferred.

6.3 Carbon nanomaterials

Carbon nanostructures, such as graphene and its derivatives, carbon nanotubes, carbon nanoparticles and carbon black are good candidates for constructing electrochemical sensors and biosensors – they possess high chemical stability, great mechanical strength and high surface-to-volume ratio, while exhibiting enhanced electron transfer and high resistance to fouling. Recent achievements with carbon nanomaterials used in electrochemical aptasensors for food analysis have been reviewed in detail in the following sections [85]. Furthermore, nano-aptasensors based on carbon NMs have been summarized in Table 5.

6.3.1 Graphene and its derivatives

The current progress with graphene (G) has revealed a capable material with greater sensing properties. Described as a carbonaceous honeycomb, graphene is an appropriate substrate for sensor applications due to its outstanding structural, biochemical and electrical properties. Graphene and its derivatives such as graphene oxide (GO) and reduced graphene oxide (rGO) are the most widely used carbon-based nanomaterials in electrochemical biosensing.

In one approach, GO was integrated with carbon-based printed electrodes with redox probe methylene blue and aptamers [36]. This sensor was used to detect AFB1 through conformational changes in the aptamer structure. Due to the catalytic behavior of GO towards the redox probe, the proposed sensor could detect AFB1 with an LOD 0.05 ng mL^{-1} [36]. A combined and hybrid nano-catalyst-based approach has also been tested using nanoceria and graphene oxide [72]. The aptamer was coupled with GO loaded onto the electrode, and the target toxin was trapped by the aptamer between the free and the ceria-tagged target, through specific aggressive means. The electrochemical response was obtained by analyzing the redox probe's interaction with the ceria tag and the response was improved by GO, which amplified the electron transfer at the electrode (Figure 15). This novel approach offered an outstanding platform to detect mycotoxins in a wide spectrum, with an LOD of 0.1 nM for OTA [72].

In another study, a graphene-sheltered aptamer with thionine label was used to develop an electrochemical aptasensor for OTA [109]. The aptamer was attached to the graphene nanosheet with strong non-covalent bonding of GO, with DNase-I-based signal amplification as the recognition element. The developed aptasensor platform could detect OTA in concentration levels down to 5.6 pg mL^{-1} [109]. In yet another interesting approach, Au NPs were attached to 2-aminothiophenol-functionalized reduced graphene oxide, and it was observed that the surface density of the Au particles played a major role in aptamer immobilization and sensitivity improvement [110]. Additionally, it helped in the development of a signal-enlarged platform for sensitive and specific screening of OTA using EIS. The developed

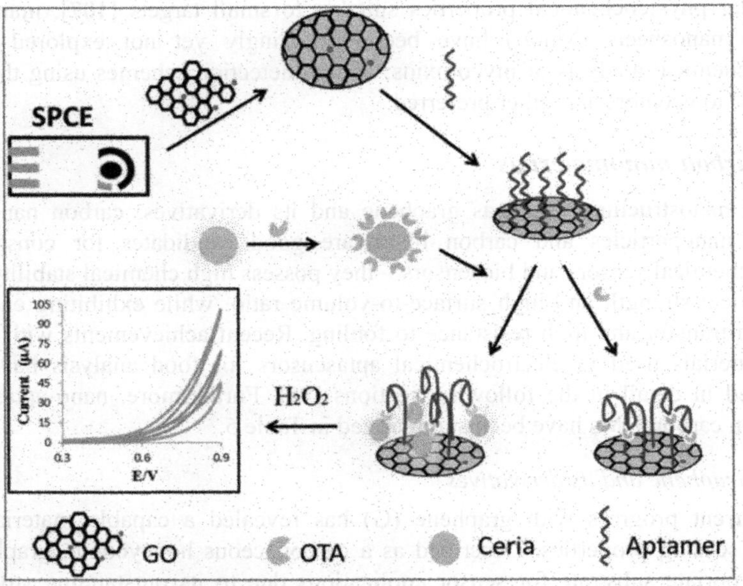

Figure 15. Non-enzymatic nanocatalyst-based electrochemical aptasensor using a nano-ceria tag and GO (schematic) [72].

sensor showed a dynamic detection range of 0.1–200 ng mL^{-1}, with an LOD of 0.03 ng mL^{-1} [110].

An aptasensor for AFB1 detection has also been proposed using anodized and porous alumina altered with graphene oxide [111]. Herein, the aptamers were covalently immobilized on alumina, and GO was coupled on the electrode surface through π-π stacking with the aptamer. The detection of the fungal toxin relied on the negative charge of GO and the aptamer. The achieved LOD was 0.13 ng mL^{-1} for AFB1. Also, the developed sensor showed good selectivity towards AFB1, offering potential for a sensitive, simple and effective method for detecting mycotoxins [111].

Geleta and colleagues reported an aptasensor based on reduced graphene oxide/molybdenum disulfide/polyaniline nanocomposite for the detection of AFB1 [112]. The rGO $MoSe_2$-PANI nanohybrids were prepared and entrapped onto glassy carbon electrodes, using chitosan. AuNPs were also added to the system to scaffold the thiolated aptamers. The aptamer-AFB1 binding event was monitored by the peak current through DPV measurement in the presence of $[Fe(CN)_6]^{3-/4-}$. The nanoplatform exhibited a wide linear detection range and a remarkably low detection limit: 0.01 fg mL^{-1} to 1.0 fg mL^{-1}, and 0.002 fg mL^{-1} respectively [112]. Overall, the synergistic interaction of AuNPs and the hybrid nanocomposite increased the active surface area and improved the electrochemical output signal of the aptasensor.

In a more simplistic platform, an electrochemical aptasensor for OTA was proposed by utilizing inherently electroactive graphene oxide nanoplatelets (GONPls) as redox labels [113]. In this case, anti-OTA aptamers were physically adsorbed onto disposable electrically printed carbon electrodes before exposure to mycotoxin.

Subsequently, incubation with GONPls served to quantify the remaining non-bound aptamers through π-π interactions (Figure 16). The proposed procedure exhibited a DPV detection range of 0.31–310 pM, and a good discrimination against ascorbic acid [113]. This was recommended as a valid signal-generating nanoprobe with all aptasensors. Nonetheless, it would require well-characterized and reproducible GO preparations.

It is worth noting that graphene structures doped with p-block elements such as nitrogen, boron, sulfur, hydrogen, oxygen and/or fluorine are widely used in electrochemical sensors and biosensors. However, research targeting mycotoxins, using aptamers, has not reported their extensive use so far.

6.3.2 Carbon nanotubes

Carbon nanotubes (CNTs) are allotropes of carbon that take the form of hollow seamless cylinders with a diameter of less than 1 nm and lengths of up to several micrometers. Researchers have shown tremendous interest in CNTs due to their unique and outstanding properties of large surface area, stability, biocompatible nature, and high electrical conductivity. Although CNTs were discovered over a decade before graphene, they are also considered as rolled-up graphene derivatives, as in single-walled carbon nanotubes (SWCNTs) and multi-walled carbon nanotubes (MWCNTs). CNTs have shown various advantages when coupled with electrochemical aptasensors, such as rapid signal response, high level of sensitivity, ease of operation, cost-effectiveness, and simplicity.

An amperometric aptamer-based biosensor has been described in the literature for OTA detection, using methylene redox probe. The developed aptasensor consisted of a gold electrode coupled with aptamers, and carbon nanotubes were used as the signal amplifier. The developed aptasensor was highly selective towards OTA and reached to an LOD of 52 pM. Subsequently, OTA detection was performed in serum and grape juice samples [61].

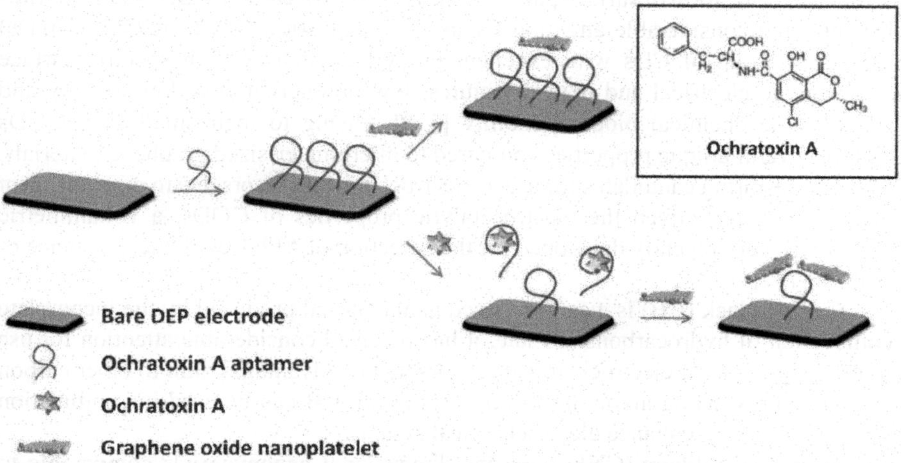

Figure 16. Application of GONPs as inherently electroactive labels for the aptasensing of OTA (schematic) [113].

More recently, a label-free aptasensor based on polyethyleneimine (PEI) functionalized with MoS_2-doped multi-walled carbon nanotubes was successfully fabricated for zearalenone detection [114]. The nanocomposite PEI-MoS_2-MWCNTs was reported as a novel nanocarrier with enhanced adsorption ability to assemble a large amount of the signal probe toluidine blue (Tb). Furthermore, platinum/gold core-shell (Pt@Au) nanoparticles were dropped on the modified electrodes to load a large amount of ZEN-binding aptamers. This resulted in a stable variation of CV responses, realizing the quantitative determination of zearalenone in the range of 0.5 pg mL^{-1}–50 ng mL^{-1}, with an LOD of 0.17 pg mL^{-1}.

To the best of the authors' knowledge, only a few reported examples describe the integration of CNTs in electrochemical aptasensors towards the detection of mycotoxins (see Table 5). Recent references suggest that interests in CNTs for the E-aptasensing of mycotoxins are still modest, against the many more detection attempts using CNT-assisted immunosensors. This is mainly due to the more convenient production methods of 2D nanostructured electrodes based on graphene and its derivatives.

6.3.3 Other carbon nanomaterials

Although graphene and its derivatives are well known in the field of biosensors, other carbonaceous nanomaterials such as carbon nanospheres and carbon nanofibers have been recently introduced in the field of electrochemical biosensing [115]. However, such a variety of carbon NMs are rarely employed in electrochemical aptasensors; even when they are, the applications are targeted towards the biomedical field rather than food analysis. Still, it can be envisaged that the developed concepts will be translated soon to the analysis of mycotoxins. To cite their use, a brief description of some emerging carbonaceous nanomaterials showing distinctive properties has been provided hereunder.

Carbon dots (CDs) are small-sized carbon NPs with a diameter of less than 10 nm and some form of surface passivation. This class of carbon-based nanomaterials has attracted considerable attention for use in diagnostics. Graphene QDs (GQDs) are a sub-group of CDs with excellent properties such as high specific surface area, unique electrical and optical features, low production cost, low toxicity and considerable chemical biocompatibility [116]. Owing to their small size, GQDs exhibit several unique properties compared to micrometer-sized graphene materials. They are inspiring candidates for use in electrochemical sensors due to more efficient electron transfer. Given the electrocatalytic properties of GQDs, a voltammetric sensor was very recently developed for the detection of AFB1, with the assistance of AuNPs [117].

Carbon black (CB) is a carbon-based nanomaterial produced by the incomplete combustion of hydrocarbons. Of late, it has received considerable attention for use in the fabrication of electrochemical sensors due to advantages akin to other carbon nanostructures, while involving lower costs [118]. CB can be used in combination with metal NPs to enhance electrochemical signals.

Carbon nanofibers (CNFs) are one-dimensional nanomaterials comparable to CNTs. Despite similarities in their cylindrical structures, CNTs are distinguished

Table 5. Electrochemical nano-aptasensors for mycotoxin detection, based on carbon nanomaterials (2010–2019).

NM	Aptasensor Assembly	Target	Detection Technique	Role of NMs	Conc. Range (ng mL^{-1})	LOD (ng mL^{-1})	Real Applications	Ref.
Graphene Oxide (GO)	SPCE/functionalized **GO**/HMDA/Aptamer-MB	AFB1	DPV	MB signal amplification via electron transfer improvement	0.05–6.0	5.00E-02	Beer, wine	[36]
	Au film electrode/porous anodized alumina nanochannels/Aptamer/**GO** stacking/AFB1	AFB1	CA	GO for steric and electronic hindrance E-detection feasibility	1–20	1.30E-01	–	[111]]
	Negatively charged SPCE/**GO**@Thionine-modified aptamer/DNaseI	OTA	DPV	Aptamer immobilization through stacking	0.01–100	5.60E-03	Wheat samples	[109]
GO nanoplatelets (GONPs)	SPCE/Aptamer/OTA/**GONPs**	OTA	DPV	Electroactive labels	0.00013–0.125	1.30E-04	–	[113]
Reduced Graphene Oxide (rGO)	GCE/aminocaproic acid/Aptamer/**rGO**	AFB1	DPV	electron transfer enhancement	0.16–1.25	2.19E-02	Pasteurized cow milk and human blood plasma	[121]
rGO and Au NPs	AuE/DNA_1-MCH/Aptamer/**AuNPs–rGO**-DNA_2	OTA	EIS	Signal amplifier via sandwich competitive assay	0.001–50	3.00E-04	Red wine	[122]
	AuE-Cys/ **rGO**-ATP-**AuNPs**/Aptamer	OTA	EIS	Signal amplification via loading capacity enhancement	0.1–200	3.00E-02	Red wine	[110]

Table 5 contd. ...

...Table 5 contd.

NM	Aptasensor Assembly	Target	Detection Technique	Role of NMs	Conc. Range (ng mL^{-1})	LOD (ng mL^{-1})	Real Applications	Ref.
Graphene Oxide and nanoceria (GO & nCe)	SPCE/**GO**/Aptamer/OTA-**nCe**	OTA	CV	GO for covalent aptamer immobilization nCe tag acts as nanocatalyst for enzyme mimicking	0.06–72.7	4.00E-02	Corn samples	[72]
Graphene & Au NPs	GCE/**AuNPs**/cDNA/ Aptamer/ **Graphene**@ thionine	FB1	CV	AuNPs for aptamer immobilization and G@ thionine nanocomposite as signal generator	0.001–0.1	1.00E-03	Wheat	[123]
Single-Walled Carbon Nanotubes (SWCNTs)	SPGE/DNA1/Aptamer/ ADN2-**SWCNTs conjugate**/ Methylene blue (MB)	OTA	DPV	Signal amplifier for an ON/Off response	0.07–18.2	2.10E-02 (Buffer) 0.054 (serum) 0.023 (grape)	Serum and grape juice	[61]
Multi-Walled Carbon Nanotubes (MWCNTs)	PEI-MoS_2-MWCNTs	ZEA	CV	Sensitivity enhancement	0.0005–50	1.70E-04	Beer	[114]

by a larger diameter and length, hollow cavities and diameter tunability. Lower diameters of CNFs enable the acceleration of electron transfer in only one direction. This can result in enhancing conductivity and decreasing the detection time [119]. Furthermore, mesoporous carbon (carbon materials with pore sizes between 2 and 50 nm) has shown a great ability towards the detection of small molecules as reported in latest electrochemical investigations [120].

6.4 Other nano-aptasensing platforms

Quantum dots have lately witnessed a rise as potential agents for assisting electrochemical aptasensing of small molecules. QDs consist of small semiconductor particles (with diameters in the range of 2–10 nm) of the elements from II–VI, III–V and IV–VI groups. They have unique optical and electronic characteristics. When incorporated into the design of a biosensor, they have three key functions: (1) act as aptamer labels, (2) be a part of the signal transducer, and (3) serve as the scaffold for bioreceptors. For example, Hao et al. described an ultrasensitive electrochemical OTA aptasensor based on the interaction of two fabricated nanocomposites: CdTe QDs-modified graphene/AuNPs and AuNP-functionalized silica-coated Fe_3O_4 NPs (Figure 17) [124]. Anti-OTA aptamers acted as linkers between the nanocomposites, previously decorated with two complementary ssDNA. Due to the high affinity between targets and aptamers, this structure would break into two parts in the

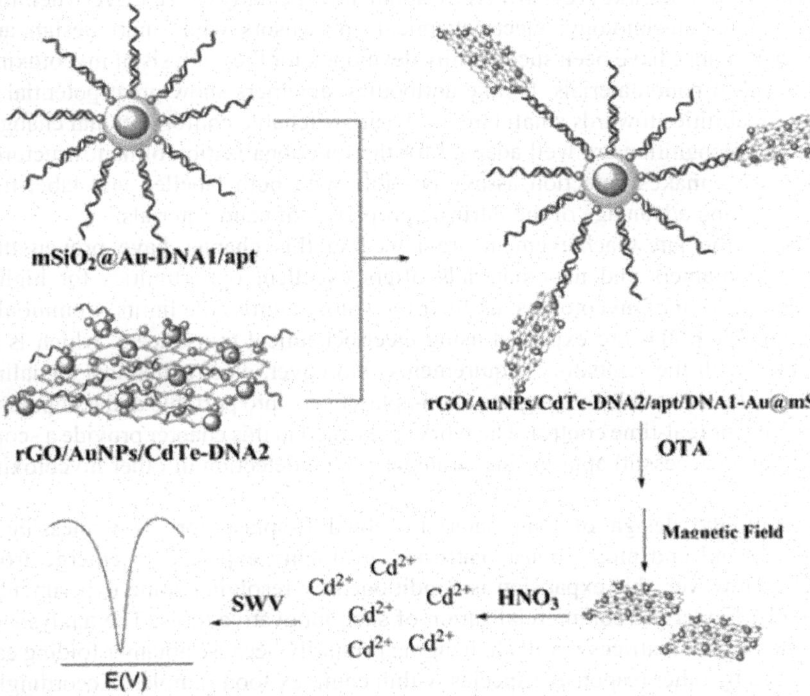

Figure 17. Aptasensor fabrication and the principle for OTA detection, using CdTe-based nanocomposite [124].

presence of OTA. Therefore, iron-based nanostructures were discarded through magnetic separation. Thereafter, CdTe QDs remaining in the supernatant liquid were dissolved in HNO_3 and the concentration of Cd^{2+} ions was detected by SWV; the concentration of Cd^{2+} ions was directly proportional to the concentration of OTA. The aptasensor exhibited a wide linear detection range of 0.2 pg mL^{-1}–4 ng mL^{-1}, and a low detection limit of 0.07 pg mL^{-1}[124].

With a methodology akin to the above, a magneto-controlled detection approach was developed for OTA and fumonisin B1 (FB1), using an electrochemical aptasensor. Therein, silica coated with CdTe/PbS QDs was utilized as the label and c-DNA was coupled with magnetic beads like a capture probe. When the target mycotoxin was available, the aptamers created the analyte binding which triggered the partial release of preloaded tags from the magnetic beads. The developed aptasensor showed detection in a wide range of 10 pg mL^{-1} to 10 ng mL^{-1} for OTA and 50 pg mL^{-1} to 50 ng mL^{-1} for FB1. Moreover, the sensor was tested in real maize samples. This method can be extended to the detection of other fungal toxins, using metal sulfide QDs, subject to the availability of specific aptamers [80].

7. Conclusion and prospects

Widespread mycotoxin contamination of food and animal feed has led researchers to investigate rapid and reliable biosensing tools for food safety monitoring. Aptasensors are constantly playing a key role in this development, especially with the overwhelming progress in nanotechnology; electrochemical aptasensors with smart design and better performance have been successfully developed for the analysis of mycotoxins, with the aid of nanomaterials. Unlike antibodies, aptamers show great potential in binding capabilities towards small targets. Their switchable conformational changes upon target recognition are well adapted for the functionalization of nanostructured supports. This makes detection assays possible with both labelled and label-free aptamers, taking advantage of the intrinsic properties of nanomaterials.

Indeed, the latest achievements reviewed in this chapter have proven that combining aptamers and nanomaterials offers excellent opportunities for highly sensitive detection of mycotoxins, on their regulatory admissible limits. Commonly, sub-nanomolar LODs are expected using electrochemical transducers, which is in accordance with the regulatory requirements. Moreover, electrochemical signaling methods make it possible to integrate nano-aptasensors into portable and equipment-free devices for real-time control. The works presented in this chapter provide a scope for extending successful aptasensing strategies to the detection of other mycotoxins as well.

Overall, the design of nanomaterial-enabled E-aptasensors is witnessing a promising trend and more robust platforms are highly expected to emerge from this field. However, this expansion is conditioned by resolving some experimental hurdles that hinder the commercialization of such innovative tools. The analysis of food toxins in real matrices is still challenging primarily because aptamer folding can be triggered by other responsive species within complex food samples. Accordingly, most of the successful applications are works of proof-of-concept, restricted to the laboratory scale and yet to reach the end users. These nano-aptasensors are usually validated by testing pretreated and buffered samples, then spiked with analytical-grade toxins. Therefore, we are still far from knowing as to when and where these novel tools will noticeably replace classic analytical tools in food analysis.

Although aptamers are now well established for some major mycotoxins (OTA and AFB1), they are yet to be developed for the remaining panel of such pathogens. Most of the aptamers targeting other mycotoxins were duly noted only during the last five years or so. In all probability, their scarce applications are not only related to their short history, but also to the difficulties in K_D determination for low-weight molecules.

Additional efforts should be deployed to study the relevant binding sites of mycotoxins among DNA aptamer strands. Considering the small size of targets, a few nucleotides can be effectively responsible for molecular recognition. Therefore, truncation of long sequences, mostly exceeding 50 nucleic bases, would probably decrease cross-reactivity and further reduce costs. New concepts integrating other novel bioreceptors with aptamers can also be explored. This is even recommended to overcome the lack of specificity and to increase the precision and sensitivity of the assay.

On the other hand, the current state-of-the-art of aptamer-conjugated nanomaterials is yet to overcome difficulties in bioconjugation chemistry, the effect of non-uniform nanostructures and the lack of functional moieties in some nanomaterials. To achieve this, researchers are rapidly improving synthesis protocols; new nanomaterials should be fully characterized before use in biosensing platforms in order to maximize their advantages. Furthermore, the toxicity of the concerned nanomaterials towards biomolecules should be also considered for preserving the assay's performance and reproducibility.

Moreover, the limited number of multiplexed aptasensors indicates that there is an urgent need to research and develop nanoplatforms for the simultaneous detection of a series of mycotoxins which can eventually occur in the same food commodity. Despite the available aptamers, multi-mycotoxin aptasensing is still immature compared to other targets. The development of microfluidic devices or lab-on-chips for rapid, sensitive, and *in-situ* analysis of mycotoxins should be particularly accelerated.

Finally, we believe that research in nanomaterial-enabled aptasensors for the detection of mycotoxins will still witness a promising trend and more robust nanosensors are highly likely and expected to result in breakthrough approaches, solving the above-mentioned limitations in this field.

References

[1] Chauhan, R., J. Singh, T. Sachdev, T. Basu and B.D. Malhotra. 2016, July. Recent advances in mycotoxins detection. Biosens. Bioelectron 81: 532–545. Doi: 10.1016/j.bios.2016.03.004.

[2] Bennett, J.W. and M. Klich. 2003, July. Mycotoxins. Clin. Microbiol. Rev. 16(3): 497–516. Doi: 10.1128/CMR.16.3.497-516.2003.

[3] Omotayo, O.P., A.O. Omotayo, M. Mwanza and O.O. Babalola. 2019, Jan. Prevalence of mycotoxins and their consequences on human health. Toxicol. Res. 35(1): 1–7. Doi: 10.5487/TR.2019.35.1.001.

[4] Park, D.L., H. Njapau and E. Boutrif. 1999. Minimizing risks posed by mycotoxins utilizing the HACCP concept. Food, Nutrition and Agriculture 23(23): 49–55 [Online]. Available: http://www.fao.org/docrep/X2100T/x2100t08.htm#TopOfPage.
[5] Eskola, M., G. Kos, C.T. Elliott, J. Hajšlová, S. Mayar and R. Krska. 2020, Sep. Worldwide contamination of food-crops with mycotoxins: Validity of the widely cited 'FAO estimate' of 25%. Crit. Rev. Food Sci. Nutr. 60(16): 2773–2789. Doi: 10.1080/10408398.2019.1658570.
[6] Alshannaq, A. and J.-H. Yu. 2017, Jun. Occurrence, toxicity, and analysis of major mycotoxins in food. Int. J. Environ. Res. Public Health 14(6): 632. Doi: 10.3390/ijerph14060632.
[7] Ostry, V., F. Malir, J. Toman and Y. Grosse. 2017, Feb. Mycotoxins as human carcinogens—the IARC Monographs classification. Mycotoxin. Res. 33(1): 65–73. Doi: 10.1007/s12550-016-0265-7.
[8] Alshannaq, A. et al. 2012. Improving public health through mycotoxin control 51(1). WHO PRESS. Doi: 10.1016/j.fct.2012.09.032.
[9] Peltomaa, R., E. Benito-Peña and M.C. Moreno-Bondi. 2018, Jan. Bioinspired recognition elements for mycotoxin sensors. Anal. Bioanal. Chem. 410(3): 747–771. Doi: 10.1007/s00216-017-0701-3.
[10] Kuiper-Goodman, T. 2004. Risk assessment and risk management of mycotoxins in food. pp. 3–31. *In*: Magan, N. and E.M. Olsen (eds.). Mycotoxins in Food Elsevier. Doi: 10.1533/9781855739086.1.3.
[11] Hussain, S.P., J. Schwank, F. Staib, X.W. Wang and C.C. Harris. 2007, Apr. TP53 mutations and hepatocellular carcinoma: Insights into the etiology and pathogenesis of liver cancer. Oncogene 26(15): 2166–2176. Doi: 10.1038/sj.onc.1210279.
[12] Bbosa, G.S., D. Kitya, A. Lubega, J. Ogwal-Okeng, W.W. Anokbonggo and D.B. Kyegombe. 2013. Review of the biological and health effects of aflatoxins on body organs and body systems. *In*: Razzaghi-Abyaneh, M. (ed.). Aflatoxins—Recent Advances and Future Prospects. Rijeka: InTech. Doi: 10.5772/51201.
[13] Bui-Klimke, T.R. and F. Wu. 2015, Nov. Ochratoxin A and human health risk: A review of the evidence. Crit. Rev. Food Sci. Nutr. 55(13): 1860–1869. Doi: 10.1080/10408398.2012.724480.
[14] Mbundi, L., H. Gallar-Ayala, M.R. Khan, J.L. Barber, S. Losada and R. Busquets. 2014. Advances in the analysis of challenging food contaminants. In Advances in Molecular Toxicology 8: 35–105. Doi: 10.1016/B978-0-444-63406-1.00002-7.
[15] Vardon, P., C. McLaughlin and C. Nardinelli. 2003. Potential Economic Costs of Mycotoxins in the United States.
[16] Wu, F. and N.J. Mitchell. 2016. How climate change and regulations can affect the economics of mycotoxins. World Mycotoxin. J. 9(5): 653–663. Doi: 10.3920/WMJ2015.2015.
[17] Debegnach, F., E. Gregori, S. Russo, F. Marchegiani, G. Moracci and C. Brera. 2017, May. Development of a LC-MS/MS method for the multi-mycotoxin determination in composite cereal-based samples. Toxins (Basel) 9(5): 169. Doi: 10.3390/toxins9050169.
[18] Köppen, R., M. Koch, D. Siegel, S. Merkel, R. Maul and I. Nehls. 2010, May. Determination of mycotoxins in foods: Current state of analytical methods and limitations. Appl. Microbiol. Biotechnol. 86(6): 1595–1612. Doi: 10.1007/s00253-010-2535-1.
[19] Evtugyn, G., V. Subjakova, S. Melikishvili and T. Hianik. 2018. Affinity biosensors for detection of mycotoxins in food. In Advances in Food and Nutrition Research 85: 263–310. Doi: 10.1016/bs.afnr.2018.03.003.
[20] Leenaars, M. and C.F.M. Hendriksen. 2005, Jan. Critical steps in the production of polyclonal and monoclonal antibodies: Evaluation and recommendations. ILAR J. 46(3): 269–279. Doi: 10.1093/ilar.46.3.269.
[21] Gray, A.C., S.S. Sidhu, P.C. Chandrasekera, C.F.M. Hendriksen and C.A.K. Borrebaeck. 2016, Dec. Animal-friendly affinity reagents: replacing the needless in the haystack. Trends Biotechnol. 34(12): 960–969. Doi: 10.1016/j.tibtech.2016.05.017.
[22] Rhouati, A., G. Bulbul, U. Latif, A. Hayat, Z.-H. Li and J. Marty. 2017, Oct. Nano-aptasensing in mycotoxin analysis: Recent updates and progress. Toxins (Basel) 9(11): 349. Doi: 10.3390/toxins9110349.
[23] Palchetti, I. and M. Mascini. 2012, Apr. Electrochemical nanomaterial-based nucleic acid aptasensors. Anal. Bioanal. Chem. 402(10): 3103–3114. Doi: 10.1007/s00216-012-5769-1.

[24] Chiorcea-Paquim, A.M. and A.M. Oliveira-Brett. 2016, Jul. Guanine quadruplex electrochemical aptasensors. Chemosensors 4(3): 13. Doi: 10.3390/chemosensors4030013.

[25] Zhuo, Z. et al. 2017, Oct. Recent advances in SELEX technology and aptamer applications in biomedicine. Int. J. Mol. Sci. 18(10): 2142. Doi: 10.3390/ijms18102142.

[26] Pfeiffer, F. and G. Mayer. 2016, Jun. Selection and biosensor application of aptamers for small molecules. Front. Chem. 4. Doi: 10.3389/fchem.2016.00025.

[27] Tomita, Y., Y. Morita, H. Suga and D. Fujiwara. 2016, Jun. DNA module platform for developing colorimetric aptamer sensors. Biotechniques 60(6): 285–292. Doi: 10.2144/000114425.

[28] Goode, J.A., J.V.H. Rushworth and P.A. Millner. 2015, Jun. Biosensor regeneration: A review of common techniques and outcomes. Langmuir 31(23): 6267–6276. Doi: 10.1021/la503533g.

[29] Evtugyn, G. and T. Hianik. 2020. Aptamer-based biosensors for mycotoxin detection. In Nanomycotoxicology, Elsevier, pp. 35–70. Doi: 10.1016/B978-0-12-817998-7.00003-3.

[30] Evtugyn, G. and T. Hianik. 2019. Electrochemical immuno- and aptasensors for mycotoxin determination. Chemosensors 7(1). Doi: 10.3390/CHEMOSENSORS7010010.

[31] Cruz-Aguado, J.A. and G. Penner. 2008. Determination of ochratoxin A with a DNA aptamer. J. Agric. Food Chem. 56(22): 10456–10461. Doi: 10.1021/jf801957h.

[32] Rhouati, A., C. Yang, A. Hayat and J.L. Marty. 2013, Nov. Aptamers: A promosing tool for ochratoxin a detection in food analysis. Toxins (Basel) 5(11): 1988–2008. Doi: 10.3390/toxins5111988.

[33] Hermann, T. and D.J. Patel. 2000, Feb. Adaptive recognition by nucleic acid aptamers. Science (1979) 287(5454): 820–825. Doi: 10.1126/science.287.5454.820.

[34] Barthelmebs, L., J. Jonca, A. Hayat, B. Prieto-Simon and J.L. Marty. 2011. Enzyme-Linked Aptamer Assays (ELAAs), based on a competition format for a rapid and sensitive detection of Ochratoxin A in wine. Food Control 22(5): 737–743. Doi: 10.1016/j.foodcont.2010.11.005.

[35] McKeague, M., R. Velu, K. Hill, V. Bardóczy, T. Mészáros and M.C. DeRosa. 2014. Selection and characterization of a novel DNA aptamer for label-free fluorescence biosensing of ochratoxin A. Toxins (Basel) 6(8): 2435–2452. Doi: 10.3390/toxins6082435.

[36] Goud, K.Y., A. Hayat, G. Catanante, S.M. Satyanarayana, K.V. Gobi and J.L. Marty. 2017, Aug. An electrochemical aptasensor based on functionalized graphene oxide assisted electrocatalytic signal amplification of methylene blue for aflatoxin B1 detection. Electrochim. Acta 244: 96–103. Doi: 10.1016/j.electacta.2017.05.089.

[37] Setlem, K., B. Mondal, S. Ramlal and J. Kingston. 2016, Dec. Immuno affinity SELEX for simple, rapid, and cost-effective aptamer enrichment and identification against aflatoxin B1. Front. Microbiol., 7. Doi: 10.3389/fmicb.2016.01909.

[38] Chryseis Le, L., J. Cruz-Aguado and G. A. Penner. 2010. DNA ligands for aflatoxins and zearalenone. PCT/CA2010/001292.

[39] Ma, X. et al. 2015, Jan. Selection, characterization and application of aptamers targeted to Aflatoxin B2. Food Control 47: 545–551. Doi: 10.1016/j.foodcont.2014.07.037.

[40] Nguyen, B.H., L.D. Tran, Q.P. Do, H. le Nguyen, N.H. Tran and P.X. Nguyen. 2013, May. Label-free detection of aflatoxin M1 with electrochemical Fe_3O_4/polyaniline-based aptasensor. Materials Science and Engineering C 33(4): 2229–2234. Doi: 10.1016/j.msec.2013.01.044.

[41] Chalyan, T. et al. 2017. Aptamer-and Fab'-functionalized microring resonators for aflatoxin M1 detection. IEEE Journal on Selected Topics in Quantum Electronics 23(2). Doi: 10.1109/JSTQE.2016.2609100.

[42] Malhotra, S., A.K. Pandey, Y.S. Rajput and R. Sharma. 2014. Selection of aptamers for aflatoxin M1 and their characterization. Journal of Molecular Recognition 27(8): 493–500. Doi: 10.1002/jmr.2370.

[43] Wu, S., N. Duan, W. Zhang, S. Zhao and Z. Wang. 2016. Screening and development of DNA aptamers as capture probes for colorimetric detection of patulin. Anal. Biochem. 508: 58–64. Doi: 10.1016/j.ab.2016.05.024.

[44] McKeague, M., C.R. Bradley, A. de Girolamo, A. Visconti, J. David Miller and M.C. de Rosa. 2010. Screening and initial binding assessment of fumonisin B 1 aptamers. Int. J. Mol. Sci. 11(12): 4864–4881. Doi: 10.3390/ijms11124864.

[45] Chen, X. et al. 2014. Selection and characterization of single stranded DNA aptamers recognizing fumonisin B1. Microchimica Acta 181(11-12): 1317–1324. Doi: 10.1007/s00604-014-1260-3.

[46] Wu, S., L. Zhang and M. Yang. 2012. Fumonisins B2 nucleic acid adapter and application thereof. Art. no. CN 102517290 A.

[47] Zhang, H.-Hui, Ya-Jun, JIN, Qing-Ri, SHAO, Chun-Yan, GUI, Hai-Luan, FANG, Wei-Huan, YANG and Meng-Hua, SONG. 2016. Development of an aptamer based technique for the detection of mycotoxin citrinin. Mycosystema 35(2): 209–216. Doi: 10.13346/j.mycosystema.140241.

[48] Liu, H., Y. Liu and S. Wu. 2012. Deoxynivalenol nucleic acid aptamer and application thereof. CN102559686A.

[49] Chen, X. et al. 2013. Selection and identification of ssDNA aptamers recognizing zearalenone. Anal. Bioanal. Chem. 405(20): 6573–6581. Doi: 10.1007/s00216-013-7085-9.

[50] Zhang, Y. et al. 2018. Selection of a DNA aptamer against zearalenone and docking analysis for highly sensitive rapid visual detection with label-free aptasensor. J. Agric. Food Chem. 66(45): 12102–12110. Doi: 10.1021/acs.jafc.8b03963.

[51] Chen, X. et al. 2014. Screening and identification of DNA aptamers against T-2 toxin assisted by graphene oxide. J. Agric. Food Chem. 62(42): 10368–10374. Doi: 10.1021/jf5032058.

[52] Krüger, A., F.M. Zimbres, T. Kronenberger and C. Wrenger. 2018. Molecular modeling applied to nucleic acid-based molecule development. Biomolecules 8(3). Doi: 10.3390/biom8030083.

[53] Hayat, A. and J.L. Marty. 2014, Jun. Aptamer based electrochemical sensors for emerging environmental pollutants. Front. Chem. 2. Doi: 10.3389/fchem.2014.00041.

[54] Raouafi, A., A. Sánchez, N. Raouafi and R. Villalonga. 2019. Electrochemical aptamer-based bioplatform for ultrasensitive detection of prostate specific antigen. Sens. Actuators B Chem. 297. Doi: 10.1016/j.snb.2019.126762.

[55] Ronkainen, N.J., H.B. Halsall and W.R. Heineman. 2010. Electrochemical biosensors. Chem. Soc. Rev. 39(5): 1747–1763. Doi: 10.1039/b714449k.

[56] Chen, W., C. Yan, L. Cheng, L. Yao, F. Xue and J. Xu. 2018. An ultrasensitive signal-on electrochemical aptasensor for ochratoxin A determination based on DNA controlled layer-by-layer assembly of dual gold nanoparticle conjugates. Biosens. Bioelectron 117: 845–851. Doi: 10.1016/j.bios.2018.07.012.

[57] Santos, A.O., A. Vaz, P. Rodrigues, A.C.A. Veloso, A. Venâncio and A.M. Peres. 2019, Jan. Thin films sensor devices for mycotoxins detection in foods: Applications and challenges. Chemosensors 7(1): 3. Doi: 10.3390/chemosensors7010003.

[58] Qing, Y. et al. 2017. Differential pulse voltammetric ochratoxin A assay based on the use of an aptamer and hybridization chain reaction. Microchimica Acta 184(3): 863–870. Doi: 10.1007/s00604-017-2080-z.

[59] He, B. and X. Dong. 2018. Aptamer based voltammetric patulin assay based on the use of ZnO nanorods. Microchimica Acta 185(10). Doi: 10.1007/s00604-018-3006-0.

[60] Kulikova, T.N., A.v. Porfireva, G.A. Evtugyn and T. Hianik. 2019. Electrochemical aptasensor with layer-by-layer deposited polyaniline for aflatoxin M1 voltammetric determination. Electroanalysis 31(10): 1913–1924. Doi: 10.1002/elan.201900274.

[61] Abnous, K., N.M. Danesh, M. Alibolandi, M. Ramezani and S.M. Taghdisi. 2017, Apr. Amperometric aptasensor for ochratoxin A based on the use of a gold electrode modified with aptamer, complementary DNA, SWCNTs and the redox marker Methylene Blue. Microchimica Acta 184(4): 1151–1159. Doi: 10.1007/s00604-017-2113-7.

[62] Tang, J., Y. Huang, Y. Cheng, L. Huang, J. Zhuang and D. Tang. 2018, Mar. Two-dimensional MoS2 as a nano-binder for ssDNA: Ultrasensitive aptamer based amperometric detection of Ochratoxin A. Microchimica Acta 185(3): 162. Doi: 10.1007/s00604-018-2706-9.

[63] Lv, E., J. Ding and W. Qin. 2018, Apr. Potentiometric aptasensing of small molecules based on surface charge change. Sens. Actuators B Chem. 259: 463–466. Doi: 10.1016/j.snb.2017.12.067.

[64] Kumar, A., M. Malinee, A. Dhiman, A. Kumar and T.K. Sharma. 2019. Aptamer technology for the detection of foodborne pathogens and toxins. In Advanced Biosensors for Health Care Applications, pp. 45–69. Doi: 10.1016/b978-0-12-815743-5.00002-0.

[65] Ocaña, C. and M. del Valle. 2018. Impedimetric aptasensors using nanomaterials. In Nanotechnology and Biosensors, Elsevier, pp. 233–267. Doi: 10.1016/B978-0-12-813855-7.00008-8.

[66] Rivas, L., C.C. Mayorga-Martinez, D. Quesada-González, A. Zamora-Gálvez, A. de La Escosura-Muñiz and A. Merkoçi. 2015, May. Label-free impedimetric aptasensor for ochratoxin-A detection using iridium oxide nanoparticles. Anal. Chem. 87(10): 5167–5172. Doi: 10.1021/acs.analchem.5b00890.

[67] Castillo, G., K. Spinella, A. Poturnayová, M. Šnejdárková, L. Mosiello and T. Hianik. 2015, Jun.Detection of aflatoxin B1 by aptamer-based biosensor using PAMAM dendrimers as immobilization platform. Food Control 52: 9–18. Doi: 10.1016/j.foodcont.2014.12.008.

[68] Fernandes, F.C.B., M.S. Góes, J.J. Davis and P.R. Bueno. 2013, May. Label free redox capacitive biosensing. Biosens. Bioelectron 50: 437–440. Doi: 10.1016/j.bios.2013.06.043.

[69] Santos, A., J.J. Davis and P.R. Bueno. 2014, Jun. Fundamentals and applications of impedimetric and redox capacitive biosensors. J. Anal. Bioanal. Tech. 0. Doi: 10.4172/2155-9872.S7-016.

[70] Ben Aissa, S., A. Mars, G. Catanante, J.L. Marty and N. Raouafi. 2019. Design of a redox-active surface for ultrasensitive redox capacitive aptasensing of aflatoxin M1 in milk. Talanta 195(August 2018): 525–532. Doi: 10.1016/j.talanta.2018.11.026.

[71] Argoubi, W., A. Rabti, S. ben Aoun and N. Raouafi. 2019. Sensitive detection of ascorbic acid using screen-printed electrodes modified by electroactive melanin-like nanoparticles. RSC Adv. 9(64): 37384–37390. Doi: 10.1039/C9RA07948C.

[72] Bulbul, G., A. Hayat and S. Andreescu. 2015. A generic amplification strategy for electrochemical aptasensors using a non-enzymatic nanoceria tag. Nanoscale 7(31): 13230–13238. Doi: 10.1039/c5nr02628h.

[73] Zhang, J., B. Liu, H. Liu, X. Zhang and W. Tan. 2013, Jun. Aptamer-conjugated gold nanoparticles for bioanalysis. Nanomedicine 8(6): 983–993. Doi: 10.2217/nnm.13.80.

[74] Azizah, N., U. Hashim, S.C.B. Gopinath and S. Nadzirah. 2016, Dec. Gold nanoparticle mediated method for spatially resolved deposition of DNA on nano-gapped interdigitated electrodes, and its application to the detection of the human Papillomavirus. Microchimica Acta 183(12): 3119–3126. Doi: 10.1007/s00604-016-1954-9.

[75] Gu, C. et al. 2019, Jun. A bimetallic (Cu-Co) Prussian Blue analogue loaded with gold nanoparticles for impedimetric aptasensing of ochratoxin A. Microchimica Acta 186(6): 343. Doi: 10.1007/s00604-019-3479-5.

[76] Kang, D., F. Ricci, R.J. White and K.W. Plaxco. 2016, Nov. Survey of redox-active moieties for application in multiplexed electrochemical biosensors. Anal. Chem. 88(21): 10452–10458. Doi: 10.1021/acs.analchem.6b02376.

[77] Hayat, A., G. Catanante and J.L. Marty. 2014, Dec. Current trends in nanomaterial-based amperometric biosensors. Sensors (Switzerland) 14(12): 23439–23461. Doi: 10.3390/s141223439.

[78] Ding, L., A.M. Bond, J. Zhai and J. Zhang. 2013, Oct. Utilization of nanoparticle labels for signal amplification in ultrasensitive electrochemical affinity biosensors: A review. Anal. Chim. Acta 797: 1–12. Doi: 10.1016/j.aca.2013.07.035.

[79] Jalalian, S.H., M. Ramezani, N.M. Danesh, M. Alibolandi, K. Abnous and S.M. Taghdisi. 2018, Oct. A novel electrochemical aptasensor for detection of aflatoxin M1 based on target-induced immobilization of gold nanoparticles on the surface of electrode. Biosens. Bioelectron 117: 487–492. Doi: 10.1016/j.bios.2018.06.055.

[80] Wang, C. et al. 2017, Mar. Magneto-controlled aptasensor for simultaneous electrochemical detection of dual mycotoxins in maize using metal sulfide quantum dots coated silica as labels. Biosens. Bioelectron 89: 802–809. Doi: 10.1016/j.bios.2016.10.010.

[81] Ting, B.P., J. Zhang, Z. Gao and J.Y. Ying. 2009. A DNA biosensor based on the detection of doxorubicin-conjugated Ag nanoparticle labels using solid-state voltammetry. Biosens. Bioelectron 25(2): 282–287. Doi: 10.1016/j.bios.2009.07.005.

[82] Shi, Z. et al. 2017. Ultra-sensitive aptasensor based on IL and Fe_3O_4 nanoparticles for tetracycline detection. Int. J. Electrochem. Sci. 12(8): 7426–7434. Doi: 10.20964/2017.08.76.

[83] Mars, A., C. Parolo, A. de la Escosura-Muñiz, N. Raouafi and A. Merkoçi. 2016, Aug. Control of electron-transfer in immunonanosensors by using polyclonal and monoclonal antibodies. Electroanalysis 28(8): 1795–1802. Doi: 10.1002/elan.201500646.

[84] Jing, P., H. Yi, S. Xue, Y. Chai, R. Yuan and W. Xu. 2015, Jan. A sensitive electrochemical aptasensor based on palladium nanoparticles decorated graphene-molybdenum disulfide flower-

like nanocomposites and enzymatic signal amplification. Anal. Chim. Acta 853(1): 234–241. Doi: 10.1016/j.aca.2014.10.003.

[85] Vasilescu, A., A. Hayat, S. Gáspár and J.L. Marty. 2018, Jan. Advantages of carbon nanomaterials in electrochemical aptasensors for food analysis. Electroanalysis 30(1): 2–19. Doi: 10.1002/elan.201700578.

[86] Goud, K.Y. et al. 2018, Dec. Progress on nanostructured electrochemical sensors and their recognition elements for detection of mycotoxins: A review. Biosens. Bioelectron 121: 205–222. Doi: 10.1016/j.bios.2018.08.029.

[87] Wei, M. and W. Zhang. 2017. A novel impedimetric aptasensor based on AuNPs-carboxylic porous carbon for the ultrasensitive detection of ochratoxin A. RSC Adv. 7(46): 28655–28660. Doi: 10.1039/c7ra04209d.

[88] Argoubi, W., M. Saadaoui, S. ben Aoun and N. Raouafi. 2015. Optimized design of a nanostructured SPCE-based multipurpose biosensing platform formed by ferrocene-tethered electrochemically-deposited cauliflower-shaped gold nanoparticles. Beilstein Journal of Nanotechnology 6(1): 1840–1852. Doi: 10.3762/bjnano.6.187.

[89] Xie, J., X. Zhang, H. Wang, H. Zheng, Y. Huang and J. Xie. 2012, Oct. Analytical and environmental applications of nanoparticles as enzyme mimetics. TrAC—Trends in Analytical Chemistry 39: 114–129. Doi: 10.1016/j.trac.2012.03.021.

[90] Shin, H.Y., T.J. Park and M.il Kim. 2015. Recent research trends and future prospects in nanozymes. J. Nanomater. 2015: 1–11. Doi: 10.1155/2015/756278.

[91] Wang, Q., H. Wei, Z. Zhang, E. Wang and S. Dong. 2018, Aug. Nanozyme: An emerging alternative to natural enzyme for biosensing and immunoassay. TrAC—Trends in Analytical Chemistry 105: 218–224. Doi: 10.1016/j.trac.2018.05.012.

[92] Evtugyn, G. et al. 2013, Aug. Electrochemical aptasensor for the determination of ochratoxin A at the Au electrode modified with Ag nanoparticles decorated with macrocyclic ligand. Electroanalysis 25(8): 1847–1854. Doi: 10.1002/elan.201300164.

[93] Karimi, A., A. Hayat and S. Andreescu. 2017, Jan. Biomolecular detection at ssDNA-conjugated nanoparticles by nano-impact electrochemistry. Biosens. Bioelectron 87: 501–507. Doi: 10.1016/j.bios.2016.08.108.

[94] Evtugyn, G. et al. 2013. Impedimetric aptasensor for ochratoxin a determination based on Au nanoparticles stabilized with hyper-branched polymer. Sensors (Switzerland) 13(12): 16129–16145. Doi: 10.3390/s131216129.

[95] Wei, M. and S. Feng. 2017. A signal-off aptasensor for the determination of Ochratoxin A by differential pulse voltammetry at a modified Au electrode using methylene blue as an electrochemical probe. Analytical Methods 9(37): 5449–5454. Doi: 10.1039/c7ay01735a.

[96] Kuang, H. et al. 2010. Fabricated aptamer-based electrochemical 'signal-off' sensor of ochratoxin A. Biosens. Bioelectron 26(2): 710–716. Doi: 10.1016/j.bios.2010.06.058.

[97] Roushani, A., M. Zare Dizajdizi, B. Rahmati and Z. Azadbakht. 2019. Development of electrochemical aptasensor based on gold nanorod and its application for detection of aflatoxin B1 in rice and blood serum sample. Nanochemistry Research 4(1): 35–42. Doi: 10.22036/ncr.2019.01.005.

[98] He, B. and X. Yan. 2019. An amperometric zearalenone aptasensor based on signal amplification by using a composite prepared from porous platinum nanotubes, gold nanoparticles and thionine-labelled graphene oxide. Microchimica Acta 186(6). Doi: 10.1007/s00604-019-3500-z.

[99] Yang, X. et al. 2014. Ultrasensitive electrochemical aptasensor for ochratoxin A based on two-level cascaded signal amplification strategy. Bioelectrochemistry 96: 7–13. Doi: 10.1016/j.bioelechem.2013.11.006.

[100] Chen, X., Y. Huang, X. Ma, F. Jia, X. Guo and Z. Wang. 2015, Jul. Impedimetric aptamer-based determination of the mold toxin fumonisin B1. Microchimica Acta 182(9-10): 1709–1714. Doi: 10.1007/s00604-015-1492-x.

[101] Zheng, W. et al. 2016, Jun. Hetero-enzyme-based two-round signal amplification strategy for trace detection of aflatoxin B1 using an electrochemical aptasensor. Biosens. Bioelectron 80: 574–581. Doi: 10.1016/j.bios.2016.01.091.

[102] Nan, M. et al. 2019, Jan. Rapid determination of ochratoxin a in grape and its commodities based on a label-free impedimetric aptasensor constructed by layer-by-layer self-assembly. Toxins (Basel) 11(2): 71. Doi: 10.3390/toxins11020071.

[103] Gu, M. et al. 2019, Aug. Development of ochratoxin aptasensor based on DNA metal nanoclusters. Nanoscience and Nanotechnology Letters 11(8): 1139–1144. Doi: 10.1166/nnl.2019.2981.

[104] Solanki, P.R., A. Kaushik, V.V. Agrawal and B.D. Malhotra. 2011, Jan. Nanostructured metal oxide-based biosensors. NPG Asia Mater. 3(1): 17–24. Doi: 10.1038/asiamat.2010.137.

[105] Ansari, A.A., A. Kaushik, P.R. Solanki and B.D. Malhotra. 2010, Feb. Nanostructured zinc oxide platform for mycotoxin detection. Bioelectrochemistry 77(2): 75–81. Doi: 10.1016/j.bioelechem.2009.06.014.

[106] He, B. and X. Dong. 2019. Hierarchically porous Zr-MOFs labelled methylene blue as signal tags for electrochemical patulin aptasensor based on ZnO nano flower. Sens. Actuators B Chem. 294: 192–198. Doi: 10.1016/j.snb.2019.05.045.

[107] Chen, Z. and M. Lu. 2016, Nov. Target-responsive aptamer release from manganese dioxide nanosheets for electrochemical sensing of cocaine with target recycling amplification. Talanta 160: 444–448. Doi: 10.1016/j.talanta.2016.07.052.

[108] Wang, C. et al. 2018, Jun. Fabrication of magnetically assembled aptasensing device for label-free determination of aflatoxin B1 based on EIS. Biosens. Bioelectron 108: 69–75. Doi: 10.1016/j.bios.2018.02.043.

[109] Sun, A.-L., Y.-F. Zhang, G.-P. Sun, X.-N. Wang and D. Tang. 2017, Mar. Homogeneous electrochemical detection of ochratoxin A in foodstuff using aptamer–graphene oxide nanosheets and DNase I-based target recycling reaction. Biosens. Bioelectron 89: 659–665. Doi: 10.1016/j.bios.2015.12.032.

[110] Qian, J., L. Jiang, X. Yang, Y. Yan, H. Mao and K. Wang. 2014, Aug. Highly sensitive impedimetric aptasensor based on covalent binding of gold nanoparticles on reduced graphene oxide with good dispersity and high density. Analyst 139(21): 5587–5593. Doi: 10.1039/C4AN01116C.

[111] Mo, R. et al. 2018, Oct. A novel aflatoxin B1 biosensor based on a porous anodized alumina membrane modified with graphene oxide and an aflatoxin B1 aptamer. Electrochem. Commun. 95: 9–13. Doi: 10.1016/j.elecom.2018.08.012.

[112] Geleta, G.S., Z. Zhao and Z. Wang. 2018. A novel reduced graphene oxide/molybdenum disulfide/polyaniline nanocomposite-based electrochemical aptasensor for detection of aflatoxin B1. Analyst 143(7): 1644–1649. Doi: 10.1039/c7an02050c.

[113] Loo, A.H., A. Bonanni and M. Pumera. 2015, May. Mycotoxin aptasensing amplification by using inherently electroactive graphene-oxide nanoplatelet labels. ChemElectroChem 2(5): 743–747. Doi: 10.1002/celc.201402403.

[114] Ma, L., L. Bai, M. Zhao, J. Zhou, Y. Chen and Z. Mu. 2019, Jul. An electrochemical aptasensor for highly sensitive detection of zearalenone based on PEI-MoS2-MWCNTs nanocomposite for signal enhancement. Anal. Chim. Acta 1060: 71–78. Doi: 10.1016/j.aca.2019.02.012.

[115] Sanati, A. et al. 2019, Dec. A review on recent advancements in electrochemical biosensing using carbonaceous nanomaterials. Microchimica Acta 186(12): 773. Doi: 10.1007/s00604-019-3854-2.

[116] Bacon, M., S.J. Bradley and T. Nann. 2014, Apr. Graphene quantum dots. Particle & Particle Systems Characterization 31(4): 415–428. Doi: 10.1002/ppsc.201300252.

[117] Gevaerd, A., C.E. Banks, M.F. Bergamini and L.H. Marcolino-Junior. 2020, Mar. Nanomodified screen-printed electrode for direct determination of aflatoxin B1 in malted barley samples. Sens. Actuators B Chem. 307: 127547. Doi: 10.1016/j.snb.2019.127547.

[118] Arduini, F., F. di Nardo, A. Amine, L. Micheli, G. Palleschi and D. Moscone. 2012, Apr. Carbon black-modified screen-printed electrodes as electroanalytical tools. Electroanalysis 24(4): 743–751. Doi: 10.1002/elan.201100561.

[119] Erdem, A., G. Congur and G. Mayer. 2015, Dec. Aptasensor platform based on carbon nanofibers enriched screen printed electrodes for impedimetric detection of thrombin. Journal of Electroanalytical Chemistry 758: 12–19. Doi: 10.1016/j.jelechem.2015.10.002.

[120] Fang, L.-X., K.-J. Huang and Y. Liu. 2015, Sep. Novel electrochemical dual-aptamer-based sandwich biosensor using molybdenum disulfide/carbon aerogel composites and Au nanoparticles for signal amplification. Biosens. Bioelectron 71: 171–178. Doi: 10.1016/j.bios.2015.04.031.

[121] Beheshti-Marnani, A., A. Hatefi-Mehrjardi and Z. Es'haghi. 2019. A sensitive biosensing method for detecting of ultra-trace amounts of AFB1 based on 'Aptamer/reduced graphene oxide' nano-bio interaction. Colloids Surf B Biointerfaces 175: 98–105. Doi: 10.1016/j.colsurfb.2018.11.087.

[122] Jiang, L. et al. 2014, Jan. Amplified impedimetric aptasensor based on gold nanoparticles covalently bound graphene sheet for the picomolar detection of ochratoxin A. Anal. Chim. Acta 806: 128–135. Doi: 10.1016/j.aca.2013.11.003.

[123] Shi, Z.Y., Y.T. Zheng, H.B. Zhang, C.H. He, W. da Wu and H. bin Zhang. 2015, May. DNA electrochemical aptasensor for detecting fumonisins B1 based on graphene and thionine nanocomposite. Electroanalysis 27(5): 1097–1103. Doi: 10.1002/elan.201400504.

[124] Hao, N., L. Jiang, J. Qian and K. Wang. 2016, Nov. Ultrasensitive electrochemical Ochratoxin A aptasensor based on CdTe quantum dots functionalized graphene/Au nanocomposites and magnetic separation. Journal of Electroanalytical Chemistry 781: 332–338. Doi: 10.1016/j.jelechem.2016.09.053.

Index